T0145246

# Advances in Intelligent Systems and Computing

Volume 840

**Series editor**

Janusz Kacprzyk, Polish Academy of Sciences, Warsaw, Poland
e-mail: kacprzyk@ibspan.waw.pl

The series "Advances in Intelligent Systems and Computing" contains publications on theory, applications, and design methods of Intelligent Systems and Intelligent Computing. Virtually all disciplines such as engineering, natural sciences, computer and information science, ICT, economics, business, e-commerce, environment, healthcare, life science are covered. The list of topics spans all the areas of modern intelligent systems and computing such as: computational intelligence, soft computing including neural networks, fuzzy systems, evolutionary computing and the fusion of these paradigms, social intelligence, ambient intelligence, computational neuroscience, artificial life, virtual worlds and society, cognitive science and systems, Perception and Vision, DNA and immune based systems, self-organizing and adaptive systems, e-Learning and teaching, human-centered and human-centric computing, recommender systems, intelligent control, robotics and mechatronics including human-machine teaming, knowledge-based paradigms, learning paradigms, machine ethics, intelligent data analysis, knowledge management, intelligent agents, intelligent decision making and support, intelligent network security, trust management, interactive entertainment, Web intelligence and multimedia.

The publications within "Advances in Intelligent Systems and Computing" are primarily proceedings of important conferences, symposia and congresses. They cover significant recent developments in the field, both of a foundational and applicable character. An important characteristic feature of the series is the short publication time and world-wide distribution. This permits a rapid and broad dissemination of research results.

More information about this series at http://www.springer.com/series/11156

Ahmad Lotfi · Hamid Bouchachia
Alexander Gegov · Caroline Langensiepen
Martin McGinnity
Editors

# Advances in Computational Intelligence Systems

Contributions Presented at the 18th UK
Workshop on Computational Intelligence,
September 5–7, 2018, Nottingham, UK

 Springer

*Editors*
Ahmad Lotfi
School of Science and Technology
Nottingham Trent University
Nottingham, UK

Caroline Langensiepen
School of Science and Technology
Nottingham Trent University
Nottingham, UK

Hamid Bouchachia
Faculty of Science and Technology
Bournemouth University
Poole, Dorset
UK

Martin McGinnity
College of Science and Technology
Nottingham Trent University
Nottingham, UK

Alexander Gegov
School of Computing
University of Portsmouth
Portsmouth, Hampshire
UK

ISSN 2194-5357          ISSN 2194-5365   (electronic)
Advances in Intelligent Systems and Computing
ISBN 978-3-319-97981-6          ISBN 978-3-319-97982-3   (eBook)
https://doi.org/10.1007/978-3-319-97982-3

Library of Congress Control Number: 2018950434

This Springer imprint is published by the registered company Springer Nature Switzerland AG
The registered company address is: Gewerbestrasse 11, 6330 Cham, Switzerland

# Preface

This volume contains the papers presented at the 18th UK Workshop on Computational Intelligence (UKCI 2018) which was held in Nottingham, UK, on 5–7 September 2018. UKCI has been the premier UK event for presenting leading research on all aspects on computational intelligence since 2001. The overall objective of UKCI is to provide a forum for the academic community and industry to share and exchange recent ideas about the theoretical and practical aspects of computational intelligence techniques.

Computational intelligence is a rapidly expanding research field, attracting a large number of scientists, engineers and practitioners working in areas such as fuzzy systems, neural networks, evolutionary computation, evolving systems and machine learning. A growing number of companies are employing computational intelligence techniques to improve previous solutions and to deal with new problems. These include evolving systems that allow high performance in spite of changes which are either external or internal to the system, thereby increasing the reusability of developed systems. This also includes smart, intelligent and autonomous systems, self-learning, self-adapting, self-calibrating and self-tuning.

UKCI 2018 has attracted 41 submissions, on areas such as fuzzy systems, neural networks, evolutionary computation, clustering and classification, machine learning, data mining, cognition and robotics, and deep learning. Each paper was reviewed by at least three members of the programme committee and additional reviewers. Based on their recommendations, 32 papers have been accepted for publication in this volume and presentation during the workshop. The presented papers in this volume are organised into five parts: (1) Search and Optimisation, (2) Modelling and Representation, (3) Learning and Adaptation, (4) Clustering and Regression and (5) Analysis and Detection.

While organised primarily as a national event, UKCI 2018 has attracted papers and participants from other countries including Japan, India, Italy, Switzerland, Portugal and Malaysia. The UKCI 2018 programme featured keynote talks by three established researchers in the field of computational intelligence—Prof. Plamen Angelov, Prof. Jon Garibaldi and Prof. Qiang Shen.

Finally, we would like to thank everyone who contributed to the success of UKCI 2018 and publication of this volume. Our special thanks to the members of the programme and organising committees, additional reviewers, the keynote speakers, the authors and the presenters of papers. We are also grateful for support received from Nottingham Trent University to host the event and for help and support received from Springer to prepare this volume.

Ahmad Lotfi
Hamid Bouchachia
Alexander Gegov
Caroline Langensiepen
Martin McGinnity

# Organization

## Programme Committee

| | |
|---|---|
| Giovanni Acampora | University of Naples Federico II, Italy |
| Peter Andras | Keele University, UK |
| Plamen Angelov | Lancaster University, UK |
| Atta Badi | University of Reading, UK |
| Abdelhamid Bouchachia | Bournemouth University, UK |
| Fei Chao | Xiamen University, China |
| Tianhua Chen | University of Huddersfield, UK |
| George Coghill | University of Aberdeen, UK |
| Sonya Coleman | Ulster University, UK |
| Simon Coupland | De Montfort University, UK |
| Keeley Crockett | Manchester Metropolitan University, UK |
| Jon Garibaldi | University of Nottingham, UK |
| Alexander Gegov | University of Portsmouth, UK |
| Hani Hagras | University of Essex, UK |
| Hongnmei He | Cranfield University, UK |
| Chris Hide | Loughborough University, UK |
| Xia Hong | University of Reading, UK |
| Thomas Jansen | Aberystwyth University, UK |
| Robert John | University of Nottingham, UK |
| Dermot Kerr | Ulster University, UK |
| Ahmed Kheiri | Lancaster University, UK |
| Caroline Langensiepen | Nottingham Trent University, UK |
| Miqing Li | University of Birmingham, UK |
| Ahmad Lotfi | Nottingham Trent University, UK |
| Trevor Martin | University of Bristol, UK |
| Martin McGinnity | Nottingham Trent University, UK |
| Daniel Neagu | University of Bradford, UK |
| George Panoutsos | The University of Sheffield, UK |

| Amir Pourabdollah | Nottingham Trent University, UK |
| Girijesh Prasad | Ulster University, UK |
| Chris Price | Aberystwyth University, UK |
| Shahin Rostami | Bournemouth University, UK |
| Steven Schockaert | Cardiff University, UK |
| Qiang Shen | Aberystwyth University, UK |
| Longzhi Yang | Northumbria University, UK |
| Shengxiang Yang | De Montfort University, UK |
| Shufan Yang | University of Glasgow, UK |
| Yingjie Yang | De Montfort University, UK |

## Additional Reviewers

| Georgina Cosma | Nottingham Trent University, UK |
| Omprakash Kaiwartya | Nottingham Trent University, UK |
| Pedro Machado | Nottingham Trent University, UK |
| Mufti Mahmud | Nottingham Trent University, UK |
| Aboozar Taherkhani | Nottingham Trent University, UK |

# Contents

# Search and Optimisation

# On the Integrity of Performance Comparison for Evolutionary Multi-objective Optimisation Algorithms

Kevin Wilson[✉] and Shahin Rostami

Computational Intelligence Research Initiative (CIRI), Bournemouth University,
Bournemouth BH12 5BB, UK
{kwilson,srostami}@bournemouth.ac.uk
http://research.bournemouth.ac.uk/project/ciri

**Abstract.** This paper proposes the notion that the experimental results and performance analyses of newly developed algorithms in the field of multi-objective optimisation may not offer sufficient integrity for hypothesis testing. The reason for this is that many implementations exist of the same optimisation algorithms, and these may vary in behaviour due to the interpretation of the developer. This is demonstrated through the comparison of three implementations of the popular Non-dominated Sorting Genetic Algorithm II (NSGA-II) from well-regarded frameworks using the hypervolume indicator. The results show that of the thirty considered comparison cases, only four indicate that there was no significant difference between the performance of either implementation.

**Keywords:** Evolutionary algorithms · Genetic algorithms
Optimisation · Hypervolume indicator

## 1  Introduction

Evolutionary Multi-objective Optimisation (EMO) algorithms are well suited to solving complex real-world problems. One strength of EMO is the ability to produce a set of multiple trade-off solutions to a multi-objective problem within a single algorithm execution. This set is referred to as an approximation-set, and is considered to be an approximation of the theoretical true optimum (Pareto-optimal front) of a problem. When proposing a new EMO algorithm, it is often expected that the proposed algorithm is applied to a multi-objective test suite according to some experimental design. The approximation-sets generated from these experiments are then subjected to some performance metrics and compared to popular or state-of-the-art algorithms as a benchmark. The typical characteristics for assessing the quality of an approximation-set have been illustrated in Fig. 1 and have been listed in the following [20]:

- **Proximity:** the distance between the approximation-set and the Pareto-optimal front.

© Springer Nature Switzerland AG 2019
A. Lotfi et al. (Eds.): UKCI 2018, AISC 840, pp. 3–15, 2019.
https://doi.org/10.1007/978-3-319-97982-3_1

- **Diversity:** the uniformity and extent of the distribution of solutions within the approximation-set.
- **Pertinence:** the relevance of the approximation-set in the presence of a Decision Maker (DM) preferences.

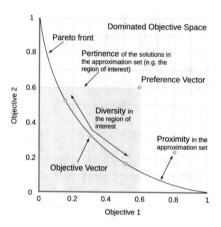

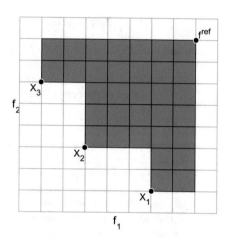

**Fig. 1.** Characteristics of an approximation-set in bi-objective space.

**Fig. 2.** An example of the hypervolume indicator in bi-objective space

Many performance metrics exist for assessing approximation-sets against these desirable characteristics [32], and these have been employed in the comparison of EMO algorithms throughout the literature [17]. These pairwise or multiple comparisons resolve whether the proposed algorithm outperforms the benchmark algorithms in respect to desired the desired characteristics on specific test-cases. This approach allows an algorithm's comparative performance to be demonstrated, and allows the verification of hypotheses and contribution.

One popular algorithm which is often used as a benchmark in the EMO literature [14] is the Non-dominated Sorting Genetic Algorithm II (NSGA-II) [4]. The NSGA-II, described in Sect. 2.2, is a powerful EMO algorithm which employs elitism, non-dominated sorting, and diversity preservation in order to produce approximation sets with good proximity and diversity.

An implementation of the NSGA-II algorithm is required to generate results for comparison. The solution is either to implement NSGA-II based on details available in the related publication, or to acquire an existing implementation. Depending on the publication, it can be time-consuming or difficult to implement an EMO algorithm, e.g. due to unusual mathematical notation or the lack of low-level description of the operators. Where the authors have not released the implementation of their algorithm (which is often the case) [13], it is possible that implementations may be available as part of existing optimisation libraries.

It should be noted that the author of NSGA-II has made the source code available on their website. However, as described below, many researchers still make use of alternative implementations provided by commonly used libraries, and in some cases, even their own implementation of the algorithm [13].

Given this situation, it would clearly be of some concern if the performance of the benchmark algorithms were subject to statistically significant variation depending on which implementation was being used, as this would impact the integrity of performance comparisons when proposing a new algorithm.

To verify this, the following testable hypotheses can be defined:

**Hypothesis 1.** *The approximation-sets produced by multiple different implementations of NSGA-II on the same test case will not always achieve significantly similar performance.*

**Hypothesis 2.** *The final approximation-sets produced by multiple different implementations of NSGA-II may achieve significantly similar performance, but the performance of the populations maintained throughout the optimisation process will vary, i.e. the rate of convergence.*

These definitions will be used to inform the experimental design, as described in Sect. 3, and are referred to again in Sect. 5 where the experimental results are discussed.

The rest of the paper is organised as follows. Section 2.1 introduces the concept of Evolutionary Multi-objective Optimisation. Section 2.2 describes the NSGA-II algorithm and discusses its importance in the field. Section 3 details the experimental design and how the algorithms were configured. Section 3.1 lists the three implementations of the NSGA-II algorithm that were used in this experiment, with details of how they have been used other experimental studies. Section 3.2 describes the hypervolume indicator, the key metric used to measure the performance of the NSGA-II algorithm in this study. The results of the experiment are summarised in Sect. 4, discussed in Sect. 5 and conclusions are drawn in Sect. 6.

## 2   Background: Evolutionary Multi-objective Optimisation and NSGA-II

### 2.1   Evolutionary Multi-objective Optimisation

A typical evolutionary algorithm will start with population of randomly generated candidate solutions. The object is to allow them to reproduce and generate another population of candidate solutions, which are closer to the theoretical optimum set of solutions, and to do this repeatedly, until the solution set is close enough to the optimum to be useful. As part of each iteration, the following steps are undertaken:

1. **Evaluation:** Each solution is evaluated and assigned a fitness level. This would involve calculating some function of its objective function value, with regard to any constraints that may exist. The optimisation is normally trying

to minimise (or in some cases maximise) this value. With multi-objective problems a number of approaches to both fitness evaluation and selection have been developed to try and deal with the problem of multiple, possibly conflicting, objectives.

2. **Selection:** A selection operator is then applied to the population. This has the job of making copies of good solutions and eliminating bad solutions from the population, whilst keeping the population size the same. This step cannot create any new solutions - that is done by the remaining two steps.

   There are several approaches to selection when dealing with multi-objective problems, including Selection by switching objectives, and Selection by Aggregation. The one used by NSGA-II is called Pareto-based Selection.

   In this approach, the solutions are sorted into a number of sets. All non-dominated solutions are labelled as Set One, and removed from consideration. The remaining non-dominated solutions are then labelled as Set Two, and removed from consideration, and so on. The fitness of a solution is determined by which set it belongs to.

3. **Crossover:** The crossover operator has the job of taking two solutions, and combining portions of them together to make two new solutions. As both portions come from solutions that have a certain level of fitness (otherwise they would not have survived the selection process) the chances are good that the crossover operation will result in solutions that also have a good level of fitness. There are a number of widely used crossover operators, including Linear [29], and Simulated Binary [15].

4. **Mutation:** The mutation operator introduces an element of random variation by changing some portion of a solution to create a new one. This ensures that the solution space is explored and as wide a range of optimum solutions as possible are identified. There are a number of widely used mutation operators, including Random [18] and Polynomial [5].

## 2.2  Non-dominated Sorting Genetic Algorithm II

The Non-dominated Sorting Genetic Algorithm II was first proposed by Deb et al. in 2000 [4]. It generates an offspring population from the parent population, and then sorts them into a set of Pareto fronts. Solutions with high dominance are included in the next generation, whether they come from the parent or offspring population, thus making it an elitist algorithm. The algorithm also attempts to maximise diversity by employing a crowding distance operator which sorts the solutions according to their distance from their neighbours, to ensure as wide a spread as possible. NSGA-II has proved to be influential - Scopus reports a citation count of 12945 for Deb's paper, at the time of writing, and it has been described by Ishibuchi [13] as "the most frequently used evolutionary multiobjective algorithm in the literature since its proposal". He goes on to say that "New EMO algorithms were almost always compared with NSGA-II for performance evaluation in 2000–2010". The execution life-cycle of NSGA-II is shown in Algorithm 1.

**Algorithm 1.** NSGA-II

1  Generate random population P of size n
2  Apply selection and variation operators to produce offspring population Q of size n
3  **repeat**
4    Combine $P$ and $Q$ to produce combined population $R$ of size 2n
5    Perform non-dominated sort of $R$ to produce set of fronts $F_1$ to $F_m$
6    x = 1
7    **while** *(enough room in $P_{new}$ for all members of $F_x$)* **do**
8      Calculate crowding distance for each solution in $F_x$
9      Copy all members of $F_x$ to $P_{new}$
10     x = x + 1
11   **end**
12   Sort $F_x$ with crowded comparison operator in descending order
13   Copy the best solutions from $F_x$ to fill the remaining spaces in $P_{new}$
14   Apply binary tournament selection (based on crowded comparison) to $P_{new}$
15   Apply crossover operators to $P_{new}$
16   Apply mutation operators to $P_{new}$
17   Operators will produce an offspring population $Q_{new}$ of size n
18   P = $P_{new}$, Q = $Q_{new}$
19 **until** *(stopping criteria satisfied);*
20 Output $P$

# 3  Experimental Design

In order to test our hypotheses, three different implementations have been compared based on their hypervolume indicator performance on a set of synthetic test functions from the ZDT [30] test suite. These test functions, their configurations, and their salient features have been listed in Table 1.

**Table 1.** Salient features of the selected ZDT synthetic test problems and their respective parameter configurations

| Problem | #Var | #Obj | Salient features | Reference vector |
|---------|------|------|------------------|------------------|
| ZDT1 | 30 | 2 | Convex front | 11,11 |
| ZDT2 | 30 | 2 | Concave front | 11,11 |
| ZDT3 | 30 | 2 | Disconnected front | 11,11 |
| ZDT4 | 10 | 2 | Convex front, many local optima | 300,300 |
| ZDT6 | 10 | 2 | Concave front, non-uniform distribution | 11,11 |

The parameters used to configure each of the considered implementations are taken from the author's original work [6] and listed in Table 2.

In order to verify Hypothesis 1, each execution has been configured to allow a maximum of 500 generations, with a sample size of 30 executions per test function. The mean hypervolume indicator values over 30 executions will be calculated and subjected to statistical comparison.

**Table 2.** Parameter configurations for all considered NSGA-II implementations

| Crossover type | Simulated binary crossover |
|---|---|
| Crossover probability | 0.9 |
| Crossover distribution Index | 20 |
| Mutation type | Polynomial |
| Mutation probability | 1/No of decision variables |
| Mutation distribution Index | 20 |
| Selection type | Binary Tournament |
| Population size | 100 |
| Generation count | 500 |

In order to verify Hypothesis 2, the three implementations of NSGA-II have also been configured to output their populations at each generation, allowing their performance to be measured throughout each execution, rather than just at the end. The performance of each will be displayed graphically and compared.

The remainder of this section describes the NSGA-II implementations considered in this experiment, and introduces the hypervolume indicator.

### 3.1 Considered Implementations

Three widely used implementations of NSGA-II are compared, in this study. Each one is provided as part of a larger library of programs. Each library also contains implementations of other similar algorithms, and other features, such as test problems. The three libraries are introduced in the following sub-sections.

**jMetal.** jMetal is an object-oriented Java-based framework, developed by Durillo and Nebro [8]. It contains implementations of many multi-objective optimisation algorithms, as well as test problems, quality indicators and other tools useful to a researcher in the field. Development was started in 2006, with a major redesign and rewrite of the code in 2014. It is publicly available and licensed under the GNU Lesser General Public License. The current version is jMetal 5.1 and it is available on a public facing code repository [19].

The study in [25] uses the NSGA-II implementation offered by jMetal, as well as their implementation of SPEA-2 and IBEA, to test the hypothesis that their hyper-heuristic (HH) approach improves upon the results produced by the traditional algorithms. They compare the performance of the three algorithms unmodified and choosing the one that returns the best results, NSGA-II, they measure its performance again, having modified it to incorporate the HH feature.

Similar experimental scenarios which use implementations from jMetal can be found in [21], where they are used to test new types of Multi-objective Evolutionary Algorithm, and in [23] where they are used to aid in the design of software products.

**Shark ML.** The Shark Machine Learning Library is an open-source C++ library which provides methods for optimisation and learning algorithms, neural networks and other machine learning techniques.

It is maintained by Igel, Heidrich-Meisner and Glasmachers and is publicly available and licensed under the GNU Lesser General Public License. The current version is 3.1.0 and it is available on a public facing code repository [12].

The study in [24] uses hybridised versions of the NSGA-II and MO-CMA-ES algorithms, as provided by the Shark Machine Learning Library, in conjunction with a partially observable Markov decision process (POMDP).

Other experimental scenarios which use the Shark ML framework can be found in [1] where they are used to develop a system for the topological optimisation of mechanical structures, and in [2] where they are used in parameter tuning for various types of algorithms, such as those used by search engines. Other examples can be found in [9] where they were used to test a new type of hypervolume indicator, and in [22] where they were used to implement a way of updating covariance matrices.

**MOEA Framework.** The MOEA Framework is a free and Open Source Java-based framework, developed by Hadka and first released in 2011 [10]. Like jMetal, it also contains implementations of a large group of multi-objective optimization algorithms, standard test problems and performance indicators.

The framework is also licensed under the GNU Lesser General Public License. The current version is 2.11 which was released Aug 16 2016 and it is available on a public facing code repository [10].

MOEA incorporates some algorithm implementations from the jMetal library, although it provides its own implementation of NSGA-II.

The study described in [7] makes use of the MOEA framework in its experimental design. The authors wish to determine the effectiveness of using a multi-objective evolutionary algorithm approach to solve the problem of how best to manage a robot. Similar experimental scenarios which use implementations from the MOEA framework can be found in [3,26].

### 3.2  Hypervolume Indicator

The HV indicator (or $s$-metric) is a performance metric for indicating the quality of a non-dominated approximation set, introduced by [31] where it is described as the *"size of the space covered or size of dominated space"*. It can be defined as [27]:

$$HV\left(f^{ref}, X\right) = \Lambda \left( \bigcup_{X_n \in X} \left[ f_1(X_n), f_1^{ref} \right] \times \cdots \times \left[ f_m(X_n), f_m^{ref} \right] \right) \qquad (1)$$

where $HV\left(f^{ref}, X\right)$ resolves the size of the space covered by an approximation set $X$, $f^{ref} \in \mathbb{R}$ refers to a chosen reference point, and $\Lambda\left(.\right)$ refers to the Lebesgue measure [16]. This has been illustrated in Fig. 2 in two-dimensional objective space (to allow for easy visualisation) with a population of three solutions.

The hypervolume (HV) indicator is appealing because it is compatible with any number of problem objectives and requires no prior knowledge of the true Pareto-optimal front. It is currently used in the field of multi-objective optimisation as both a proximity and diversity performance metric and also in the decision making process [11].

Unlike dominance-based criteria which require only two solutions for performing a comparison (which can be used on an ordinal scale), a reference vector is required to calculate the HV indicator value (i.e. it requires the objective to be measured on an interval scale). When used for pairwise or multiple comparison of optimisation algorithms, this reference vector must be the same, otherwise the resulting HV indicator values are not comparable. The reference vectors used in these experiments are listed in Table 1.

## 4 Results

Average hypervolume values over 30 runs for each test problem are shown in Fig. 3. The hypervolume indicator results for the worst, mean, and best execution at generation 500 and generation 100 have been presented in Table 3.

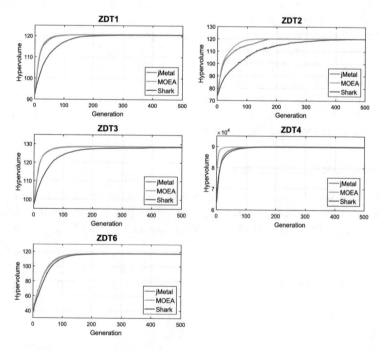

**Fig. 3.** Mean hypervolume indicator results for the ZDT test problems, 500 generations, for NSGA-II implementations taken from jMetal, MOEA Framework, and Shark ML.

The Wilcoxon signed-rank nonparametric test [28] has been employed to test for statistical significance, and the results of this are shown in Table 4. For each pair of implementations, the $p$ value is shown, and a symbol is used to show relative performance. A '+' indicates that the first implementation outperformed the second, a '−' indicates that the second outperformed the first, and a '=' indicates that there was no significant difference between the first and second.

## 5  Discussion

The MOEA Framework implementation outperforms the implementations from jMetal and Shark ML on 26 of the 30 results listed in Table 3, with the fastest overall rate of convergence. The greatest difference in performance can be seen early in the optimisation process, e.g. before 200 generations (Fig. 3). This suggests that the number of generations used in an experiment will significantly impact the integrity of the results, i.e. if an experiment is using Shark ML's implementation of NSGA-II as a benchmark, the comparison would not be the same as if they were to use jMetal's implementation of NSGA-II.

Hypothesis 1 is concerned with the final approximation sets produced by each of the implementations. Although the magnitude of the variance is smaller in the later generations, even in the final approximation set, 13 of the 15 pairwise comparisons rejected the null hypothesis. This shows a significant difference in the hypervolume quality of each implementation's final approximation set.

**Table 3.** Mean hypervolume results from 30 executions of three versions of NSGA-II on the ZDT test suite. The boldface values indicate superior performance

| Problem | jMetal | | | MOEA | | | Shark | | |
|---|---|---|---|---|---|---|---|---|---|
| | Worst | Mean | Best | Worst | Mean | Best | Worst | Mean | Best |
| *gen* = 100 | | | | | | | | | |
| ZDT1 | 118.8605 | 120.2570 | 120.4908 | **120.3037** | **120.4232** | **120.5114** | 114.7115 | 117.0696 | 118.6003 |
| ZDT2 | 109.9355 | 114.9227 | 119.7335 | **110.4201** | **119.1809** | **119.8377** | 99.3786 | 105.1081 | 113.5910 |
| ZDT3 | 125.1561 | 128.1568 | 128.5589 | **128.1541** | **128.4240** | **128.5620** | 121.2054 | 124.1729 | 126.9816 |
| ZDT4 | 89532.74 | 89712.37 | 89856.94 | **89538.86** | **89815.22** | **89999.52** | 88683.47 | 89403.65 | 89722.09 |
| ZDT6 | 111.8097 | **113.7393** | **114.7459** | 111.3142 | 112.9072 | 114.3293 | 108.3006 | 110.8531 | 112.8922 |
| Problem | jMetal | | | MOEA | | | Shark | | |
| | Worst | Mean | Best | Worst | Mean | Best | Worst | Mean | Best |
| *gen* = 500 | | | | | | | | | |
| ZDT1 | 120.5031 | 120.6304 | 120.6550 | **120.6595** | **120.6601** | **120.6608** | 120.6234 | 120.6510 | 120.6601 |
| ZDT2 | 119.3920 | 120.1841 | 120.3202 | **120.3260** | **120.3268** | **120.3275** | 119.3120 | 120.2408 | 120.3249 |
| ZDT3 | 128.2403 | 128.7063 | 128.7711 | **128.7745** | **128.7748** | **128.7751** | 125.5630 | 128.4439 | 128.7729 |
| ZDT4 | 89983.72 | 89996.25 | 89999.61 | 89619.57 | 89818.45 | **89999.66** | **89992.71** | **89998.54** | 89999.53 |
| ZDT6 | 116.7795 | 117.4678 | 117.4982 | **117.4874** | **117.4963** | **117.5044** | 117.4800 | 117.4900 | 117.5005 |

The results suggest that MOEA Framework's NSGA-II implementation offers the best performance, despite being configured with the same parameters and operators. It outperformed the other two implementations on all of the test problems except ZDT4, where it performed the worst.

**Table 4.** Results showing pairwise comparison of the mean hypervolume values using the Wilcoxon signed-ranks non-parametric test

| Problem | Generation 100 | | | | | | Generation 500 | | | | | |
|---|---|---|---|---|---|---|---|---|---|---|---|---|
| | jMetal/MOEA | | MOEA/Shark | | jMetal/Shark | | jMetal/MOEA | | MOEA/Shark | | jMetal/Shark | |
| | p-value | | p-value | | p-value | | p-value | | p-value | | p-value | |
| ZDT1 | 6.8714e−02 | = | 1.7344e−06 | + | 1.7344e−06 | + | 1.7344e−06 | − | 1.7344e−06 | + | 3.6094e−03 | − |
| ZDT2 | 1.2506e−04 | − | 1.7344e−06 | + | 2.6033e−06 | + | 1.7344e−06 | − | 1.7344e−06 | + | 4.0483e−01 | = |
| ZDT3 | 1.5886e−01 | = | 1.7344e−06 | + | 1.7344e−06 | + | 1.7344e−06 | − | 1.7344e−06 | + | 1.7138e−01 | = |
| ZDT4 | 8.9443e−04 | − | 1.9209e−06 | + | 4.2857e−06 | + | 1.9209e−06 | + | 1.9209e−06 | − | 1.1748e−02 | − |
| ZDT6 | 4.5336e−04 | + | 3.1817e−06 | + | 1.9209e−06 | + | 1.6566e−02 | − | 1.4773e−04 | + | 4.9498e−02 | − |

The Shark implementation was second best overall, given that it performed best on ZDT4, and second best on three of the test problems (ZDT1, ZDT2 and ZDT6), with only its performance on ZDT3 being the worst out of the three.

The jMetal implementation was the worst performer on three of the test problems, and came second on ZDT3 (behind MOEA) and ZDT4 (behind Shark).

This provides experimental confirmation of Hypothesis 1 - that the approximation sets produced by multiple different implementations of NSGA-II on the same test case will not always achieve significantly similar performance.

Hypothesis 2 is concerned with the rate of convergence of each implementation. Figure 3 clearly illustrates that, at generation 100 of the optimisation process, for all considered test functions, the Shark ML implementation of NSGA-II offers the worst performance, with regards to the mean hypervolume indicator quality achieved by each population. Table 3 lists results indicating that Shark ML also achieved the lowest hypervolume indicator quality for the worst performing and best performing populations for each problem.

The relative performance of the three implementations at generation 100 is also different when compared with generation 500. The MOEA implementation is still the best performer, outperforming the other implementations on all of the test problems except ZDT6, where it comes second after jMetal. The jMetal implementation is second best on all of the other test problems, which leaves Shark as the worst performer on all 5 test problems.

Table 4 shows that there is a statistically significant difference in 13 of the 15 pairwise comparisons in the generation 100 results. The two comparisons which do support the null hypothesis are between jMetal and MOEA, on the test problems ZDT1 and ZDT3.

It is clear from the results listed in Table 4, and from the graphs shown in Fig. 3 that the rate of convergence of the three implementations is different, in some cases markedly so. Shark has a lower rate of convergence in all of the test problems, the difference being most marked in ZDT2, but almost as large in ZDT1 and ZDT3. jMetal and MOEA converge at a similar rate in ZDT1, ZDT3 and ZDT6, but in ZDT4 the convergence rate of jMetal and Shark seem to be similar, but both are slower than MOEA.

However, although Shark has the slowest rate of convergence in all of the test problems, it actually results in the the best final approximation set in the case of ZDT1, and the second best in three of the other test problems.

Any discussion of which implementation is best will involve a tradeoff between the following factors - the speed at which the algorithm can converge towards an approximation set which is close enough to the Pareto front to be useful, and the desired quality of the final approximation set.

# 6    Conclusions

In this study, three implementations of the NSGA-II algorithm were configured with the same parameters and operators, and applied to the ZDT test suite in a three-way comparison. The hypervolume indicator was used as a performance metric to identify the worst, mean, and best performance of populations generated by each implementation for each test function. The implementations should not have offered significantly different performance, however, the null hypothesis was rejected in 26 of the 30 comparisons.

MOEA Framework's implementation of NSGA-II outperformed those from jMetal and Shark ML - this shows that, when comparing a newly proposed algorithm to two implementations of NSGA-II, it is possible for the algorithm to both outperform NSGA-II, and be outperformed by NSGA-II, simultaneously.

It seems increasingly important that researchers should publish source code for their algorithm, and specify which implementation (and version) of a benchmark algorithm is used in their comparisons. The difference in performance can emerge for many reasons, including the interpretation of the pseudo-code or algorithm listing, if the author does not publish their source code, and the features of a particular programming language e.g. the accuracy of floating point numbers.

Further work in this research direction would benefit the field of Evolutionary Computation, as there are many new algorithms proposed in the literature and the majority of them are compared against existing algorithms as a benchmark. It is also worth noting that differences in the implementation of test functions and performance metrics may also exist, which would also impact the integrity of performance comparison.

# References

1. Aulig, N., Olhofer, M.: Neuro-evolutionary topology optimization of structures by utilizing local state features. In: Proceedings of the 2014 Annual Conference on Genetic and Evolutionary Computation, GECCO 2014, pp. 967–974. ACM (2014)
2. Branke, J., Elomari, J.A.: Meta-optimization for parameter tuning with a flexible computing budget. In: Proceedings of the 14th Annual Conference on Genetic and Evolutionary Computation, GECCO 2012, pp. 1245–1252. ACM (2012)
3. Cocaña-Fernández, A., Sánchez, L., Ranilla, J.: Improving the eco-efficiency of high performance computing clusters using EECluster. Energies $9(3)$, 197 (2016)

4. Deb, K., Agrawal, S., Pratap, A., Meyarivan, T.: A fast elitist non-dominated sorting genetic algorithm for multi-objective optimisation: NSGA-II. In: Parallel Problem Solving from Nature - PPSN VI, 6th International Conference, Paris, France, 18-20 September 2000, Proceedings, pp. 849–858 (2000)
5. Deb, K., Goyal, M.: A combined genetic adaptive search (geneas) for engineering design. Comput. Sci. Inform. **26**, 30–45 (1996)
6. Deb, K., Pratap, A., Agarwal, S., Meyarivan, T.: A fast and elitist multiobjective genetic algorithm: NSGA-II. IEEE Trans. Evol. Comput. **6**(2), 182–197 (2002)
7. Desjardins, B., Falcon, R., Abielmona, R., Petriu, E.: A multi-objective optimization approach to reliable robot-assisted sensor relocation. In: 2015 IEEE Congress on Evolutionary Computation (CEC), pp. 956–964 (2015)
8. Durillo, J.J., Nebro, A.J.: jMetal: a java framework for multi-objective optimization. Adv. Eng. Softw. **42**(10), 760–771 (2011)
9. Friedrich, T., Bringmann, K., Voß, T., Igel, C.: The logarithmic hypervolume indicator. In: Proceedings of the 11th Workshop Proceedings on Foundations of Genetic Algorithms, FOGA 2011, pp. 81–92. ACM (2011)
10. Hadka, D.: MOEA - a free and open source java framework for multiobjective optimization (2015). https://github.com/MOEAFramework/MOEAFramework
11. Helbig, M., Engelbrecht, A.P.: Performance measures for dynamic multi-objective optimisation algorithms. Inf. Sci. **250**, 61–81 (2013)
12. Igel, C., Heidrich-Meisner, V., Glasmachers, T.: The shark machine learning library (2013). https://github.com/Shark-ML/Shark
13. Ishibuchi, H., Imada, R., Setoguchi, Y., Nojima, Y.: Performance comparison of NSGA-II and NSGA-III on various many-objective test problems. In: 2016 IEEE Congress on Evolutionary Computation (CEC), pp. 3045–3052 (2016)
14. Ishibuchi, H., Tsukamoto, N., Nojima, Y.: Evolutionary many-objective optimization: a short review. In: 2008 IEEE Congress on Evolutionary Computation (IEEE World Congress on Computational Intelligence), pp. 2419–2426 (2008)
15. Kumar, K.D.A.: Real-coded genetic algorithms with simulated binary crossover. Complex Syst. **9**, 431–454 (1995)
16. Lebesgue, H.: Intégrale, longueur, aire. Annali di matematica pura ed applicata **7**(1), 231–359 (1902)
17. Li, M., Yang, S., Liu, X.: A performance comparison indicator for pareto front approximations in many-objective optimization. In: Proceedings of the 2015 Annual Conference on Genetic and Evolutionary Computation, GECCO 2015, pp. 703–710. ACM (2015)
18. Michalewicz, Z., Hartley, S.J.: Genetic algorithms + data structures = evolution programs. Math. Intell. **18**(3), 71 (1996)
19. Nebro, A.: jMetal: a framework for multi-objective optimization with metaheuristics (2014). https://github.com/jMetal/jMetal
20. Purshouse, R.C.: On the evolutionary optimisation of many objectives. University of Sheffield, Sheffield (2003)
21. Rostami, S., Neri, F.: Covariance matrix adaptation Pareto archived evolution strategy with hypervolume-sorted adaptive grid algorithm. Integr. Comput. Aided Eng. **23**(4), 313 (2016)
22. Rostami, S., Shenfield, A.: A multi-tier adaptive grid algorithm for the evolutionary multi-objective optimisation of complex problems. Soft Comput. **21**, 1–17 (2016)
23. dos Santos Neto, P.d.A., Britto, R., Rabêlo, R.d.A.L., Cruz, J.J.d.A., Lira, W.A.L.: A hybrid approach to suggest software product line portfolios. Appl. Soft Comput. **49**, 1243–1255 (2016)

24. Soh, H., Demiris, Y.: Evolving policies for multi-reward partially observable markov decision processes (MR-POMDPs). In: Proceedings of the 13th Annual Conference on Genetic and Evolutionary Computation, GECCO 2011, pp. 713–720. ACM (2011)
25. Strickler, A., Prado Lima, J.A., Vergilio, S.R., Pozo, A.T.R.: Deriving products for variability test of feature models with a hyper-heuristic approach. Appl. Soft Comput. **49**, 1232–1242 (2016)
26. Svensson, M.K.: Using evolutionary multiobjective optimization algorithms to evolve lacing patterns for bicycle wheels. Master's thesis, NTNU-Trondheim (2015)
27. Voß, T., Hansen, N., Igel, C.: Improved step size adaptation for the MO-CMA-ES. In: Proceedings of the 12th Annual Conference on Genetic and Evolutionary Computation, GECCO 2010, pp. 487–494. ACM (2010)
28. Wilcoxon, F.: Individual comparisons by ranking methods. Biometrics Bull. **1**(6), 80–83 (1945)
29. Wright, A.H., et al.: Genetic algorithms for real parameter optimization. Found. Genet. Algorithms **1**, 205–218 (1991)
30. Zitzler, E., Deb, K., Thiele, L.: Comparison of multiobjective evolutionary algorithms: empirical results. Evol. Comput. **8**(2), 173–195 (2000)
31. Zitzler, E., Thiele, L.: An evolutionary algorithm for multiobjective optimization: The strength pareto approach. Citeseer, Swiss Federal Institute of Tech (1998)
32. Zitzler, E., Thiele, L., Laumanns, M., Fonseca, C.M., Fonseca, V.G.d.: Performance assessment of multiobjective optimizers: an analysis and review. IEEE Trans. Evol. Comput. **7**(2), 117–132 (2003)

# The Influence of Age Assignments on the Performance of Immune Algorithms

Alessandro Vitale[1], Antonino Di Stefano[1], Vincenzo Cutello[2], and Mario Pavone[2]($\boxtimes$)

[1] Department of Electric, Electronics and Computer Science, University of Catania, V.le A. Doria 6, 95125 Catania, Italy
[2] Department of Mathematics and Computer Science, University of Catania, V.le A. Doria 6, 95125 Catania, Italy
{cutello,mpavone}@dmi.unict.it

**Abstract.** How long a B cell remains, evolves and matures inside a population plays a crucial role on the capability for an immune algorithm to jump out from local optima, and find the global optimum. Assigning the right age to each clone (or offspring, in general) means to find the proper balancing between the exploration and exploitation. In this research work we present an experimental study conducted on an immune algorithm, based on the clonal selection principle, and performed on eleven different age assignments, with the main aim to verify if at least one, or two, of the top 4 in the previous efficiency ranking produced on the one-max problem, still appear among the top 4 in the new efficiency ranking obtained on a different complex problem. Thus, the NK landscape model has been considered as the test problem, which is a mathematical model formulated for the study of *tunably rugged fitness landscape*. From the many experiments performed is possible to assert that in the elitism variant of the immune algorithm, two of the best age assignments previously discovered, still continue to appear among the top 3 of the new rankings produced; whilst they become three in the no elitism version. Further, in the first variant none of the 4 top previous ones ranks ever in the first position, unlike on the no elitism variant, where the previous best one continues to appear in *1st* position more than the others. Finally, this study confirms that the idea to assign the same age of the parent to the cloned B cell is not a good strategy since it continues to be as the worst also in the new efficiency ranking.

## 1 Introduction

The Immune algorithms are an established and successful computational methodology, which take inspiration from the information processing mechanism of the living things, and from their dynamics. What makes the immune system really challenging from a computational perspective is its ability in recognize, distinguish, detect, and remember (memory) foreign entities to the living

© Springer Nature Switzerland AG 2019
A. Lotfi et al. (Eds.): UKCI 2018, AISC 840, pp. 16–28, 2019.
https://doi.org/10.1007/978-3-319-97982-3_2

organism, as well as its capability in learning, and self-regulation [6]. As in any evolutionary algorithms, also in the immune algorithms the success's key for having efficient performances is given by an appropriate, and proper balancing between the exploration and exploitation mechanisms: the first is carried out by the perturbation and recombination operators, whilst the last one is obtained through the selection process adopted. What makes, however, more efficient and accurate the search process, and at the same time helps to learn information as more as possible, is given by the right maturation time of each solution. Indeed, how long a solution stays, evolves, and matures inside a population is strictly related to a good balancing of the exploration/exploitation processes, and, therefore, plays a decisive role in the performances of any evolutionary algorithm.

In a previous work published in [5], a study on how much time an individual needs to stay into the population to properly explore the search space has been conducted. This study has been performed on an Immune Algorithm (IA), and the outcomes obtained have been summarized in an efficiency ranking of the several age assignment types considered. Further, this ranking was performed and produced on the one-max problem [4,12], one of the classical toy problems used for understanding the dynamics, variants and search's ability of any stochastic algorithm [2]. In light of this, in this paper we present the same experimental study but tackling a different problem, such as the *NK-model* [8] that is a mathematical model able to describe a *"tunably rugged"* fitness landscape, with the primary goal to verify if the top 4 in the previous ranking, or at least 2 of them, still appear among the top 4 in the new efficiency ranking produced on this model. If so, we give a reliable age's assignment that provides, with high probability, efficient and robust performances in discrete search spaces to can use especially in uncertainty environments. Thus, the same IA proposed in [1,11] was developed, whose core components are the cloning, hypermutation and aging operators, and eleven (11) age assignment types have been considered for being studied. The *Elitism* and *No Elitism* versions were investigated as well. Inspecting the new outcomes it is possible to assert that (a) in the elitism version, two of the top 4 of the previous ranking appear among the top 3 of the new ranking, but none of them rank in the first position; (b) in the no elitism variant, instead, three of the previous top 4 are still among the first three positions of the new efficiency ranking, and, further, the best of the previous ranking continues to be still the best one; finally, (c) the worst of the previous ranking still continues to be in the last position in the new one.

## 2   The NK Landscape Model

For validating and generalizing the outcomes obtained, it is needed to avoid that the algorithm is tailored to a specific problem, keeping it instead unaware on the knowledge of the application domain. In this way, the efficiency ranking produced will be suitable and applicable also on other problems. At this end, the NK-model was considered for validating the experiments performed, and the outcomes produced. This model was developed in [8,9] as a powerful analytic tool able to model and represent the effects of epistatic interactions in population genetic. The basic idea behind this model is that in complex systems with many components $(N)$, the functional contribution of each component is affected by the interaction of one, or more $(K)$ parts of the system. The multidimensional space in which each component is represented by a dimension of the space, to which is assigned a fitness level with respect to a specific property, constitutes the *fitness landscape*, whose topology depends on the interactions and on the degree of interdependence of the functional contribution of the various components. This interdependence degree influences strictly the smoothness or ruggedness of the fitness landscape: if a component is affected by a variety of other parts, then the landscape produced will be quite *rugged* [10]. The NK-model represents, a mathematical model that captures the central factors for an ensemble of tunable rugged fitness landscapes, via changes between overall size of the landscape and the number of its local *hills and valleys*. It is based on two parameters: (1) the number $N$ of components; and (2) a parameter $K$ that measures the richness of interactions among components. Increasing this last parameter means to increase the number of peaks and valleys, and thus raising the ruggedness of the corresponding fitness landscape: moving from single peaks and smooths, to multi-peaks and fully uncorrelated. Any component is assumed to be in a binary state, 0 or 1, and the fitness contribution of such a component depends on its own state, and the one of those $K$ other components explicitly linked to it.

Formally, given a bit string $s = (s_1, s_2, \ldots, s_N)$ of length $N$, it assigns a fitness contribution $\phi_i$ to each locus $s_i$ of the $N$ residue chain such that $\phi_i$ depends on $s_i$ and $K$ other bits $(0 \le K \le N-1)$. There are $2^{(k+1)}$ combinations of states of the $K+1$ loci that determine the fitness contribution of each locus. For each of this combination is randomly selected a variable in $[0, 1]$ as its fitness contribution. The total fitness of the string $s$ is, then, defined as the average of the fitness contributions of each part, and the ones of $K$ neighbors that affect upon it:

$$F_{NK}(s) = \frac{1}{N} \sum_{i=1}^{N} \phi_i(s_i, s_{i+1}, s_{i+2}, \ldots, s_{i+K}) \tag{1}$$

The goal in dealing with the NK-model is to *maximize* the equation above. Note that the tuning of the parameter $K$ alters how rugged the landscape should be, and then affords us a tunable rugged fitness landscape. Indeed, for $K = 0$ each site is independent of all other ones, and this generate the smoothest landscape; whilst, for $K = N-1$ the fitness contribution of each site depends on all of other ones, producing then most rugged landscapes, with very many local optima. Solving the NK-model was proved to be NP-complete problem [13]. This model, thanks to its properties, allows us to keep the study, and its validation within the discrete domain, as done in [5], and, in the same time, testing it on a different rugged fitness landscape level.

# 3   The Immune Algorithm

In order to achieve the prefixed purposes in this research work, we have faithfully developed the immune inspired algorithm proposed in [3,5], and whose main features are given by the three operators: (1) cloning, which generates a new population centered on the higher affinity values, (2) inversely proportional hypermutation, which explores the neighborhood of each point in the search space, and (3) aging, which introduces diversity in the population and avoid to get trapped into local optima. This algorithm is based on four main parameters (all user-defined), such as population size ($d$);number of clones to be produced ($dup$); mutation rate ($\rho$); and maximum number of generations allowed to a B cell to stay into the population ($\tau_B$). Its pseudo-code is showed in Algorithm 1.

---

**Algorithm 1.** The Immune Algorithm $(d, dup, \rho, \tau_B)$

---

$t \leftarrow 0$;
$P^{(t)} \leftarrow$ Initialize_Population($d$);
Evaluate_Fitness($P^{(t)}$);
**repeat**
    Increase_Age($P^{(t)}$);
    $P^{(clo)} \leftarrow$ Cloning $(P^{(t)}, dup)$;
    $P^{(hyp)} \leftarrow$ Hypermutation($P^{(clo)}, \rho$);
    Evaluate_Fitness($P^{(hyp)}$);
    $(P_a^{(t)}, P_a^{(hyp)}) \leftarrow$ Aging($P^{(t)}, P^{(hyp)}, \tau_B$);
    $P^{(t+1)} \leftarrow (\mu + \lambda)$-Selection($P_a^{(t)}, P_a^{(hyp)}$);
    $t \leftarrow t + 1$;
**until** (stop criterion is satisfied)

---

The immune algorithm begins with the inizialization of the population $P^{(t=0)}$ of size $d$ by generating random solutions in the binary domain ($\{0, 1\}$); after that, for each B cell ($x \in P^{(t)}$), i.e. a candidate solution, the fitness value is computed via the Evaluate_Fitness($P^{(t)}$) procedure (lines 3 and 8 in Algorithm 1). Inside the evolutionary process the three main operators are performed together to a selection operator, which attempts to exploit as better as possible the information gained during the evolution. The main loop (i.e. the evolution) terminates once a fixed stop criterion is satisfied, that is when it is reached the maximum number of fitness function evaluations allowed ($T_{max}$).

The first immune operator applied is the *cloning operator*, which simply copies $dup$ times each solution (i.e. each B cell) producing an intermediate population $P^{(clo)}$ of size ($d \times dup$). Once a clone is produced to this is assigned an age that determines its lifetime into the population. Indeed, based on this assignment, the B cell will live until its age will be greater than $\tau_B$. For doing so, at the beginning of each time step $t$, the age of all B cells in the population is increased by one (line 5 in Algorithm 1). The assignment of the age to each

B cell, together with the aging operator, has the purpose to reduce premature convergences, and keep a right diversity into the population. Then, what age to assign to each clone (or individual) plays a crucial role on the performances of any evolutionary algorithm. Therefore, in order to analyse and confirm if the previous efficiency ranking produced is still valid for this new model, even if in different order, we have conducted the same age assignment study proposed in [5], evaluating the overall performances in terms of performance, convergence and success. Eleven different age assignment types for each clone has been considered in this study:

1. (type0) age zero (0);
2. (type1) random age chosen in the range $[0, \tau_B]$;
3. (type2) random age chosen in the range $[0, \frac{2}{3}\tau_B]$. This option guarantees to each B cell to evolve at least for a minimal number of generations (in the worst case $\frac{1}{3}\tau_B$);
4. (type3) random age chosen in the range $[0, inherited]$, where with *inherited* we indicate the same age of the parent. With this option, in the worst case the clone will have the same age of its parent;
5. (type4) random age chosen in the range $[0, \frac{2}{3} inherited]$. In this way for each offspring is guaranteed a lower age than the parent;
6. to each clone is assigned the same age of the parent (inherited), but if after $M$ mutations performed on the clone its fitness value improves, then its age is updated as follows:
   (a) (type5) zero;
   (b) (type6) randomly chosen in the range $[0, \tau_B]$;
   (c) (type7) randomly chosen in the range $[0, \frac{2}{3}\tau_B]$;
   (d) (type8) randomly chosen in the range $[0, inherited]$;
   (e) (type9) randomly chosen in the range $[0, \frac{2}{3} inherited]$;
7. (type10) same age of parent less one ($inherited - 1$).

We would to highlight, and stress once again, that what we want to prove with this research study is not to get the same and identical efficiency ranking previously produced, but rather verify if the previous best 4, or just some of these, in overall still appear among the top positions in the new ranking. In this way, we may provide one or more age assignments to be considered, for having reliable and efficient performances, especially when it is tackled uncertainty environments.

The *hypermutation operator* acts on each solution of population $P^{(clo)}$ performing $M$ mutations, whose number is determined by an *inversely proportional law*, that is the higher is the fitness function value, the lower is the number of mutations performed on the B cell. Interestingly that this perturbation operator is not based on any mutation probability. In particular, the number of mutations $M$ to perform over a clone $\boldsymbol{x}$ is determined by $\alpha = e^{-\rho \hat{f}(\boldsymbol{x})}$, where $\alpha$ represents the mutation rate, and $\hat{f}(\boldsymbol{x})$ the fitness function value normalized in $[0, 1]$. The number M of mutations is then determined by $M = \lfloor (\alpha \times \ell) + 1 \rfloor$, with $\ell$ the length of the B cell. In this way, at least one mutation is guaranteed on each B cell; and this happens exactly when the solution is very close to the optimal one. For each clone, the hypermutation operator randomly choose a bit, and it inverts its value (from 0 to 1, or viceversa), and this is repeated without redundancy for M times. At the end, all hypermutated clones produce a new population, labelled $P^{(hyp)}$. During the normalization of the fitness function into the range $[0, 1]$, the best current fitness is decreased by a threshold $\theta$, and it is used in place of the global optima, because usually it is often unknown. In this way, no *a priori* knowledge about the problem is used.

*Aging operator* acts as last immune operator with the main goal to produce enough diversity into the population in order to avoid premature convergences and then getting trapped into local optima. It eliminates the old B cells from the populations $P^{(t)}$ and $P^{(hyp)}$: every B cell is allowed to stay in the population for a fixed number of generations $\tau_B$; then as soon as a B cell becomes older than $\tau_B$ it is removed from the population of belonging independently from its fitness value. There exists a variant of this operator, called elitism version, which makes an exception on the best solution found so far: it is always kept in the population, even it is older than $\tau_B$. In this experimental study, both variants of the aging operator have been taken into account: elitism and no elitism. An analysis on the benefits and efficiency of the aging operator can be found in [7].

After the three immune operators have been performed, a new population $P^{(t+1)}$ for the next iteration is produced by the merging of the best B cells via the $(\mu + \lambda)$-*Selection operator*, which, then, selects the best $d$ survivors to the aging step from the populations $P_a^{(t)}$ and $P_a^{(hyp)}$. Such operator, with $\mu = d$ and $\lambda = (d \times dup)$, reduces the offspring B cell population of size $\lambda \geq \mu$ to a new parent population of size $\mu = d$. The selection operator identifies the $d$ best elements from the offspring set and the old parent B cells, thus guaranteeing monotonicity in the evolution dynamics. Nevertheless, due to the aging operator, it could happen that only $d_1 < d$ B cells survived; in this case, the selection operator randomly generates new $d - d_1$ B cells.

## 4    Results and Discussion

In this section all outcomes obtained in the experimental study conducted are presented, whose main aim is to investigate if the best age assignments previously produced are still valid in solving a new model, working always in discrete domain (bit strings), but facing a more rugged fitness landscape. For these experiments we have considered the NK-model that is a mathematical model based on two parameters ($N$ and $K$), from which is possible to produce tunable rugged landscapes. In particular, tuning the parameter $K$ allows to alter the roughness of the landscape: from the smoothest produced for $K = 0$, to the roughest one with many local optima for $K = N - 1$. For our test case we have considered $N = 100$, and $K = \frac{N}{2} = 50$. In this way, a more rugged fitness landscape is produced with many local optima, which makes surely harder the search process for the global optima but, on the other hand, allow us to better validate the outcomes obtained. In Fig. 1 is showed the fitness landscape produced for $N = 12$, and $K = 6 = \frac{N}{2}$. It was considered $N = 12$ due to the high number of the all binary strings ($2^N$) that affect the given landscape. It gives anyway an idea about the hardness of the landscape of the problem to be solved.

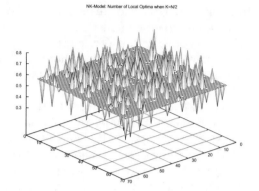

**Fig. 1.** Rugged fitness landscape produced by setting $N = 12$ and $K = N/2$.

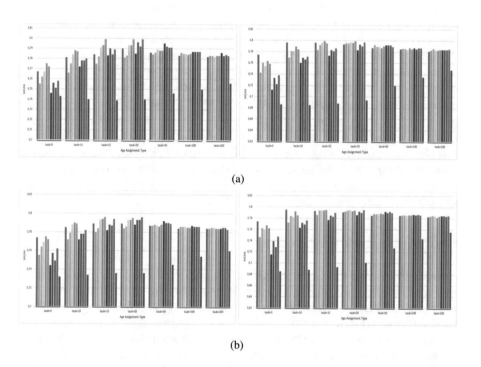

**Fig. 2.** Mean of the best solutions versus age types. Results are obtained by both immune algorithm versions: elitism (upper plots) and without elitism (bottom plots). Plot (a) shows the results obtained by setting $d = 50$ and $dup = 2$; whilst the ones of plot (b) using the parameters $d = 100$ and $dup = 2$.

In order to validate all results, and produce the new efficiency ranking, we used the same experimental protocol proposed in [5], that is: $d = \{50, 100\}$; $dup = \{2, 5, 10\}$; $\tau_B = \{5, 10, 15, 20, 50, 100, 200\}$; $T_{max} = 10^5$; and each experiment was performed on 100 independent runs. Because the parameter $\rho$ is related only to the problem dimension $(N)$, after several tests, it was fixed to 5.9 for all experiments. Further, all experiments were performed for both versions of IA: *elitism*, and *no elitism* variants. Since the optimal solution is unknown, we have considered as evaluation measures, in order, respectively: (i) mean of the best solutions found on 100 independent runs; (ii) best solution found; and (iii) standard deviation $(\sigma)$. Of course, seeing the high number of experiments performed, in this section we report figures and tables of the more meaningful results. Note that each plot in Figs. 2 and 3, and for each group of columns inside them (i.e. $\tau_B$ values), shows the age assignments in increasing order, as suggested in Sect. 3, from left to right (type0 leftmost, and type10 the rightmost one).

Figure 2, plots (a) and (b), shows the results produced by the developed immune algorithm on the 11 age assignment test cases. Each plot reports the *mean* of the best solutions found on the 100 independent runs by varying the $\tau_B$ parameter, and for both variants of the algorithm: with elitism (upper plots) and without elitism (bottom plots). In particular, the plot (a) is obtained by setting $d = 50$ and $dup = 2$; whilst plot (b) using the parameters $d = 100$ and $dup = 2$. Inspecting plot (a) of Fig. 2, is possible to see how at least 2 of the previous best

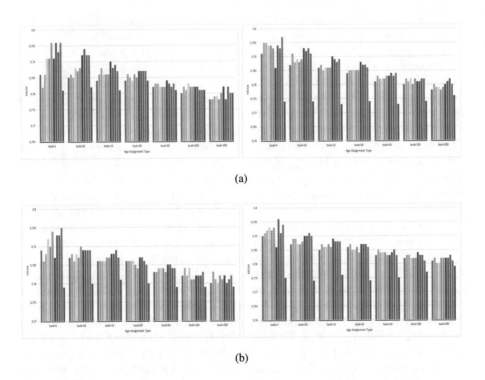

(a)

(b)

**Fig. 3.** Mean of the best solutions versus age types. Results are obtained by both immune algorithm versions: elitism (upper plots) and without elitism (bottom plots). Plot (a) shows the results obtained by setting $d = 50$ and $dup = 10$; whilst the ones of plot (b) using the parameters $d = 100$ and $dup = 10$.

4 age assignments (type0, type4, type3 and type2, in the order) often appear among the first positions, although not always as the best ones. This appears clearer if we analyse the no elitism version (bottom plot). In particular, at low values of $\tau_B$ all 4 best previous ones appear among the new 4–5 top positions, and sometimes also in the same order of the previous ranking. It is interesting to note from both plots (a) how for low values of $\tau_B$ the performances at various age types are feelingly different; whilst at the increasing of $\tau_B$ the performances are roughly equivalents. In both variants, the overall best performances, in terms of mean of best solutions found, are obtained with $\tau_B = 20$. What instead is clear in both plots, and confirms what asserted in the previous efficiency ranking, is that the last age assignment type proposed (type10) still continues to be the worst. Same analysis may be done also for the plot (b) in Fig. 2. In addition to the previous analysis is possible to assert for both versions that all previous 4 top age assignment types appear always among the 4–5 best (reflecting the same order in some way), except for $\tau_B = 50$. Also in these experiments, the best performances in the overall are obtained by setting $\tau_B = 20$, besides, the last age assignment option (type10), that is to assign the same age of parent minus 1, continues to be a bad choice.

Plots (a) and (b) in Fig. 3 report, instead, the performances of both variants of the immune algorithm on all age assignment test cases when $dup = 10$, and by varying the population size. The upper plots show the results obtained with the elitism version, whilst on the bottom ones the variant without elitism. Analysing the plots (a), it is possible to see for these experiments none of the top 4 previous ones appears in steady way among the new best positions, and when they appear they are in sparse order. This is due to the high number of clones (10 for each B cell) that require a greater diversity into the population, which is better produced by the age assignment types 5 to 9. This statement is also confirmed by having better performances at low $\tau_B$ values. However, if we consider only the rank, and possible equivalent ranks, then at least one of the previous best 4 is almost always among the 4–5 best in the new ranking. The plots (b) show a similar behavior to previous ones, but at the increasing of $\tau_B$ values it becomes more pronounced the presence of the previous top age types among the new best ones. Also in these experiments, the type10 appears as the worst age assignment.

Table 1 reports the efficiency ranking produced for each pair of values $(d, dup)$. In boldface is highlighted when one of the previously best 4 [5] is ranked in the top 4 positions of the new ranking. Looking to the elitism version for $d = 100$, we find type3 and type 4 among the first 4 ranked for $dup = \{2, 5\}$, whilst only the type3 to $4th$ place for $dup = 10$. A different efficiency ranking is, instead, produced for $d = 50$, where for $dup = 2$ the type3 and type4 are ranked in $2nd$ and $3rd$ position respectively; whilst, for the last $dup$ value, none of the previously top 4 achieve the first positions. This different behavior may be explained because by setting the population size to 50 the algorithm shows a stronger form of elitism that limits the effect of the age assignments. This means, in a nutshell, that the age type chosen alone is not able to produce a right diversity into the population. Inspecting the no elitism variant is clear how,

**Table 1.** Efficiency ranking produced by each pairs $(d, dup)$.

| Type | $d = 50$ | | | $d = 100$ | | |
|---|---|---|---|---|---|---|
| | $dup = 2$ | $dup = 5$ | $dup = 10$ | $dup = 2$ | $dup = 5$ | $dup = 10$ |
| | Elitism ranking | | | | | |
| 0 | 6 | 8 | 10 | 5 | 8 | 9 |
| 1 | 10 | 10 | 9 | 10 | 10 | 7 |
| 2 | 8 | 9 | 8 | 7 | 9 | 9 |
| 3 | 3 | 5 | 6 | 3 | 3 | 4 |
| 4 | 2 | 6 | 7 | 2 | 4 | 8 |
| 5 | 1 | 1 | 5 | 1 | 1 | 4 |
| 6 | 9 | 7 | 2 | 9 | 7 | 4 |
| 7 | 5 | 3 | 1 | 6 | 4 | 2 |
| 8 | 7 | 4 | 3 | 8 | 6 | 3 |
| 9 | 4 | 2 | 3 | 4 | 2 | 1 |
| 10 | 11 | 11 | 11 | 11 | 11 | 11 |
| | No elitism ranking | | | | | |
| 0 | 1 | 7 | 10 | 1 | 4 | 10 |
| 1 | 7 | 6 | 4 | 7 | 8 | 3 |
| 2 | 4 | 3 | 8 | 4 | 3 | 7 |
| 3 | 5 | 4 | 6 | 5 | 6 | 9 |
| 4 | 2 | 2 | 9 | 2 | 1 | 6 |
| 5 | 3 | 1 | 6 | 3 | 2 | 8 |
| 6 | 10 | 10 | 5 | 10 | 10 | 5 |
| 7 | 8 | 8 | 3 | 8 | 7 | 1 |
| 8 | 9 | 9 | 2 | 9 | 9 | 2 |
| 9 | 6 | 5 | 1 | 6 | 5 | 4 |
| 10 | 11 | 11 | 11 | 11 | 11 | 11 |

for $dup = \{2, 5\}$, three over four of the previous ranking appear still among the top 4, as type0, type2, and type4. In particular, for small $dup$ values type0 is always the best option to choose; type4 is always the second best; whilst type3 fluctuates between $3rd$ and $4th$ position. If, instead, the number of clones is increased ($dup = 10$), then, an opposite ranking to the one showed in [5] is produced, where the previous age types at the bottom of the ranking are now in the best ranks. This is because they are able to produce more diversity than the others, which is a key point when working with a high number of solutions. From these single efficiency rankings, what emerges clearly is that in any ranking the type10 is always the worst. This confirms once again the outcome obtained in the previous work.

After analysing the single experiments, and individual efficiency rankings, the goal of this research work is to understand which age assignment types show best performances in the overall (i.e. independently of the parameter values used), and if among these appears one or more of the top 4 previous ones. At this end, in Table 2 it is reported the summarized and overall outcomes. Thus, once computed the single rankings for each set value of $d$ and $dup$, in Table 2 we summarize everything, reporting the number of times that each age assignment appears among the top $k$ ($\varphi_k$) over 6 overall experiments (by varying $d = 50, 100$ and $dup = 2, 5, 10$), and the statistic success rate in using that age assignment type. This last was computed only with respect to the top 4, which means that, inspecting all experiments, a given age type appears $SR_{top4}$ times among the best 4. Obviously, the higher the percentage is, more robust and efficient are the performances using the relative age assignment type. Further, always in Table 2, $\varphi_5$ indicates how many times the given age assignment appears among the top 5; $\varphi_4$ among top 4; and so on. From this table, analysing the elitism version, is possible to note how the best age assignment in the previous ranking (type0) appears only one time among the top 5, and never among the first 4; whilst the types 3 and 4, third and second position of the previous ranking, appear in these new experiments among the best 3, respectively 3 and 2 times. Need to be highlight as type3 is among top 3 for 3 times over 6, but never appear in the first two positions, unlike of type4 that it ranks in the first two positions 2 times over 6 experiments. Focusing only on the top, none of the previous best 4 is able to outperform the other age assignment types. In particular, type 5 is the age type that shows the best performances, winning 4 times over 6. In general, it is possible to assert that the type5, type7 and type9, the only ones to appear as top, represent an assignment type that introduce diversity into the population, but in the same time leaves enough time for the maturation of all promising solutions. Inspecting the no elitism version, we achieve a different behavior of the performances with respect the other variant (of course, as we expected), where, instead, the previous best one still continues to be the top (2 over 6) also in this case study, together with the type4 (the previous 2nd) that appears in the first rank 1 times. Looking to the first 4 ranks, is possible to see as 3 of the previous best are still among the top 4 in the new ranking (type4, type2 and type0), together to the assignment type5. It is worth noting though that although type0 is not the best among the top 4, neither among the top 2, it is instead the one that appear in the first position more times than the others.

Finally, keeping in mind the goals of this research work, and following this conducted analysis, it is possible to conclude that, for the elitism variant of the developed immune algorithm, 2 of the previous top 4 (type3 and type4) still appear among the top 4 in the new ranking, and, furthermore, the type4 is again among the best 2, placing in the first two positions 2 times over 6. For the no elitism variant, instead, three of the previous top 4 (type4, type2 and type0) continue to appear in the new top 3 (of course in different order); but if we inspect only the top, we may state that the best previous (type0) continues to be it also in this new case study. What emerges clearly, and in common between the two variants of the immune algorithm, are the bad performances produced by the age assignment type10, as highlighted also in the previous efficiency ranking.

**Table 2.** Summary of the overall outcomes, where $\varphi_k$ reports the number of times that each age assignment appears among the top $k$, and $SR_{top4}$ the statistic success rate that indicates the percentage of how many times the relative age type appears among the best 4.

| Type | Elitism | | | | | | No elitism | | | | | |
|---|---|---|---|---|---|---|---|---|---|---|---|---|
| | $\varphi_5$ | $\varphi_4$ | $\varphi_3$ | $\varphi_2$ | $\varphi_1$ | $SR_{top4}$ | $\varphi_5$ | $\varphi_4$ | $\varphi_3$ | $\varphi_2$ | $\varphi_1$ | $SR_{top4}$ |
| 0 | 1 | 0 | 0 | 0 | 0 | 0% | 3 | 3 | 2 | 2 | 2 | 50% |
| 1 | 0 | 0 | 0 | 0 | 0 | 0% | 2 | 2 | 1 | 0 | 0 | 33.33% |
| 2 | 0 | 0 | 0 | 0 | 0 | 0% | 4 | 4 | 2 | 0 | 0 | 66.67% |
| 3 | 5 | 4 | 3 | 0 | 0 | 66.67% | 3 | 1 | 0 | 0 | 0 | 16.67% |
| 4 | 3 | 3 | 2 | 2 | 0 | 50.00% | 4 | 4 | 4 | 4 | 1 | 66.67% |
| 5 | 6 | 5 | 4 | 4 | 4 | 83.33% | 4 | 4 | 4 | 2 | 1 | 66.67% |
| 6 | 2 | 2 | 1 | 1 | 0 | 33.33% | 2 | 0 | 0 | 0 | 0 | 0% |
| 7 | 5 | 4 | 3 | 2 | 1 | 66.67% | 2 | 2 | 2 | 1 | 1 | 33.33% |
| 8 | 3 | 3 | 2 | 0 | 0 | 50% | 2 | 2 | 2 | 2 | 0 | 33.33% |
| 9 | 6 | 6 | 4 | 3 | 1 | 100% | 4 | 2 | 1 | 1 | 1 | 33.33% |
| 10 | 0 | 0 | 0 | 0 | 0 | 0% | 0 | 0 | 0 | 0 | 0 | 0% |

## 5   Conclusions

Starting from the well-know assertion that a right balancing between exploration and exploitation is a successful key for any evolutionary algorithm, in this research work we present an experimental study focused on understanding the right maturation of each solution for having a careful search process, and good information learning. Since this study follows a previous one, our main goal is investigate if the best age assignments previously obtained are still validate when is tackled a new and different problem.

An Immune Algorithm (IA) was developed for tackling and solving the NK-model, which is a mathematical model able to capture the fitness interactions producing a tunable rugged fitness landscape. Using this model we are keeping the study on discrete search space (binary strings), as done in the previous study, but with more roughness of the fitness landscape. In order to lead a proper study the tunable parameter $(K)$, with whom is altered the roughness of the search landscape, was set to 50% of the length of the given bit string $(N)$, producing in this way a hard landscape with many local optima.

Eleven (11) age assignment test cases have been studied and analysed with the aim to understand which one shows the best performances; produce an efficiency ranking; but, primarily, investigate if one or more best age types in the previous ranking are still among the top ranks in the new one. An overall of 924 experiments have been performed, and from their analysis we may assert that: (i) for the elitism version, two of the best previous ones (type3 and type4) still appear among the top 3, with, moreover type4 ranked in 2*nd* position 2 times over 6 in the new efficiency ranking; (ii) always for elitism version, looking only the top, any of the best previous ones reach never the 1*st* position, and this is likely due to the lower diversity they produce with respect the other age assignment types; (iii) for the no elitism variant, instead, three of the previous top 4 (type4, type2 and type0) still appear among the best 3 in the new efficiency ranking; (iv) for this last variant, the best of the previous ranking (type0), although it is never the best among the top 3, it is the one that however appears in 1*st* position more than the other age assignment types studied; and, finally, (v) type10 continues to be the worst option as already proved in the previous study.

# References

1. Conca, P., Stracquadanio, G., Greco, O., Cutello, V., Pavone, M., Nicosia, G.: Packing equal disks in a unit square: an immunological optimization approach. In: 1st International Workshop on Artificial Immune Systems (AIS), pp. 1–5. IEEE Press (2015)
2. Cutello, V., De Michele, A.G., Pavone, M.: Escaping local optima via parallelization and migration. In: VI International Workshop on Nature Inspired Cooperative Strategies for Optimization (NICSO). Studies in Computational Intelligence, vol. 512, pp. 141–152 (2013)
3. Cutello, V., Morelli, G., Nicosia, G., Pavone, M., Scollo, G.: On discrete models and immunological algorithms for protein structure prediction. Natural Comput. **10**(1), 91–102 (2011)
4. Cutello, V., Narzisi, G., Nicosia, G., Pavone, M.: Clonal selection algorithms: a comparative case study using effective mutation potentials. In: 4th International Conference on Artificial Immune Systems (ICARIS). LNCS, vol. 3627, pp. 13–28 (2005)
5. Di Stefano, A., Vitale, A., Cutello, V., Pavone, M.: How long should offspring lifespan be in order to obtain a proper exploration? In: 2016 IEEE Symposium Series on Computational Intelligence (SSCI), pp. 1–8, INSPEC number 16670548 (2016)
6. Fouladvand, S., Osareh, A., Shadgar, B., Pavone, M., Sharafi, S.: DENSA: an effective negative selection algorithm with flexible boundaries for selfspace and dynamic number of detectors. Eng. Appl. Artif. Intell. **62**, 359–372 (2017)
7. Jansen, T., Zarges, C.: On benefits and drawbacks og aging strategies for randomized search heuristics. Theor. Comput. Sci. **412**, 543–559 (2011)
8. Kauffman, S.A., Levin, S.: Towards a genarl theory of adaptive walks on rugged landscapes. J. Theor. Biol. **128**(1), 11–45 (1987)
9. Kauffman, S.A., Weinberger, E.D.: The NK model of rugged fitness landscapes and its application to maturation of the immune response. J. Theor. Biol. **141**(2), 211–245 (1989)
10. Levinthal, D.A.: Adaptation on rugged landscapes. Manage. Sci. **43**(7), 934–950 (1997)
11. Pavone, M., Narzisi, G., Nicosia, G.: Clonal selection - an immunological algorithm for global optimization over continuous spaces. J. Global Optim. **53**(4), 769–808 (2012)
12. Schaffer, J.D., Eshelman, L.J.: On crossover as an evolutionary viable strategy. In: 4th International Conference on Genetic Algorithms, pp. 61–68 (1991)
13. Weinberger, E.: NP-completeness of Kauffman's NK model, a tuneably rugged fitness landscape. Santa Fe Institute Working Paper, 96-02-003 (1996)

# Evolutionary Constraint in Artificial Gene Regulatory Networks

Alexander P. Turner$^{(\boxtimes)}$ , George Lacey , Annika Schoene ,
and Nina Dethlefs

Hull University, Cottingham Road, Hull HU6 7RX, UK
`alexander.turner@hull.ac.uk`

**Abstract.** Evolutionary processes such as convergent evolution and rapid adaptation suggest that there are constraints on how organisms evolve. Without constraint, such processes would most likely not be possible in the time frame in which they are seen. This paper investigates how artificial gene regulatory networks (GRNs), a connectionist architecture designed for computational problem solving may too be constrained in its evolutionary pathway. To understand this further, GRNs are applied to two different computational tasks and the way their underlying genes evolve over time is observed. From this, rules about how often genes are evolved and how this correlates with their connectivity within the GRN are deduced. By generating and applying these rules, we can build an understanding of how GRNs are constrained in their evolutionary path, and build measures to exploit this to improve evolutionary performance and speed.

**Keywords:** Artificial gene regulatory networks
Evolutionary dynamics · Computational optimisation

## 1 Introduction

For a species to adapt over time to a specific environment, it must be able to incorporate meaningful change into its genome whilst selecting against maladaptive outcomes. There are certain regions within the DNA of organisms which code for functions which are essential to the survival of the organism, and are under strong purifying selection. Processes such as redundancy mean that essential processes within an organism are often coded by multiple genes [2,19]. This reduces the likelihood of that function being removed due to a specific mutation. Conversely, there are sections of the DNA which are unlikely to provide any change to the organism regardless of if they are mutated, and such regions are often found to be void of modularity and redundancy [18]. In between these two examples is an evolutionary sweet spot, where mutations can cause meaningful changes to an organism without the risk of lethality. Hence, it is likely that there are constraints onto how evolution can progress and the direction it will take based upon an organisms genetic structure, and which areas allow for

© Springer Nature Switzerland AG 2019
A. Lotfi et al. (Eds.): UKCI 2018, AISC 840, pp. 29–40, 2019.
https://doi.org/10.1007/978-3-319-97982-3_3

useful variability [22]. If such constraints can be known, it may be possible to inform predictions about how a species will evolve, which is the theory behind evolutionary constraint.

Bio-inspired computational techniques have been designed to mimic biological systems *insilico* for decades [6,7,9,11,12,14]. There have been two main reasons for doing this. Firstly, and most predominately is to develop computational systems which are intelligent and are more able to behave like biological systems. The front runner in this field is artificial neural networks [9]. The second is to build computational models to improve the understanding of biology. Previously, conclusions drawn about biological processes were typically derived from comparatively expensive wet-lab techniques [1,3,5]. Computational methods have helped to improve the time frame in which certain biological experiments can be conducted, whilst alleviating a range of ethical concerns [15]. Some models are able to both solve computational problems, whilst being able to provide evidence to support biological theory [13].

In this work we consider how an artificial gene regulatory network (GRN), a type of bio-inspired computational model which is based upon gene regulation can be used to quantify constraint during evolution based upon its topology [4]. This will be done by analysing the connectivity of the underlying genes within the GRN over its evolutionary path to ascertain if certain levels of connectivity between the genes are more likely to provide beneficial mutations. Based upon these findings, this work will then create a targeted evolutionary approach which focuses on mutating genes which are more likely to cause useful meaningful change without loss of functionality within the GRN. There are two main reasons for using a GRN in this capacity. Firstly, genes can evolve a wide range of behaviors, allowing the emergent GRN to possess a wide range of functionality, akin to genetic networks in biology [14,17]. Secondly, the GRN is particularly robust to structural perturbations [14]. This allows for variation within the networks in a similar vein to a biological organism, a key property in evolvable systems [16,17]. The GRN will be applied to two tasks, the inverted pendulum task and a pattern generating task, both of which are modified to run indefinitely to promote evolution over long time-spans. If a task is solved, the task will randomly select new parameters and the network will have to adapt to this new task again. In order for the GRN to be applied to a task, a simple evolutionary algorithm will be used, which allows for the evolution of the GRN to be closely monitored throughout experimentation.

## 2   Biological Evolution

Evolution can be considered the heritable change to a species over generations. The backbone of evolution is DNA, which holds the code responsible for describing the primary structure of proteins in an organism, as well as certain regulatory information. DNA can be organised into genes, which specify a unit of hereditary information. The process of evolution is necessary to allow species to adapt over time, allowing organisms to thrive in varying environments. It is an evolutionary advantage to be able to allow for positive genetic variation whilst reducing lethality due to such variation.

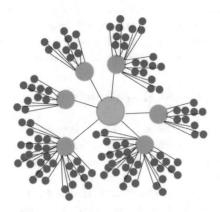

**Fig. 1.** An illustration of a hub, abstracted from [20] which shows how genes are interacting. If we are to look at this hub from a connectivity point of view, the blue center node is connected to all the green nodes, and indirectly to all the purple nodes. Hence, if the blue node were to mutate, the entire hub could fail to function. If for example one of the purple nodes was mutated, it would be unlikely to cease the functionality of the hub or cause a significant change. The green nodes sit in between the blue and purple ones, a mutation would mostly likely not cause the hub to fail, but would probably cause significant change in its functionality.

The ability for a species to be able to accept change yet maintain its function can be described as robustness [17], of which there are three key concepts. The first is modularity, which is the process of keeping functioning genes and genetic behaviors within a module meaning that the failure of a single module can be prevented from causing the failure of an entire organism. The second is redundancy, which as mentioned earlier specifies that a functioning unit is encoded by multiple genes. This means that more important functioning units are not coded for in exactly one place, and multiple mutations would be required to cause negative effects. The third is decoupling, and is the idea that the phenotype of an organism is the product of an indirect representation of that organism or functionality [17]. This provides separation from the function or an organism and low level genetic encoding, providing robustness and yet the opportunity to maintain variation in the species.

## 2.1 Constraint

Evolutionary constraint is a hypothesis which suggests that the pathway of evolution is most likely to occur in certain directions [23,24]. Indeed, such processes as convergent evolution and rapid adaptation are difficult to explain without evolutionary constraint. Take Fig. 1 as an abstracted example of the evolution of a hub. If the center node is mutated, it will likely have a large effect on the hub, which although could be positive, could also be negative. If one of the outside nodes is mutated, it is much less likely to cause an effect on the hub as its connectivity is much lower. Evolutionary constraint proposes that there is a sweet spot between these two examples, such as the green nodes, which can produce meaningful change without risking non-functionality. The sweet spot guides the path of evolution allowing for processes such as convergent evolution [22–24].

# 3   The Artificial Gene Regulatory Network

The artificial gene regulatory network (GRN) is a computational model designed for problem solving which takes inspiration from gene regulation in nature. The GRN consists of a set of genes (nodes), each containing a parameterised regulatory function, typically a variable sigmoid function. Each gene has an expression level, which can either be used to interact with a task, or to update the expression of other genes. In this work, the expression level of a gene is calculated using the sigmoid function in Eq. (1), where $s$ (sigmoid slope) $\in [0, 20]$, $b$ (sigmoid offset) $\in [-1, 1]$. Each gene in the network has a number of connections and a weight, which are used as inputs to update a given genes' expression levels. This can be seen in Eq. 2, which specifies that a genes expression level is the weighted sum of the expression levels of the genes in which it is connected, and then this value is passed through that genes' regulatory function and this is the expression level for that gene (Eq. 1). The GRN is the emergent property of the behavior of its underlying genes.

$$f(n) = (1 + e^{-sx-b})^{-1} \tag{1}$$

$$x = \sum_{J=0}^{n} i_j w_j \tag{2}$$

Formally, this GRN architecture can be defined by the tuple $\langle G, L, \mathsf{In}, \mathsf{Out} \rangle$, where:

$G$ is a set of genes $\{n_0 \ldots n_{|N|} : n_i = \langle a_i, I_i, W_i \rangle\}$ where:
  $a_i : \mathbb{R}$ is the activation level of the gene.
  $I_i \subseteq G$ is the set of inputs used by the gene.
  $W_i$ is a set of weights, where $0 \leq w_i \leq 1$, $|W_i| = |I_i|$.
$L$ is a set of initial activation levels, where $|L_N| = |N|$.
$\mathsf{In} \subset G$ is the set of genes used as external inputs.
$\mathsf{Out} \subset G$ is the set of genes used as external outputs.

For the GRN to interact with a task, a set of inputs relating to the state of the task are mapped onto the genes of the network. This is done by setting the expression level of a gene to a value representing the state of the task. Other genes in the network can then use the expression values of the genes which are mapped to the inputs to update their own expression values using the processes shown above. The GRN can then execute by updating the expression values for all genes in the network. The expression value(s) of a sub set of genes are then mapped back to the task. These steps are repeated at each time step in the task.

## 4    Optimisation

In this work we want to focus on the evolutionary process and observe each step of the process to understand how it is functioning. Moreover, we wish to look at a single GRN throughout the evolutionary process rather than a population of them. To achieve this we use a 1+1 evolutionary algorithm which is similar to a hill climbing heuristic [8]. As the genes constitute a GRN, a single mutation of a network will focus on a given gene, for which its regulatory function, its connectivity and its weight are all able to be mutated. It would be possible to use a more classical evolutionary algorithm with a large population which would probably yield stronger objective performance of the GRN, however the focus on this work is on understanding evolutionary constraint in the GRN, and the objective performance of the networks is of lower priority. The algorithm detailing the optimisation process can be seen in Algorithm 1.

---

**Algorithm 1.** Optimising a GRN

---
1: $P \leftarrow$ new random GRN
2: **for** *number of evaluations* **do**
3:        CLONE P AS Q
4:        MUTATE(Q)
5:        EVALUATE(Q)
6:        **if** Q.fitness $>=$ P.fitness **then**
7:                P = Q
8:        **end if**
9: **end for**

---

## 5    Task Definitions

To focus on the evolutionary process, we have adapted two tasks to be in a never ending computational loop. These tasks can be solved in a classical fashion, but whenever they are solved the parameters of the task will change, which requires the GRNs to be continuously evolving. This is to provide an environment for the GRNs to adapt evolvable characteristics. Each of the tasks will have a set number of time-steps, and the objective performance of a GRN will be the amount of times it completes a given task. For each experiment, the GRN will contain 20 genes, and this will not be changed during the optimization process. The mutation process, as specified in Algorithm 1 will mutate 3 genes at every single time step.

In the first set of experiments the mutations will be randomly selected, so that at each time step, 3 genes are selected randomly for mutation, in which any one of their parameters will be modified. In the second set of experiments, the mutations will be guided according to the distributions discovered in the first set of experiments, which will have provided rules and correlations as to which genes are most likely to induce positive change of the GRN. 50 runs will be conducted for the coupled inverted pendulums tasks, and 100 for the pattern generation task as it is less computationally expensive.

## 5.1   Coupled Inverted Pendulums

The inverted pendulum task [10] in this instance consists of a cart mounted to a 1-dimensional track within a finite space, with a pendulum hanging vertically downwards from the cart. The boundaries of the space must not be exceeded by the cart. The objective of the task is to move the cart in such a way that the pendulum swings from below the cart and can be balanced in equilibrium above it. This task was deigned as an efficient proxy for evolving decentralised robotic controllers, and has served as a benchmark for many applications [21]. The inputs and outputs from the task can be seen in Table 1.

**Table 1.** Sensory inputs used for the inverted pendulums task. The values are re-scaled to [0, 1] before they are used as inputs to a network.

| ID | Sensor name | System to sensor mapping |
|----|-------------|--------------------------|
| $S_0$ | Pendulum Angle 0 | $\phi \in [0, 0.5\pi] \rightarrow [127, 0]$, 0 else |
| $S_1$ | Pendulum Angle 1 | $\phi \in [1.5\pi, 2\pi] \rightarrow [0, 127]$, 0 else |
| $S_2$ | Pendulum Angle 2 | $\phi \in [0.5\pi, \pi] \rightarrow [127, 0]$, 0 else |
| $S_3$ | Pendulum Angle 3 | $\phi \in [\pi, 1.5\pi] \rightarrow [0, 127]$, 0 else |
| $S_4$ | Proximity 0 | Distance left $\rightarrow [0, 127]$ |
| $S_5$ | Proximity 1 | Distance right $\rightarrow [0, 127]$ |
| $S_6$ | Cart Velocity 0 | $v \in [-2, 0] \rightarrow [127, 0]$, 0 else |
| $S_7$ | Cart Velocity 1 | $v \in [0, 2] \rightarrow [0, 127]$, 0 else |
| $S_8$ | Angular Velocity 0 | $w \in [-5\pi, 0] \rightarrow [127, 0]$, 0 else |
| $S_9$ | Angular Velocity 1 | $w \in [0, 5\pi] \rightarrow [0, 127]$, 0 else |
| $A_i$ | Actuators 0 | $A_i \in [0, 127]$, for $i \in 0,1$ |
| $u$ | Motor Control 0 | $2(A_0/127 - A_1/127) \rightarrow [0, 1]$ |

A single evolutionary run of the inverted pendulum task will consist of 1'000'000 evaluations, and at each point if the task is solved, the gravitational constant within the system will be randomly modified from 9.81 initially to a number between 7 and 12. This changes the dynamics of the tasks to an extent where it is likely that the GRN will have to evolve new behaviors to learn how to solve it again, yet maintains the overall order of the system. This is aimed at evolving the GRN to be able to solve many different variants of the same task, and to improve its evolutionary robustness.

## 5.2    Pattern Generation

The pattern generation task consists of 10 randomly selected Boolean values which the GRN has to emulate in order. This is done by executing the network over 10 time steps, and generating an output at each time step. If the output is above 0.5, this represents a true value, otherwise it is false. If the outputs from the network matches the Boolean values, the task is reset, and a new set of Boolean values is generated which must then be matched. The Boolean values required to solve the task are never given to the GRN, so the network has to enter a dynamical regime which is capable of recreating these values. For a single run, 10'000'000 evaluations will be conducted.

# 6    Results

The results showing the number of mutations depending on the connectivity of a given gene for each task can be seen in Figs. 2 and 3. Both of these graphs show that there is a similar characteristic regardless of which task the GRN is being optimised for. If we look at the mean mutations for a given gene (left graph of Figs. 2 and 3) it can be seen that genes which have 2 connections are more positively mutated than genes with any other number of connections. The distributions show that for the inverted pendulum task (Fig. 2) the median number of positive mutations for genes with a single connection was slightly higher than that for a gene with two connections. Both tasks show that after 2 connections, as the number of connections increase genes are less likely to get mutated and yield a positive outcome for the GRN.

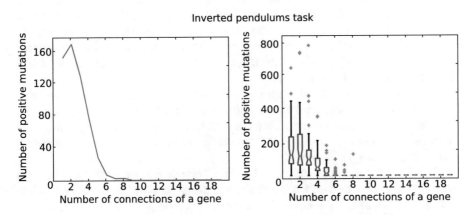

**Fig. 2.** The data for inverted pendulums task (mean left, distributions right). Given a gene with a certain number of connections, how often was that gene positively mutated. The results show that genes with 2 connections achieved the most positive mutations.

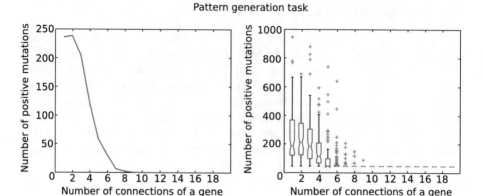

**Fig. 3.** The data for the pattern generation task (mean left, distributions right). Given a gene with a certain number of connections, how often was that gene positively mutated. Similar to the results for the inverted pendulums task, the results show that genes with 2 connections achieved the most positive mutations.

The raw data in Figs. 2 and 3 doesn't take into account the distribution of gene connectivity within the networks. This is important, as if there are more genes with 2 connections in the networks by default (which may be a property of the networks and their parameters) then they are more likely to be positively mutated for the simple reason that there are more of them. The data in Figs. 4 and 5 modifies the raw data in Figs. 2 and 3 by taking into account how many genes there are with a given number of connections in each run. What can be seen by doing this is that the general trend of the raw data (Figs. 2 and 3) still persists. There is a clear trend in both tasks for the mutation of genes with a given number of connections. For the inverted pendulum task, the mutation of genes with 2 connections resulted in proportionality the most positive benefit for the network, and in the pattern generation task, mutating genes with 3 connections was the most beneficial.

It is possible that the raw data was affected by individual run bias, where a single run may have produced a disproportionate number of mutations, favoring a specific gene and thus could affect the overall results. When the results were normalised for each run, so no one run could produce such a bias, the trends seen in Figs. 4 and 5 still persisted.

This information suggests that because certain connectivity of genes are favored in particular situations, there are genes which are more likely to yield positive mutations for the GRN than others. This suggests that we can predict based upon the distributions seen, the likelihood of certain genes being positively mutated over others, suggesting that multiple GRNs develop similar evolutionary patterns, which is the basis of convergent evolution.

Inverted pendulum task

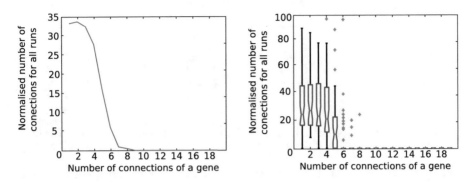

**Fig. 4.** The data from the inverted pendulum task (Fig. 2) divided the number of genes with a given connection which exist within the GRN. This is to prevent bias associated with genes of a certain connectivity being more common in a GRN.

Pattern Generation Task

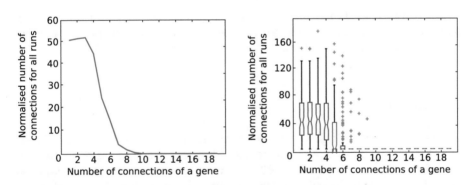

**Fig. 5.** The data from the pattern generation task (Fig. 3) divided the number of genes with a given connection which exist within the GRN. This is to prevent bias associated with genes of a certain connectivity being more common in a GRN.

## 6.1  Applying Mutational Distributions

We apply the distributions in Figs. 2 and 3 on a new set of experiments, where the genes are mutated according to their connectivity of these distributions. The results for this can be seen in Fig. 6. It can be seen that if these distributions are followed for both tasks, the GRNs evolve to solve the task more frequently, indicating that it is a positive evolutionary strategy. The difference between using random mutations and mutating genes according to the distribution is statistically significant for both tasks (For the pendulums the significance was 2.037e−5, and for the pattern generation task the significance was 0.0422. Both

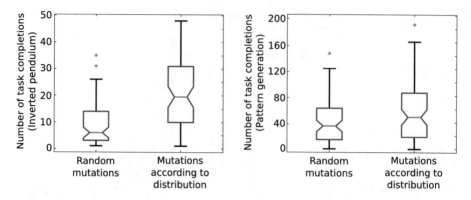

**Fig. 6.** The results of the inverted pendulum task and the pattern generation tasks when the distributions from Figs. 2 and 3 are used throughout evolution respectively. The graph shows that modifying the genes according to a distribution yields significantly better (2.037e−5) results for the inverted pendulum task, and significantly better results for the pattern generation task (0.0422). Both used the Wilcoxon rank-sum test.

used the Wilcoxon rank-sum test). This suggests that there are evolutionary pathways in GRNs which are more likely to yield positive results than random mutations alone, suggesting that certain properties of convergent evolution are present in the networks.

## 7   Conclusions

In this paper we investigated how the connectivity of genes within artificial gene regulatory networks (GRNs) influences the likelihood that a mutation with either be beneficial or detrimental. We looked at two different experiments, the inverted pendulum task and the pattern matching task. Both showed that genes with a certain connectivity (2 and 3 connections) are more likely to yield improvements to the network when mutated. We then showed that there are performance benefits when such distributions are used to guide the evolutionary process, focusing on genes which are most likely to provide beneficial mutations according to their connectivity. This significantly improved the performance of the GRNs for both tasks.

This work shows that for a given GRN and task there may be a more likely evolutionary path depending on the topology of the network and the connectivity of its genes. Although convergent evolution is a multifaceted topic in genetics, this work shows that a computational version of convergent evolution is present in GRNs, and exploiting it speeds up the evolutionary process. In future work it is possible that GRNs could be used as a platform to better understand convergent evolution in nature.

# References

1. Pigliucci, M.: Is evolvability evolvable? Nat. Rev. Genet. **9**(1), 75–82 (2008)
2. Li, J., Yuan, Z., Zhang, Z.: The cellular robustness by genetic redundancy in budding yeast. PLoS Genet. **6**(11), e1001187 (2010)
3. Tokuriki, N., Tawfik, D.S.: Protein dynamism and evolvability. Science **324**(5924), 203–207 (2009)
4. Pavlicev, M., Wagner, G.P.: A model of developmental evolution: selection, pleiotropy and compensation. Trends Ecol. Evol. **27**(6), 316–322 (2012)
5. Graves, C.J., Ros, V.I., Stevenson, B., Sniegowski, P.D., Brisson, D.: Natural selection promotes antigenic evolvability. PLoS Pathog **9**(11), e1003766 (2013)
6. Mitchell, M.: An Introduction to Genetic Algorithms. MIT Press, Cambridge (1998)
7. Michalewicz, Z.: Genetic Algorithms + Data Structures = Evolution Programs. Springer, Heidelberg (2013)
8. Droste, S., Jansen, T., Wegener, I.: On the analysis of the (1+1) evolutionary algorithm. Theor. Comput. Sci. **276**(1–2), 51–81 (2002)
9. LeCun, Y., Bengio, Y., Hinton, G.: Deep learning. Nature **521**(7553), 436–444 (2015)
10. Hamann, H., Schmickl, T., Crailsheim, K.: Coupled inverted pendulums: a benchmark for evolving decentral controllers in modular robotics. In: Proceedings of the 13th Annual Conference on Genetic and Evolutionary Computation, pp. 195–202. ACM (2011)
11. Dowsland, K.A., Thompson, J.M.: Simulated annealing. In: Handbook of Natural Computing, pp. 1623–1655. Springer (2012)
12. Castro, L.N.D., Timmis, J.: Artificial Immune Systems: A New Computational Intelligence Approach. Springer Science & Business Media (2002)
13. Schwab, J.D., Siegle, L., Kühlwein, S.D., Kühl, M., Kestler, H.A.: Stability of signaling pathways during aging-a boolean network approach. Biology **6**(4), 46 (2017)
14. Turner, A.P., Caves, L.S., Stepney, S., Tyrrell, A.M., Lones, M.A.: Artificial epigenetic networks: Automatic decomposition of dynamical control tasks using topological self-modification (2016)
15. Ribba, B., Grimm, H.P., Agoram, B., Davies, M.R., Gadkar, K., Niederer, S., van Riel, N., Timmis, J., van der Graaf, P.H.: Methodologies for quantitative systems pharmacology (QSP) models: design and estimation. CPT Pharmacometrics Syst. Pharmacol. **6**(8), 496–498 (2017)
16. Kirschner, M.: Beyond Darwin: evolvability and the generation of novelty. BMC Biol. **11**(1), 1 (2013)
17. Kitano, H.: Biological robustness. Nat. Rev. Genet. **5**(11), 826–837 (2004)
18. Palazzo, A.F., Gregory, T.R.: The case for junk DNA. PLoS Genet. **10**(5), e1004351 (2014)
19. Pennisi, E.: Encode project writes eulogy for junk DNA (2012)
20. Zotenko, E., Mestre, J., O'Leary, D.P., Przytycka, T.M.: Why do hubs in the yeast protein interaction network tend to be essential: reexamining the connection between the network topology and essentiality. PLoS Comput. Biol. **4**(8), e1000140 (2008)
21. Silva, F., Duarte, M., Correia, L., Oliveira, S.M., Christensen, A.L.: Open issues in evolutionary robotics. Evol. Comput. **24**(2), 205–236 (2016)

22. Valero, K.W.: Aligning functional network constraint to evolutionary outcomes. bioRxiv (2018)
23. Pervouchine, D.D., Djebali, S., Breschi, A., Davis, C.A., Barja, P.P., Dobin, A., Tanzer, A., Lagarde, J., Zaleski, C., See, L.-H., et al.: Enhanced transcriptome maps from multiple mouse tissues reveal evolutionary constraint in gene expression. Nat. Commun. **6**, 5903 (2015)
24. Foote, A.D., Liu, Y., Thomas, G.W.C., Vinař, T., Alföldi, J., Deng, J., Dugan, S., van Elk, C.E., Hunter, M.E., Joshi, V., et al.: Convergent evolution of the genomes of marine mammals. Nat. Genet. **47**(3), 272 (2015)

# Exploiting Tournament Selection for Efficient Parallel Genetic Programming

Darren M. Chitty[(✉)]

Department of Computer Science, University of Bristol,
Merchant Venturers Bldg, Woodland Road, Bristol BS8 1UB, UK
darrenchitty@googlemail.com

**Abstract.** Genetic Programming (GP) is a computationally intensive technique which is naturally parallel in nature. Consequently, many attempts have been made to improve its run-time from exploiting highly parallel hardware such as GPUs. However, a second methodology of improving the speed of GP is through efficiency techniques such as subtree caching. However achieving parallel performance and efficiency is a difficult task. This paper will demonstrate an efficiency saving for GP compatible with the harnessing of parallel CPU hardware by exploiting tournament selection. Significant efficiency savings are demonstrated whilst retaining the capability of a high performance parallel implementation of GP. Indeed, a 74% improvement in the speed of GP is achieved with a peak rate of 96 billion GPop/s for classification type problems.

**Keywords:** Genetic Programming · HPC · Computational Efficiency

## 1 Introduction

Genetic Programming (GP) [9] is widely known as being highly computationally intensive. This is due to candidate GP programs being typically evaluated using an interpreter which is an inefficient method of running a program due to the use of a conditional statement at each step in order to ascertain which instruction to execute. Moreover, GP is a population based technique with generally large population sizes of candidate GP programs. As such, there have been many improvements to the execution speed of GP from a compiled approach or direct machine code, exploiting parallel computational hardware or finding efficiencies within the technique such as caching common subtrees. Recent advances in computational hardware have led to multi-core architectures such that very fast parallel implementations of GP have been implemented but a drawback is that although they are fast, they are not efficient. Techniques such as subtree caching are difficult to use with these parallel implementations. Searching for common subtrees can incur a significant time cost slowing execution speeds.

© Springer Nature Switzerland AG 2019
A. Lotfi et al. (Eds.): UKCI 2018, AISC 840, pp. 41–53, 2019.
https://doi.org/10.1007/978-3-319-97982-3_4

This paper introduces an efficiency saving that can be made by exploiting the characteristics of tournament selection which can be easily implemented within a CPU based parallel GP approach. Consequently, a considerable performance gain can be made in the execution speed of GP. The paper is laid out as follows: Sect. 2 describes GP and prior methods of improving the speed and efficiency of the technique. Section 3 introduces a tournament selection strategy embedded with a highly parallel GP model and demonstrates efficiency savings with regards classification tasks. Finally, Sect. 4 demonstrates significant enhancements by consideration of evaluated solutions surviving intact between generations.

## 2  Background

Genetic Programming (GP) [9] solves problems by constructing programs using the principles of evolution. A population of candidate GP programs are maintained and evaluated as to their respective effectiveness against a target objective. New populations are generated using the genetic operators of selection, crossover and mutation. Selection is usually conducted with tournament selection where a subset of GP programs compete to be selected as parents based on their fitness. The evaluation of a GP program is typically achieved by *interpreting* it against a set of fitness cases. GP is computationally intensive as a result of using an interpreter, maintaining typically large populations of programs and often, using a large volume of fitness cases such as for classification tasks.

Recently, with the advent of multi-core CPUs and many core GPUs, the focus has been on creating highly parallel implementations of GP achieving speedups of several hundred [1–5]. However, prior to the move to parallel architectures the primary method of improving the speed of GP was through efficiency savings. A simple methodology is to reduce the number of fitness cases by dynamic sampling of the more difficult instances [7]. A further selection strategy known as Limited Error Fitness (LEF) was investigated by Gathercole and Ross for classification problems whereby an upper bound on the number of permissible misclassifications was used to terminate evaluations [8]. Maxwell implemented a time rationing approach where each GP program was evaluated for a fixed time [10]. Tournament selection was then performed and the fittest candidate GP program was declared the winner. Smaller programs could evaluate more fitness cases. Teller used a similar technique known as the *anytime* approach [14].

With regards tournament selection Teller and Andre introduced a technique known as the Rational Allocation of Trials (RAT) [15] whereby population members are evaluated on a small subset of individuals. A prediction model is then used to test if further evaluations should be made to establish the *winners* of tournaments with a 36x speedup but only small populations are considered and regression type problems. Park et al. implemented a methodology whereby the fitness of the best candidate GP program within the current population is tracked and when evaluating a GP program, if the accumulated fitness becomes worse than this best found then the evaluation is terminated and the fitness approximated [11]. Speedups of four fold were reported with little degradation in overall fitness. Finally, Poli and Langdon ran GP backwards by generating tournaments for the whole GP process and working backwards detecting offspring that are not sampled by tournament selection and not evaluating them or their parents if all offspring are not sampled [13]. However, most of these methods do not accurately reflect tournament winners thereby changing the evolutionary path.

## 3  Improved Efficiency Through Tournament Selection

The typical methodology of GP is to completely evaluate a population of GP programs to ascertain their fitness. A new population is then constructed using the GP operators of selection, crossover and mutation. The selection process typically chooses two *parents* from the current population to generate two offspring programs using crossover and mutation. The most widely used selection operator within GP is tournament selection whereby $t$ GP programs are randomly selected from the current population to form a *tournament*. The program within this *tournament* with the best fitness is then selected as a parent.

However, tournament selection can be exploited by considering an alternative evaluation methodology. Rather than evaluating each candidate GP program over every single fitness case before moving onto the next candidate, consider the opposite approach whereby all programs are evaluated upon a single fitness case before considering the next fitness case. Using this approach, the fitness levels of candidate GP programs slowly build up and facilitates comparisons between programs during the evaluation process. Consequently, if a set of tournaments is generated prior to the evaluation stage it is possible to ascertain which candidate GP programs within a given tournament reach a stage whereby they cannot possibly *win*. Moreover, if a candidate GP program is deemed unable to *win* any of the tournaments it is involved in before all the fitness cases have been evaluated upon, there is clearly no reason to continue to evaluate it. Consequently an efficiency saving can be realised using this approach. This technique can be described as *smart sampling* whereby only the minimum fitness cases necessary to establish *losers* of tournaments is required.

**Example:** Consider a classification problem whereby the fitness metric is the sum of correct classifications with a tournament size of two to select a potential parent. Using ten fitness cases the first GP program of the tournament correctly classifies the first six cases and the second incorrectly classifies them. Effectively the second GP program in the tournament cannot possibly *win* as the maximum correct classifications it can now achieve is four from the remaining fitness cases. If this second GP program is not involved in any other tournament then there is no value in continuing to evaluate it for the remaining fitness cases as it will definitely not be selected as a parent. Consequently, a 40% efficiency saving can be achieved on the evaluation of the second GP program within the tournament.

To realise this potential efficiency saving a new GP evaluation model is proposed whereby at each generation a set of tournaments composed of randomly selected candidate GP programs from the population are generated prior to the evaluation. Using a system whereby two selected programs generate two offspring, this set consists of population size $n$ tournaments each of which contain tournament size $t$ randomly selected programs. Each fitness case is then evaluated by every member of the population before the next fitness case is considered. Before a program is evaluated on the given fitness case it is checked to establish if it is has effectively *lost* all of the tournaments it is involved in and if so labelled as a *loser* and not to be further evaluated. The efficient tournament selection work suggested here ensures the same candidate GP programs *win* tournaments as with standard GP. Algorithm 1 provides a high level overview of the efficient tournament selection model.

---

**Algorithm 1.** Efficient Tournament Selection

1: initialise population
2: **for** number of generations **do**
3:     generate set of tournaments for generation of next population
4:     **for** each fitness case **do**
5:         **for** each population member **do**
6:             **if** population member has not lost all its tournaments **then**
7:                 evaluate population member on given fitness case and update fitness
8:             **end if**
9:         **end for**
10:     **end for**
11:     generate new population using tournament winners
12: **end for**

---

**Algorithm 2.** Tournament Check

1: **for** each tournament given population member is involved in **do**
2:     identify current best fitness of population members in given tournament
3:     **if** fitness of given population member plus the potential additional fitness from the remaining fitness cases is lower than this best fitness **then**
4:         given population member has already *lost* this tournament
5:     **end if**
6: **end for**
7: **if** given population member cannot possibly *win* any of its tournaments **then**
8:     stop evaluating population member
9: **end if**

---

To determine if a candidate GP program has *lost* all of the tournaments it is involved in, each tournament needs to be checked once the current fitness case has been evaluated upon by all candidate GP programs in the population. Thus, each candidate GP program maintains the subset of tournaments it is involved in. For each of these tournaments, the currently best performing candidate GP program in the tournament is identified and its fitness compared with the fitness of the candidate GP program under consideration. If using a given fitness metric it can be ascertained that the GP program under consideration cannot beat this best then it is designated as having *lost* this tournament. If a program is deemed to have *lost* all of the tournaments it is involved in then it is designated as requiring no further evaluation. A high level description of tournament checking is shown in Algorithm 2. Using this approach all the candidate GP programs that would *win* the tournaments using a standard implementation of GP will still *win* their respective tournaments. Consequently, there is no disturbance to the evolutionary path taken by GP but an efficiency saving can be achieved.

### 3.1 Efficient Tournament Selection and a Fast GP Approach

Given that the goal of efficient tournament selection is to reduce the computational cost of GP and hence improve the speed, it is only natural that the technique should be able to operate within a fast parallel GP model. Integrating the efficient tournament selection model with a GPU implementation would prove difficult without compromising the speed of the approach as communication across GPU cores evaluating differing GP programs is difficult. The alternative platform is a CPU based parallel GP [3] which introduced a two dimensional stack approach to parallel GP demonstrating significantly improved execution times. A multi-core CPU with limited parallelism was used with the two dimensional stack model to exploit the cache memory and reduce interpreter overhead. In fact, this model actually operates similarly to efficient tournament selection, GP programs are evaluated in parallel over blocks of fitness cases. Once all the candidate GP programs have been evaluated on a block of fitness cases, the next block of fitness cases is considered. Using blocks of fitness cases provided the best utilisation of cache memory and hence the best speed.

Subsequently, the efficient tournament selection model is implemented within this two dimensional stack GP model using a CPU and instead of evaluating candidate GP programs on a single fitness case at a time, they are evaluated on a larger block of fitness cases. Once the block of fitness cases has been evaluated upon then candidate GP programs that may have *lost* all of their respective tournaments can be identified. A block of 2400 fitness cases is used which was identified by Chitty [3] as extracting the best performance from the cache memory and efficiency from reduced reinterpretation of candidate GP programs.

### 3.2 Initial Results

In order to evaluate if efficient tournament selection can provide a computational saving it will be tested against three classification type problems. The first two are the Shuttle and KDDcup classification problems available from the Machine Learning Repository [6] consisting of 58,000 and 494,021 fitness cases respectively. The GP function set for these problems consists of $*$, $/$, $+$, $-$, $>$, $<$, $==$, AND, OR, IF and the terminal values the input features or a constant value between −20,000.0 and 20,000.0. The third problem is the Boolean 20-multiplexer problem [9] with the goal to establish a rule which takes address bits and data bits and correctly outputs the value of the data bit which the address bits specify. The function set consists of AND, OR, NAND, NOR and the terminal set consists of A0-A3, D0-D15. There are 1048576 fitness cases which can be reduced using bit level parallelism such that each bit of a 32 bit variable represents a differing fitness case reducing fitness cases to 32768.

The results for these experiments were generated using an i7 2600 Intel processor running at 3.4 GHz with four processor cores each able to run two threads of execution independently. The algorithms used were compiled using Microsoft Visual C++. Table 1 provides the GP parameters that were used throughout the work presented in this paper. Each experiment was averaged over 25 runs for a range of differing tournament sizes to demonstrate how efficiency can change dependant on the selection pressure.

**Table 1.** GP parameters used throughout results presented in this paper

| Population Size: 4000 | Maximum Generations: 50 |
|---|---|
| Maximum Tree Depth: 50 | Maximum Tree Size: 1000 |
| Probability of Crossover: 0.50 | Probability of Mutation: 0.50 |

In order to use efficient tournament selection a fitness metric is required which establishes if a given candidate GP program cannot *win* a tournament. This metric is described as comparing the classification rates of the best performing GP program in a tournament with the rate of the candidate GP program under consideration. If the performance of the best is greater than that of the program under consideration whilst also assuming that the all the remaining fitness cases are correctly classified then the program under consideration cannot possibly *win*. This can also be described as the candidate GP program under consideration being *mathematically* unable to *win* the given tournament.

Total efficiency saving is measured as the number of fitness cases not evaluated by each GP program multiplied by their size divided by the sum of the size of all GP programs evaluated multiplied by the number of fitness cases. It should be noted that this work is concerned with efficiency in the training phase of GP and not classification accuracy. These rates are provided merely as a demonstration that the same results are achieved between techniques.

Table 2 demonstrates the performance of a standard GP approach using the 2D stack model [3] and the efficient tournament selection model for a range of tournament sizes. Note that for two of the problem instances, as the tournament size increases, the average GP tree size similarly increases which obviously increases the execution time of GP. Also note that there is no deviation from the classification accuracy from both approaches as would be expected as there has been no deviation from the evolutionary path. However, in all cases, an efficiency saving has been observed. The greatest efficiency saving is made with the lowest levels of tournament size as a result of the non-sampled issue whereby some members of the population are not involved in any tournaments [12]. If a candidate GP program is not involved in any tournaments then there is no value in evaluating it. Additionally, in cases of low selection pressure, a GP program is likely to be involved in few tournaments and thus a poor solution can quickly *lose* all its tournaments. Also note that as the tournament size increases to ten or greater the efficiency savings begin to improve once more. This effect is due to an increased probability of a highly fit GP program being in any given tournament making it easier to identify weak solutions at an earlier stage.

**Table 2.** Results from a comparison between a standard implementation of GP and the efficient tournament selection method.

| Problem | Tournament Size | Class Accuracy (%) | Average Tree Size | Standard GP Execution Time (s) | Efficient Tournament Selection GP | | |
|---|---|---|---|---|---|---|---|
| | | | | | Efficiency Saving (%) | Execution Time (s) | Speedup |
| Shuttle | 3 | 83.96 ± 3.67 | 43.02 | 13.27 ± 3.02 | 10.07 ± 1.39 | 12.31 ± 2.66 | 1.078x |
| | 4 | 87.92 ± 5.34 | 46.57 | 14.04 ± 4.49 | 8.18 ± 1.84 | 13.22 ± 4.05 | 1.062x |
| | 5 | 88.10 ± 6.47 | 42.07 | 12.72 ± 2.17 | 7.84 ± 1.87 | 12.10 ± 1.97 | 1.051x |
| | 6 | 89.23 ± 6.19 | 44.49 | 13.59 ± 3.53 | 8.37 ± 1.62 | 12.86 ± 3.25 | 1.057x |
| | 7 | 88.93 ± 6.42 | 43.64 | 13.20 ± 4.32 | 7.81 ± 2.56 | 12.53 ± 3.86 | 1.053x |
| | 8 | 91.44 ± 5.99 | 41.34 | 12.64 ± 3.14 | 8.36 ± 2.24 | 12.00 ± 2.81 | 1.053x |
| | 9 | 92.02 ± 6.68 | 44.56 | 13.50 ± 2.62 | 9.25 ± 2.03 | 12.75 ± 2.47 | 1.059x |
| | 10 | 90.53 ± 6.24 | 41.98 | 12.83 ± 3.70 | 8.44 ± 2.32 | 12.14 ± 3.31 | 1.057x |
| | 20 | 90.10 ± 6.75 | 47.20 | 14.44 ± 5.69 | 8.65 ± 2.83 | 13.67 ± 5.26 | 1.057x |
| | 30 | 86.43 ± 7.36 | 47.72 | 14.39 ± 5.69 | 8.26 ± 2.91 | 13.73 ± 5.34 | 1.048x |
| KDDcup | 3 | 93.71 ± 6.84 | 52.03 | 130.11 ± 28.87 | 12.57 ± 0.65 | 113.83 ± 25.16 | 1.143x |
| | 4 | 95.96 ± 5.37 | 54.03 | 136.38 ± 28.21 | 11.37 ± 1.07 | 121.61 ± 24.41 | 1.121x |
| | 5 | 95.52 ± 5.29 | 55.04 | 138.10 ± 35.76 | 11.01 ± 1.66 | 123.79 ± 32.55 | 1.116x |
| | 6 | 95.58 ± 5.78 | 47.65 | 120.61 ± 25.67 | 11.50 ± 1.52 | 107.92 ± 23.42 | 1.118x |
| | 7 | 92.71 ± 8.09 | 50.78 | 127.83 ± 34.22 | 11.25 ± 1.98 | 114.51 ± 30.50 | 1.116x |
| | 8 | 96.26 ± 5.24 | 51.91 | 129.11 ± 37.06 | 11.74 ± 1.32 | 114.97 ± 32.58 | 1.123x |
| | 9 | 96.19 ± 5.36 | 56.98 | 143.57 ± 50.50 | 11.64 ± 2.28 | 128.27 ± 45.40 | 1.119x |
| | 10 | 95.40 ± 6.28 | 58.16 | 149.34 ± 63.38 | 12.42 ± 1.58 | 132.02 ± 55.25 | 1.131x |
| | 20 | 95.91 ± 5.37 | 64.90 | 162.10 ± 87.00 | 13.80 ± 2.31 | 141.25 ± 75.01 | 1.148x |
| | 30 | 94.64 ± 7.12 | 65.95 | 163.46 ± 79.64 | 12.68 ± 2.51 | 143.81 ± 70.15 | 1.137x |
| 20-Mult. | 3 | 61.46 ± 0.88 | 187.28 | 53.07 ± 7.81 | 6.46 ± 0.11 | 49.85 ± 7.25 | 1.065x |
| | 4 | 63.44 ± 1.18 | 159.25 | 46.91 ± 8.19 | 3.76 ± 0.18 | 45.63 ± 7.87 | 1.028x |
| | 5 | 64.85 ± 1.35 | 154.58 | 46.46 ± 6.87 | 2.99 ± 0.23 | 45.62 ± 6.69 | 1.018x |
| | 6 | 66.61 ± 1.78 | 156.98 | 46.86 ± 8.38 | 2.91 ± 0.22 | 46.12 ± 8.22 | 1.016x |
| | 7 | 67.02 ± 1.75 | 141.64 | 42.69 ± 6.65 | 3.04 ± 0.28 | 41.97 ± 6.51 | 1.017x |
| | 8 | 67.36 ± 2.06 | 159.75 | 47.06 ± 13.01 | 2.99 ± 0.36 | 46.19 ± 12.94 | 1.019x |
| | 9 | 68.38 ± 1.74 | 132.88 | 41.41 ± 6.98 | 3.24 ± 0.32 | 40.60 ± 7.03 | 1.020x |
| | 10 | 68.61 ± 2.25 | 128.93 | 40.19 ± 7.99 | 3.38 ± 0.40 | 39.50 ± 7.90 | 1.017x |
| | 20 | 71.17 ± 2.06 | 135.81 | 41.37 ± 8.85 | 4.36 ± 0.46 | 40.09 ± 8.70 | 1.032x |
| | 30 | 72.61 ± 2.22 | 157.60 | 45.52 ± 10.29 | 4.81 ± 0.60 | 43.92 ± 9.86 | 1.036x |

The KDDcup classification problem demonstrates the greatest efficiency savings with up to a 13.8% saving. The multiplexer problem demonstrates the lowest efficiency saving as a result of the lower accuracy achieved being a more difficult problem. Consider that as the fitness improves within the population during the evolutionary process, identifying weak candidate GP programs becomes easier to achieve at an earlier stage thereby increasing efficiency savings. The speedups observed are less than the efficiency savings as a result of the computational cost associated with repeatedly establishing if candidate GP programs have not *lost* any tournament they are involved in. Clearly the use of the efficient tournament selection technique has provided a boost in the performance of GP with a minor speedup in all cases with a maximum of 15% achieved.

## 4     Consideration of Previously Evaluated Individuals

The previous section demonstrated that speedups in GP can be achieved using the efficient tournament selection technique whilst not affecting the outcome of the GP process but the gains are rather limited as at least 50% of the fitness cases need to be evaluated before *losers* of tournaments can be identified. The fitness metric is boolean in nature in that a fitness case is either correctly classified or not so a GP program cannot have *mathematically lost* a tournament whilst 50% of fitness cases remain. However, the crossover and mutation parameters used both had a probability of 0.5. Consequently, approximately 25% of candidate GP programs will survive intact into the next generation and not reevaluating these will in itself provide an efficiency saving of approximately 25%. More importantly, these GP programs are previous tournament *winners* whereby their complete fitness is known. This will make it possible to identify *losers* of tournaments at an earlier stage in the evaluation process. Additionally, as *winners* of tournaments, they are likely highly fit and leaving little margin for error for other tournament contenders.

**Example:** Consider a tournament size of two where the first GP program has survived intact into the next generation and been previously evaluated correctly classifying eight of the ten fitness cases. If the second candidate GP program incorrectly classifies more than two of the fitness cases then it cannot possibly *win* the tournament. So if the first three fitness cases are incorrectly classified the solution has *lost* the tournament and an efficiency saving of 70% can be achieved.

To test this theory the standard approach to GP is rerun as a benchmark but this time not reevaluating candidate GP programs which survive intact into the next generation. Additionally, candidate GP programs which are not involved in any tournament, the non-sampled issue, are also not evaluated. The results are shown in Table 3 with an expected 25% reduction in execution speed through this efficiency for standard GP. The comparison results from the efficient tournament selection method are also shown in Table 3. Efficiency savings are now considerably higher than those achieved in Table 2 even taking into account the efficiency savings achieved by not reevaluating intact candidate GP programs. Indeed, the additional efficiency is now as much as 30%. Subsequently, it can be considered that having previously evaluated candidate GP programs within tournaments makes it easier to establish *losers* of tournaments hence increasing the efficiency savings. Furthermore, it can also be observed that in cases of higher selection pressure the efficiency savings increase further. Indeed, in the case of the KDDcup classification problem, efficiency savings of 60% are achieved. The reason for this is that GP programs that survive intact into the next generation have won greatly competitive tournaments and are thus highly fit making it easier to establish *losers* of tournaments. Note that there is still no change in the evolutionary path when using efficient tournament selection.

**Table 3.** Results from a comparison between a standard implementation of GP and the efficient tournament selection method taking into account not reevaluating non-modified or evaluating non-selected candidate GP programs.

| Problem | Tournament Size | Class Accuracy (%) | Average Tree Size | Standard GP Execution Time (s) | Efficient Tournament Selection GP | | |
|---|---|---|---|---|---|---|---|
| | | | | | Efficiency Saving (%) | Execution Time (s) | Speedup |
| Shuttle | 3 | 83.96 ± 3.67 | 43.02 | 10.12 ± 2.08 | 35.73 ± 1.77 | 9.45 ± 1.86 | 1.070x |
| | 4 | 87.92 ± 5.34 | 46.57 | 10.84 ± 3.26 | 35.38 ± 2.50 | 10.00 ± 2.77 | 1.084x |
| | 5 | 88.10 ± 6.47 | 42.07 | 9.97 ± 1.62 | 35.92 ± 2.73 | 9.15 ± 1.23 | 1.090x |
| | 6 | 89.23 ± 6.19 | 44.49 | 10.68 ± 2.62 | 37.37 ± 2.57 | 9.59 ± 2.27 | 1.114x |
| | 7 | 88.93 ± 6.42 | 43.64 | 10.37 ± 3.19 | 37.24 ± 4.21 | 9.33 ± 2.57 | 1.112x |
| | 8 | 91.44 ± 5.99 | 41.34 | 10.00 ± 2.31 | 38.99 ± 4.10 | 8.80 ± 1.80 | 1.137x |
| | 9 | 92.02 ± 6.68 | 44.56 | 10.64 ± 1.88 | 41.23 ± 4.17 | 9.14 ± 1.73 | 1.164x |
| | 10 | 90.53 ± 6.24 | 41.94 | 10.15 ± 2.70 | 40.42 ± 4.68 | 8.78 ± 2.14 | 1.156x |
| | 20 | 90.10 ± 6.75 | 47.20 | 11.40 ± 4.18 | 46.56 ± 8.49 | 9.11 ± 3.28 | 1.252x |
| | 30 | 86.43 ± 7.36 | 47.72 | 11.48 ± 4.19 | 45.03 ± 8.94 | 9.51 ± 3.44 | 1.207x |
| KDDcup | 3 | 93.71 ± 6.84 | 52.03 | 94.84 ± 20.17 | 38.42 ± 0.90 | 83.98 ± 17.62 | 1.129x |
| | 4 | 95.96 ± 5.37 | 54.03 | 102.27 ± 20.34 | 38.90 ± 1.35 | 87.08 ± 16.51 | 1.174x |
| | 5 | 95.52 ± 5.29 | 55.04 | 104.58 ± 26.29 | 39.69 ± 2.15 | 87.44 ± 22.24 | 1.196x |
| | 6 | 95.58 ± 5.78 | 47.65 | 91.98 ± 18.90 | 41.24 ± 2.38 | 75.23 ± 15.59 | 1.223x |
| | 7 | 92.71 ± 8.09 | 50.78 | 97.57 ± 24.97 | 41.80 ± 3.05 | 78.99 ± 20.32 | 1.235x |
| | 8 | 96.26 ± 5.24 | 51.91 | 98.74 ± 27.43 | 44.01 ± 2.00 | 76.99 ± 20.86 | 1.283x |
| | 9 | 96.19 ± 5.36 | 56.98 | 109.29 ± 37.46 | 45.67 ± 3.80 | 83.15 ± 28.72 | 1.314x |
| | 10 | 95.40 ± 6.28 | 58.16 | 113.60 ± 47.04 | 48.06 ± 2.92 | 82.96 ± 33.30 | 1.369x |
| | 20 | 95.91 ± 5.37 | 64.90 | 123.26 ± 64.26 | 58.76 ± 5.72 | 73.39 ± 37.33 | 1.680x |
| | 30 | 94.64 ± 7.12 | 65.66 | 123.74 ± 59.28 | 57.98 ± 7.10 | 76.15 ± 36.99 | 1.625x |
| 20-Mult. | 3 | 61.46 ± 0.88 | 185.11 | 38.17 ± 5.43 | 31.49 ± 0.17 | 37.73 ± 5.33 | 1.012x |
| | 4 | 63.44 ± 1.18 | 159.25 | 35.14 ± 5.98 | 29.73 ± 0.20 | 34.60 ± 5.85 | 1.016x |
| | 5 | 64.85 ± 1.35 | 154.58 | 35.19 ± 5.05 | 29.45 ± 0.31 | 34.43 ± 4.90 | 1.022x |
| | 6 | 66.61 ± 1.78 | 156.98 | 35.63 ± 6.24 | 29.66 ± 0.33 | 34.61 ± 6.02 | 1.029x |
| | 7 | 67.02 ± 1.75 | 141.64 | 32.58 ± 4.95 | 30.00 ± 0.49 | 31.46 ± 4.80 | 1.036x |
| | 8 | 67.36 ± 2.06 | 159.75 | 35.77 ± 9.81 | 30.13 ± 0.59 | 34.46 ± 9.58 | 1.038x |
| | 9 | 68.38 ± 1.74 | 132.88 | 31.54 ± 5.23 | 30.69 ± 0.54 | 30.19 ± 5.10 | 1.045x |
| | 10 | 68.61 ± 2.25 | 128.93 | 30.73 ± 6.01 | 31.04 ± 0.67 | 29.19 ± 5.78 | 1.053x |
| | 20 | 71.17 ± 2.06 | 135.81 | 31.64 ± 6.55 | 34.28 ± 1.03 | 28.84 ± 6.21 | 1.097x |
| | 30 | 72.61 ± 2.22 | 157.60 | 34.71 ± 7.56 | 35.42 ± 1.54 | 31.11 ± 6.80 | 1.116x |

In terms of the effective speed of GP from using these tournament selection efficiency savings, the greatest increase in speed has been achieved for the KDDcup problem with a 1.68x performance gain when using larger tournament sizes. Indeed, for all problem instances, the best performance gains are observed from higher selection pressure. The higher the selection pressure, the greater probability that highly fit candidate GP programs survive intact into the next generation.

## 4.1   Elitism

Given that having candidate GP programs that have survived intact into subsequent generations has been shown to make it easier to correctly identify *losers* of tournaments it could be considered that using the elitism operator would have a similar effect. Elitism involves the best subset of candidate GP programs in a given population being copied intact into the next generation meaning the best solutions are never lost from the population. As previously, these *elitist* candidate GP programs do not need to be reevaluated thus providing a basic efficiency saving. Indeed, Table 4 shows the results generated from standard GP using an elitism operator of 10% of the population and also not evaluating non-modified or non-sampled candidate GP programs. From these results it can be seen that

**Table 4.** Results from using standard GP with 10% elitism and not reevaluating non-modified or evaluating non-selected individuals

| Problem | Tournament Size | Classification Accuracy (%) | Average Tree Size | Execution Time (s) | GPop/s (bn) | Efficiency Saving (%) |
|---|---|---|---|---|---|---|
| Shuttle | 3 | 87.60 ± 5.81 | 30.64 | 7.10 ± 1.82 | 50.88 | 36.48 ± 0.63 |
| | 4 | 89.86 ± 5.30 | 33.71 | 7.86 ± 2.15 | 50.08 | 34.61 ± 0.61 |
| | 5 | 89.09 ± 5.07 | 32.71 | 7.78 ± 1.75 | 49.26 | 33.76 ± 0.52 |
| | 6 | 89.40 ± 6.65 | 36.38 | 8.38 ± 2.28 | 50.36 | 33.61 ± 0.53 |
| | 7 | 91.73 ± 4.64 | 37.36 | 8.64 ± 2.37 | 50.36 | 33.33 ± 0.57 |
| | 8 | 86.90 ± 6.84 | 36.01 | 8.33 ± 2.51 | 50.24 | 33.18 ± 0.59 |
| | 9 | 88.89 ± 6.81 | 32.91 | 7.68 ± 1.97 | 50.05 | 33.05 ± 0.82 |
| | 10 | 89.20 ± 6.74 | 38.49 | 8.73 ± 2.76 | 50.90 | 33.17 ± 0.78 |
| | 20 | 89.70 ± 7.11 | 38.46 | 8.98 ± 2.57 | 49.82 | 33.08 ± 0.80 |
| | 30 | 90.31 ± 6.94 | 38.75 | 9.01 ± 2.68 | 49.02 | 33.07 ± 0.68 |
| KDDcup | 3 | 93.79 ± 7.67 | 28.98 | 53.70 ± 12.28 | 53.59 | 35.68 ± 0.71 |
| | 4 | 94.76 ± 7.01 | 38.07 | 69.67 ± 21.00 | 53.85 | 34.09 ± 0.35 |
| | 5 | 96.45 ± 4.12 | 38.15 | 68.80 ± 21.40 | 54.86 | 33.33 ± 0.20 |
| | 6 | 92.25 ± 8.72 | 36.75 | 67.17 ± 16.15 | 54.28 | 32.73 ± 0.64 |
| | 7 | 94.39 ± 6.93 | 39.47 | 73.35 ± 32.31 | 53.20 | 32.94 ± 0.38 |
| | 8 | 95.40 ± 5.30 | 42.71 | 77.21 ± 30.06 | 54.15 | 32.81 ± 0.46 |
| | 9 | 94.43 ± 7.12 | 42.64 | 75.39 ± 26.44 | 55.95 | 32.67 ± 0.96 |
| | 10 | 96.47 ± 4.05 | 43.20 | 76.85 ± 25.67 | 55.64 | 32.86 ± 0.34 |
| | 20 | 96.20 ± 5.33 | 51.49 | 92.24 ± 34.83 | 55.21 | 33.08 ± 0.37 |
| | 30 | 95.00 ± 6.15 | 50.33 | 92.05 ± 50.41 | 54.57 | 32.97 ± 0.60 |
| 20-Mult. | 3 | 65.22 ± 1.17 | 117.25 | 26.07 ± 3.44 | 958.04 | 36.43 ± 0.34 |
| | 4 | 66.75 ± 1.24 | 128.45 | 27.86 ± 5.29 | 973.35 | 34.44 ± 0.33 |
| | 5 | 67.86 ± 1.74 | 121.00 | 26.95 ± 4.93 | 951.26 | 33.71 ± 0.29 |
| | 6 | 67.87 ± 1.73 | 113.12 | 25.95 ± 4.06 | 923.90 | 33.41 ± 0.34 |
| | 7 | 69.31 ± 1.48 | 113.86 | 26.25 ± 3.71 | 923.68 | 33.44 ± 0.19 |
| | 8 | 69.45 ± 1.67 | 115.16 | 26.40 ± 5.50 | 914.43 | 33.33 ± 0.38 |
| | 9 | 69.73 ± 2.01 | 125.68 | 27.67 ± 6.01 | 953.56 | 33.50 ± 0.26 |
| | 10 | 70.40 ± 1.57 | 128.21 | 27.52 ± 6.89 | 974.39 | 33.54 ± 0.16 |
| | 20 | 73.09 ± 1.97 | 126.28 | 27.30 ± 5.09 | 974.77 | 33.59 ± 0.19 |
| | 30 | 73.80 ± 2.70 | 140.39 | 29.78 ± 6.34 | 990.59 | 33.66 ± 0.19 |

an expected approximate 32% efficiency saving is now achieved in cases of high selection pressure whereby non-sampling is not an issue. It should be noted that the use of the elitism operator has had an effect on the classification accuracy achieved leading to slight differences when compared to the results in Table 2.

It should be expected from the earlier results that the elitism operator will improve the savings of the efficient tournament selection model further and as such the experiments are rerun using 10% elitism with the results shown in Table 5. It should be firstly noted that the classification accuracy and average tree sizes differ slightly to that those of the standard GP approach as shown in Table 4, a divergence in the evolutionary process has occurred. The reason behind this is that a highly fit candidate GP program capable of being selected by the elitism operator was not selected because it *lost* all the tournaments it was involved in and hence its continued evaluation terminated resulting in a lower fitness level and thereby no longer qualifying for selection by elitism. Normally, it would be selected even though it would not *win* any tournaments. However, the effect on classifier accuracy is minimal and in some cases the accuracy is actually improved. Efficiency savings of 40–65% are achieved for all problem instances

**Table 5.** Results from using efficient tournament selection with 10% elitism and not reevaluating non-modified or evaluating non-selected individuals

| Problem | Tournament Size | Classification Accuracy (%) | Average Tree Size | Execution Time (s) | GPop/s (bn) | Efficiency Saving (%) | Speedup |
|---|---|---|---|---|---|---|---|
| Shuttle | 3 | 87.17 ± 5.14 | 30.55 | 6.60 ± 1.07 | 54.68 | 44.26 ± 2.21 | 1.076x |
| | 4 | 90.09 ± 5.06 | 33.53 | 7.14 ± 1.78 | 54.82 | 44.32 ± 2.51 | 1.100x |
| | 5 | 88.77 ± 4.85 | 32.50 | 6.89 ± 1.72 | 54.94 | 45.04 ± 2.74 | 1.129x |
| | 6 | 88.96 ± 6.76 | 36.90 | 7.40 ± 2.14 | 57.43 | 46.08 ± 3.72 | 1.133x |
| | 7 | 91.36 ± 4.24 | 35.23 | 6.95 ± 1.45 | 58.96 | 48.59 ± 4.08 | 1.244x |
| | 8 | 87.47 ± 7.09 | 36.72 | 7.33 ± 2.15 | 58.26 | 46.83 ± 4.49 | 1.135x |
| | 9 | 88.94 ± 7.06 | 32.03 | 6.46 ± 1.52 | 57.57 | 48.55 ± 6.13 | 1.187x |
| | 10 | 89.21 ± 6.75 | 38.25 | 7.25 ± 2.26 | 61.17 | 49.77 ± 4.61 | 1.204x |
| | 20 | 89.70 ± 7.11 | 38.46 | 7.14 ± 2.16 | 63.24 | 54.26 ± 8.86 | 1.257x |
| | 30 | 90.31 ± 6.94 | 38.75 | 7.23 ± 1.95 | 61.20 | 53.11 ± 7.67 | 1.247x |
| KDDcup | 3 | 95.41 ± 6.31 | 33.42 | 50.71 ± 12.55 | 64.41 | 47.69 ± 1.58 | 1.059x |
| | 4 | 95.30 ± 6.03 | 32.17 | 48.72 ± 8.02 | 65.89 | 49.20 ± 1.69 | 1.430x |
| | 5 | 94.37 ± 7.02 | 38.41 | 54.49 ± 18.42 | 69.87 | 50.38 ± 2.60 | 1.263x |
| | 6 | 94.42 ± 7.01 | 38.48 | 53.65 ± 12.75 | 71.95 | 52.26 ± 3.13 | 1.252x |
| | 7 | 95.75 ± 5.34 | 41.19 | 55.36 ± 17.52 | 74.18 | 54.09 ± 3.69 | 1.325x |
| | 8 | 95.94 ± 5.20 | 39.11 | 50.52 ± 13.79 | 76.67 | 56.07 ± 3.15 | 1.528x |
| | 9 | 95.63 ± 6.30 | 40.34 | 50.38 ± 20.19 | 80.33 | 58.17 ± 4.74 | 1.496x |
| | 10 | 96.14 ± 5.36 | 41.67 | 50.64 ± 17.67 | 81.53 | 59.24 ± 5.23 | 1.518x |
| | 20 | 96.20 ± 5.33 | 51.49 | 52.94 ± 18.57 | 95.86 | 65.50 ± 4.28 | 1.742x |
| | 30 | 95.00 ± 6.15 | 50.33 | 54.58 ± 27.23 | 91.24 | 64.68 ± 5.89 | 1.686x |
| 20-Mult. | 3 | 65.29 ± 1.15 | 118.13 | 25.52 ± 4.35 | 977.80 | 38.45 ± 0.42 | 1.021x |
| | 4 | 66.55 ± 1.43 | 129.45 | 27.37 ± 5.50 | 997.36 | 37.09 ± 0.36 | 1.018x |
| | 5 | 68.11 ± 1.79 | 122.27 | 25.98 ± 4.22 | 994.67 | 36.98 ± 0.29 | 1.037x |
| | 6 | 67.87 ± 1.59 | 115.87 | 24.97 ± 3.46 | 980.32 | 37.15 ± 0.62 | 1.039x |
| | 7 | 69.44 ± 1.59 | 117.88 | 25.29 ± 4.40 | 984.98 | 37.81 ± 0.60 | 1.038x |
| | 8 | 69.67 ± 1.69 | 115.97 | 25.00 ± 5.24 | 972.22 | 38.24 ± 0.75 | 1.056x |
| | 9 | 69.82 ± 1.86 | 123.76 | 25.78 ± 5.88 | 1007.83 | 38.56 ± 0.66 | 1.073x |
| | 10 | 70.43 ± 1.57 | 127.35 | 25.48 ± 6.42 | 1048.23 | 39.25 ± 0.79 | 1.080x |
| | 20 | 73.09 ± 1.97 | 126.28 | 24.44 ± 4.74 | 1089.81 | 41.82 ± 1.31 | 1.117x |
| | 30 | 73.80 ± 2.70 | 140.39 | 26.45 ± 5.71 | 1115.85 | 42.50 ± 1.60 | 1.126x |

with a peak occurring for the KDDcup classification problem and a high level of selection pressure whereby highly fit solutions are more influential. Indeed, for all problem instances, the greatest efficiency savings are once again achieved for high selection pressure. It should be noted though that when using increased tournament sizes, the average execution time tends to be greater as a result of larger GP trees being considered. Thus, a tradeoff needs to be considered between efficiency savings and the increase in the size of the average GP tree.

From the results observed in Table 4, using the elitism operator should provide an additional efficiency of approximately 7%. Observing the efficiency savings in Table 5 and comparing with the previous results in Table 3, efficiency savings have improved from between 8% and 12%. Thereby, it can be considered that use of the elitism operator has further benefited identification of candidate GP programs who *mathematically* cannot *win* tournaments at an earlier point in the evaluation phase. In terms of performance compared to a standard implementation of GP which uses elitism and does not reevaluate non-modified candidate GP programs, classifier training on the KDDcup problem occurs up to 1.74x faster. A brief mention of Genetic Programming Operations per Second (GPop/s) should be made with a maximum effective GP speed of 96 billion GPop/s achieved for the KDDcup classification problem and 1116 billion GPop/s for the multiplexer problem benefiting from an extra 32x bitwise parallelism.

## 5    Conclusions

In this paper a methodology for evaluating candidate GP programs is presented which provides significant computational efficiency savings even when embedded within a high performance parallel model. This methodology exploits tournament selection in that it is possible to identify *losers* of tournaments before evaluation of all the fitness cases is complete. Essentially, by evaluating all GP programs on subsets of fitness cases before moving to the next subset, comparisons can be made between solutions and hence *losers* of tournaments can be identified and early termination of the evaluation of these solutions achieved. This approach was shown to provide minor efficiency savings and hence runtime speedups. However, this paper discovers that the true advantage of the technique arises when solutions survive intact into the next generation. These solutions have *won* tournaments so are highly fit with a known fitness enabling much earlier detection of *losers* of tournaments especially when using high selection pressure. Efficiency savings of up to 65% and subsequent speedups in execution speed of up to 1.74x were demonstrated with a peak rate of 96 billion GPop/s achieved by GP running on a multi-core CPU. Further work needs to consider alternative fitness metrics such as the Mean Squared Error (MSE) for regression problems, combining the technique with sampling approaches and alternative methods for correctly predicting tournament *losers* earlier.

# References

1. Augusto, D.A., Barbosa, H.J.: Accelerated parallel genetic programming tree evaluation with OpenCL. J. Parallel Distrib. Comput. **73**(1), 86–100 (2013)
2. Cano, A., Zafra, A., Ventura, S.: Speeding up the evaluation phase of GP classification algorithms on GPUs. Soft Comput. **16**(2), 187–202 (2012)
3. Chitty, D.M.: Fast parallel genetic programming: multi-core CPU versus many-core GPU. Soft Comput. **16**(10), 1795–1814 (2012)
4. Chitty, D.M.: Improving the performance of GPU-based genetic programming through exploitation of on-chip memory. Soft Comput. **20**(2), 661–680 (2016)
5. Chitty, D.M.: Faster GPU-based genetic programming using a two-dimensional stack. Soft Comput. **21**(14), 3859–3878 (2017)
6. Frank, A., Asuncion, A.: UCI machine learning repository (2010)
7. Gathercole, C., Ross, P.: Dynamic training subset selection for supervised learning in genetic programming. In: International Conference on Parallel Problem Solving from Nature, pp. 312–321. Springer (1994)
8. Gathercole, C., Ross, P.: Tackling the boolean even N parity problem with genetic programming and limited-error fitness. Genet. Program. **97**, 119–127 (1997)
9. Koza, J.R.: Genetic programming (1992)
10. Maxwell, S.R.: Experiments with a coroutine execution model for genetic programming. In: Proceedings of the First IEEE Conference on Evolutionary Computation. IEEE World Congress on Computational Intelligence, pp. 413–417. IEEE (1994)
11. Park, N., Kim, K., McKay, R.I.: Cutting evaluation costs: an investigation into early termination in genetic programming. In: 2013 IEEE Congress on Evolutionary Computation (CEC), pp. 3291–3298. IEEE (2013)
12. Poli, R., Langdon, W.B.: Backward-chaining evolutionary algorithms. Artif. Intell. **170**(11), 953–982 (2006)
13. Poli, R., Langdon, W.: Running genetic programming backwards. In: Yu, T., Riolo, R., Worzel, B. (eds.) Genetic Programming Theory and Practice III, Genetic Programming, vol. 9, pp. 125–140. Springer (2006)
14. Teller, A.: Genetic programming, indexed memory, the halting problem, and other curiosities. In: Proceedings of the 7th Annual Florida Artificial Intelligence Research Symposium, pp. 270–274 (1994)
15. Teller, A., Andre, D.: Automatically choosing the number of fitness cases: the rational allocation of trials. Genet. Program. **97**, 321–328 (1997)

# Modelling and Representation

# Solving Partial Differential Equations with Bernstein Neural Networks

Sina Razvarz[1], Raheleh Jafari[2(✉)], and Alexander Gegov[3]

[1] Departamento de Control Automatico, CINVESTAV-IPN (National Polytechnic Institute), Mexico City, Mexico
srazvarz@yahoo.com
[2] Centre for Artificial Intelligence Research (CAIR), University of Agder, Grimstad, Norway
Jafari3339@yahoo.com
[3] School of Computing, University of Portsmouth, Buckingham Building, Portsmouth PO13HE, UK
alexander.gegov@port.ac.uk

**Abstract.** In this paper, a neural network-based procedure is suggested to produce estimated solutions (controllers) for the second-order nonlinear partial differential equations (PDEs). This concept is laid down so as to produce a prevalent approximation on the basis of the learning method which is at par with quasi-Newton rule. The proposed neural network contains the regularizing parameters (weights and biases), that can be utilized for making the error function least. Besides, an advanced technique is presented for resolving PDEs based on the usage of Bernstein polynomial. Numerical experiments alongside comparisons show the fantastic capacity of the proposed techniques.

**Keywords:** Neural network · Bernstein polynomial
Partial Differential Equations

## 1 Introduction

Exact solution of differential equation plays a noteworthy role in the fitting seizing of qualitative characters of several processes as well as occurrences at par with several zones of natural science. Exact solutions authorize researchers for designing and carrying out experiments by developing valid natural conditions for determining these parameters or functions. However, obtaining the exact solutions of the PDEs is complicated and is case specific.

Several techniques have been proposed in literature in order to resolve different types of PDEs. In [1] the Homotopy perturbation technique is utilized to obtain the solution of PDEs with variable coefficients. In [2] the resolving of the PDEs requires two-dimensional differential transformation techniques. In [3] the modified technique of simplest equation is employed for extracting exact analytical solutions of nonlinear PDEs. In [4] an iteration technique in order to solve both linear as well as nonlinear wave equations is analyzed. In the following, some numerical solutions are laid down that have been suggested by other researchers. In [5], implementation of a spreadsheet

© Springer Nature Switzerland AG 2019
A. Lotfi et al. (Eds.): UKCI 2018, AISC 840, pp. 57–70, 2019.
https://doi.org/10.1007/978-3-319-97982-3_5

program produces the numerical solution of the hyperbolic equation. Some researchers also generate an array solution that includes the value of the solution at a selected group of points [6].

Evje et al. [7] utilized an explicit monotone difference technique in order to estimate the entropy solutions related to degenerate parabolic equation. Bulbul et al. [8] proposed a Taylor polynomial estimation for the solution of hyperbolic type PDEs having constant coefficients. In [9], the researchers analyzed double non-traveling wave solutions of two systems associated with nonlinear PDEs. In [10] convolution quadrature is employed to the time-domain boundary integral formulation related to the wave equation having non-zero initial conditions. In [11], Martinez worked on a linear wave equation having a boundary damping term. Catania et al. For prevailing results related to the feedback control of the wave equation we suggest [12] and for open-loop control of the wave equation we refer [13]. However, the above discussed techniques are very complicated. Since the solutions associated with PDE are considered to be uniformly continuous, also the problems linked to the viable sets are usually compact, neural networks are best suited candidates in order to estimate viability problems [14].

Neural network finds its application in the fields of mathematics, chemistry, physics, and numerous applications [15–20]. They have become recently popular for solving PDEs. In [21] a feed-forward neural network is laid down in order to resolve an elliptic PDE in 2D. Another methodology is proposed in [22] for resolving a class of first-order PDEs on the basis of Multilayer neural networks. In [23] an unsupervised neural network is suggested in order to solve the nonlinear Schrodinger equation. In [24] the solutions of vibration control problems by utilizing artificial neural networks is discussed. In [25] a controlled heat problem up to three decimal digits accuracy is resolved by utilizing feed forward neural networks.

In this paper, a methodology based on neural networks is proposed in order to solve the strongly degenerate parabolic equations. The trial solution related to the PDE is stated as an addition of two parts. The primary part suffices the initial and boundary conditions, and does not have adjustable parameters. The secondary part includes a neural network having adjustable parameters (weights and biases). Further-more, a superiorly modified technique is laid down in order to solve a wave equation in concerned with the application of Bernstein neural networks, that contains an excellent attributes of Bernstein polynomial. The Bernstein polynomial extracts its position in the theory of estimation considering the fact that it has good uniform approximation capability for continuous functions. These polynomials are suitable to produce a smooth estimation for equal distance knots [26]. The implementation of Bernstein polynomials is suggested in this paper, because it is extensively simple to apply. Also it is continuously differentiable due to the nature of theoretical contents.

The rest of this paper is organized as follows. Section 2 provides a summarized description of PDEs. An innovative method for resolving PDEs based on the neural networks is proposed in this section. Also, an advanced method is supplied for solving PDEs based on the application of a Bernstein neural network named as dynamic model. Experiments, simulation results, and comparisons are completed and discussed in Sect. 3. Section 4 concludes the paper.

## 2 Nonlinear System Modeling with Partial Differential Equations

**Definition 1 (Second-order nonlinear PDE).** The second-order singular nonlinear PDE can be illustrated by utilizing the equation mentioned below

$$\frac{\partial^2 v(x,t)}{\partial t^2} + \frac{2}{t}\frac{\partial v(x,t)}{\partial t} = F\left(x, v(x,t), \frac{\partial v(x,t)}{\partial x}, \frac{\partial^2 v(x,t)}{\partial x^2}\right) \qquad (1)$$

in which $t$ and $x$ are independent variables, $v$ is the dependent variable, $F$ is a nonlinear function of $x, v, v_x$ and $v_{xx}$, also the initial conditions for the PDE (1) are illustrated as below

$$v(x,0) = f(x), \quad v_t(x,0) = g(x)$$

**Definition 2 (Strongly degenerate parabolic equations).** The strongly degenerate parabolic equation is described as

$$v_t + Q(v)_x = A(v)_{xx}, (x,t) \in \Pi_T := [0,1] \times (0,T), \quad T > 0 \qquad (2)$$

with boundary conditions

$$v(x,0) = g_0(x), \quad v(0,t) = f_0(t), \quad v(1,t) = f_1(t) \qquad (3)$$

in which the integrated diffusion coefficient $A$ is demonstrated by

$$A(v) = \int_0^v a(s)ds, \quad a(v) \geq 0, a \in L^\infty([0,1]) \cap L^1([0,1]) \qquad (4)$$

The function $a$ is permitted to disappear on $v$-intervals of positive length, on which Eq. (2) degenerates to a first-order scalar conservation law. Therefore, Eq. (2) is named as strongly degenerate.

**Definition 3 (wave equation).** The Cauchy problem related to the wave equation in one dimension is defined as

$$\frac{\partial^2 v(x,t)}{\partial t^2} = c^2 \frac{\partial^2 v(x,t)}{\partial x^2}, \quad (x,t) \in [0,a] \times [0,b] \qquad (5)$$

with

$$v(x,0) = \phi(x), \quad v_t(x,0) = \psi(x)$$

where $a$ as well as $b$ are constants. In the above mentioned equation, the parameter $c$ is termed as the speed of light.

## 2.1    Controller Design with Neural Networks Approximation

Here, we construct a neural network for resolving the strongly degenerate parabolic equation that obtains the solution of differential equations in a closed analytic and differentiable form (Fig. 1). The relation between the input-output of each unit in the suggested neural architecture is described as follows

- Input units

$$o_1^1 = x, o_2^1 = t$$

- Hidden units

$$o_j^2 = F\left(b_j + w_j^1 x + w_j^2 t\right), \quad j = 1, \ldots, m$$

- Output unit:

$$N(x,t) = \sum_{j=1}^{m} \left(W_j o_j^2\right)$$

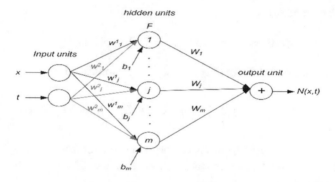

**Fig. 1.**  Neural network equivalent to strongly degenerate parabolic equations

The activation function of the hidden units in this neural network is $F(r) = \frac{2}{1+e^{-2r}} - 1$ (tan-sigmoid function). The trial solution associated with (2) is portrayed as

$$v_m(x,t) = (1-x)f_0(t) + xf_1(t) + (1-t)\{g_0(x) - [(1-x)g_0(0) + xg_0(1)]\} \\ + x(1-x)tN(x,t)$$

where

$$N(x,t) = \sum_{j=1}^{m} \left(W_j F\left(b_j + w_j^1 x + w_j^2 t\right)\right)$$

The least mean square error is obtained for $(x,t) = (x_i, t_j)$ as below

$$E_{i,j} = \frac{1}{2}\left(M_{i,j}\right)^2$$

where

$$M_{i,j} = \frac{\partial v_m(x,t)}{\partial t}\Big|_{\substack{x = x_i \\ t = t_j}} + \frac{\partial Q(v_m(x,t))}{\partial x}\Big|_{\substack{x = x_i \\ t = t_j}} - \frac{\partial^2 A(v_m(x,t))}{\partial x^2}\Big|_{\substack{x = x_i \\ t = t_j}}$$

In order to adjust the parameters we utilize Newton's rule. The standard self-learning process works as below

$$W_q(r+1) = W_q(r) - \mu(r)\frac{\partial E_{i,j}}{\partial w_q} + \gamma\left[W_q(r) - W_q(r-1)\right] \tag{6}$$

Now, the explicit technique of Eq. (6) is illustrated as

$$\begin{bmatrix} W_1 \\ \vdots \\ W_m \end{bmatrix}_{r+1} = \begin{bmatrix} W_1 \\ \vdots \\ W_m \end{bmatrix}_r - \frac{\left(\nabla E_{i,j}(W)_r\right)^T \nabla E_{i,j}(W)_r}{\left(\nabla E_{i,j}(W)_r\right)^T Q_r \nabla E_{i,j}(W)_r} \nabla E_{i,j}(W)_r + \gamma \begin{bmatrix} \Delta W_1 \\ \vdots \\ \Delta W_m \end{bmatrix}_{r-1} \tag{7}$$

where

$$\nabla E_{i,j}(W) = \left(\frac{\partial E_{i,j}}{\partial W_1}, \ldots, \frac{\partial E_{i,j}}{\partial W_m}\right)^T$$

and

$$Q = \begin{bmatrix} \frac{\partial^2 E_{i,j}}{\partial W_1^2} & \frac{\partial^2 E_{i,j}}{\partial W_1 \partial W_2} & \cdots & \frac{\partial^2 E_{i,j}}{\partial W_1 \partial W_m} \\ \frac{\partial^2 E_{i,j}}{\partial W_2 \partial W_1} & \frac{\partial^2 E_{i,j}}{\partial W_2^2} & \cdots & \frac{\partial^2 E_{i,j}}{\partial W_2 \partial W_m} \\ \cdots & \cdots & \cdots & \cdots \\ \frac{\partial^2 E_{i,j}}{\partial W_m \partial W_1} & \frac{\partial^2 E_{i,j}}{\partial W_m \partial W_2} & \cdots & \frac{\partial^2 E_{i,j}}{\partial W_m^2} \end{bmatrix}$$

The chain rule for differentiation can be illustrated as

$$\frac{\partial E_{i,j}}{\partial W_q} = \frac{\partial E_{i,j}}{\partial M_{i,j}} \cdot \frac{\partial M_{i,j}}{\partial W_q} = M_{i,j}\frac{\partial M_{i,j}}{\partial W_q}$$

The above learning procedure can be extended to other network parameters ($w_q^1$, $w_q^2$ and $b_q$) in a same way.

## 2.2    Controller Design with Bernstein Neural Networks Approximation

We carry forward our strategy for resolving wave equations by utilizing two patterns of Bernstein neural networks. Let us take into consideration the Cauchy problem (5), where the solution v relies on both spatial as well as temporal variables $x$ and $t$ respectively. The trial solution associated with (5) in the form of the Bernstein neural network is portrayed as

$$v_m(x,t) = \phi(x) + t\psi(x) + t\left[B(x,t) - B(x,0) - \frac{\partial B(x,0)}{\partial t}\right]$$

where $B(x,t)$ is the bivariate Bernstein polynomial series of solution function $v(x,t)$, termed as

$$B(x,t) = \sum_{i=0}^{n}\sum_{j=0}^{m}\binom{n}{i}\binom{m}{j}\frac{x^i(a-x)^{n-i}}{a^n}\frac{t^j(b-t)^{m-j}}{b^m}q_{i,j}(x,t), \quad n,m \in N$$

$q_{i,j}$ is the coefficient. We can state the above relation as

$$B(x,t) = \sum_{i=0}^{n}\sum_{j=0}^{m}\beta_{i,j}x^i(a-x)^{n-i}t^j(b-t)^{m-j}q_{i,j}(x,t),$$
$$n,m \in N, \beta_{i,j} = \frac{1}{a^n b^m}\binom{n}{i}\binom{m}{j} \tag{8}$$

where $\binom{n}{i} = \frac{n!}{i!(n-i)!}$ and $\binom{m}{j} = \frac{m!}{j!(m-j)!}$

Taking into account the follow relations

$$\frac{\partial^2 v_m(x,t)}{\partial x^2} = \phi''(x) + t\psi''(x) + t\left[\frac{\partial^2 B(x,t)}{\partial x^2} - \frac{\partial^2 B(x,0)}{\partial x^2} - \frac{\partial^2 \partial B(x,0)}{\partial x^2 \partial t}\right]$$

and

$$\frac{\partial^2 v_m(x,t)}{\partial t^2} = 2\frac{\partial B(x,t)}{\partial t} + t\frac{\partial^2 B(x,t)}{\partial t^2}$$

replacing the above relations in the origin problem (5) gives us the following differential equation

$$2\frac{\partial B(x,t)}{\partial t} + t\frac{\partial^2 B(x,t)}{\partial t^2} = c^2\left(\phi''(x) + t\psi''(x) + t\left[\frac{\partial^2 B(x,t)}{\partial x^2} - \frac{\partial^2 B(x,0)}{\partial x^2} - \frac{\partial^2 \partial B(x,0)}{\partial x^2 \partial t}\right]\right) \tag{9}$$

$$(x,t) \in [0,a] \times [0,b]$$

For simplicity the above relation can be justified as mentioned below

$$\sum_{i=0}^{n} \sum_{j=0}^{m} \xi_{i,j}(x,t) q_{i,j}(x,t) = \zeta(x,t), \quad (x,t) \in [0,a] \times [0,b] \tag{10}$$

Where

$$\xi_{i,j}(x,t) = 2\beta_{i,j}x^i(a-x)^{n-i}\left(jt^{j-1}(b-t)^{m-j}-(m-j)t^j(b-t)^{m-j-1}\right)$$
$$+ t\beta_{i,j}x^i(a-x)^{n-i}\left(j(j-1)t^{j-2}(b-t)^{m-j}-2j(m-j)t^{j-1}(b-t)^{m-j-1}\right.$$
$$+ (m-j)(m-j-1)t^j(b-t)^{m-j-2}\right) + c^2 t\beta_{i,j}\left(i(i-1)x^{i-2}(a-x)^{n-i}\right.$$
$$- 2i(n-i)x^{i-1}(a-x)^{n-i-1} + (n-i)(n-i-1)x^i(a-x)^{n-i-2}\right)t^j(b-t)^{m-j}$$
$$- c^2 t\beta_{i,j}\left(i(i-1)x^{i-2}(a-x)^{n-i}-2i(n-i)x^{i-1}(a-x)^{n-i-1}\right.$$
$$+ (n-i)(n-i-1)x^i(a-x)^{n-i-2}\right)\left(jt^{j-1}(b-t)^{m-j}-(m-j)t^j(b-t)^{m-j-1}\right)$$

and

$$\zeta(x,t) = c^2\left(\phi''(x) + t\psi''(x) + t\frac{\partial^2 B(x,0)}{\partial x^2}\right)$$

The Bernstein neural network (8) is shown in Fig. 2.

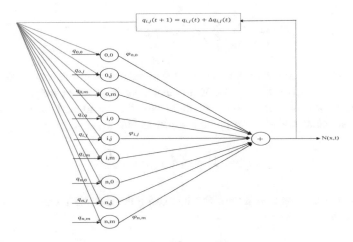

**Fig. 2.** Dynamic Bernstein neural network

In the above architecture the mathematical symbol is defined as

$$\varphi_{i,j} = \sum_{i=0}^{n} \sum_{j=0}^{m} \binom{n}{i} \binom{m}{j} \frac{x^i (a-x)^{n-i}}{a^n} \frac{t^j (b-t)^{m-j}}{b^m}, \quad n, m \in N$$

The relation between the input-output of each unit in the suggested neural architecture is described as follows

- Input unit

$$o_{i,j} = q_{i,j} \quad i = 0, \ldots, n, \quad j = 0, \ldots, m$$

- Output unit

$$N(x,t) = \varphi_{i,j} o_{i,j}$$

Now, an appropriate numerical methodology should be capable enough to supply a suitable tool in order to measure and compute the preciseness of the extracted solution. Laying down the cost function $E_{i,j}$ over the model parameters makes it a good predictor. The least mean square error is stated to be as one of the most talked usable cost function associated with $E_{i,j}$. Assume that $0 \leq x \leq 1$, the rectangle $[0,1] \times [0,T]$ is divided into $nn'$ mesh points $(x_i, t_j) = ((i-1)h, (j-1)h'), h = \frac{1}{n-1}, h' = \frac{T}{n'-1}$, $(i = 1, \ldots, n; j = 1, \ldots, n')$. Therefore for comparing the exact solution with its extracted one, the least mean square error is utilized that is described as mentioned below

$$E_{i,j} = \frac{1}{2} (\sum_{i=0}^{n} \sum_{j=0}^{m} \xi_{i,j}(x,t) q_{i,j}(x,t) - \zeta(x,t))^2$$

We utilize Newton's rule as mentioned in (12) in order to adjust the parameters in such a manner that the network error attains minimal over the space of weights setting. The initial parameter $q_{i,j}$ is chosen on a random basis in order to start the procedure. The illustrated standard self-learning process works as below

$$q_{i,j}(r+1) = q_{i,j}(r) - \mu(r) \frac{\partial E_{i,j}}{\partial q_{i,j}}$$

where $\mu$ is the training rate $\mu > 0$. For increasing the training process, a momentum term is added up as follows

$$q_{i,j}(r+1) = q_{i,j}(r) - \mu(r) \frac{\partial E_{i,j}}{\partial q_{i,j}} + \gamma [q_{i,j}(r) - q_{i,j}(r-1)] \tag{11}$$

where $\gamma > 0$. The index $r$ is the iteration number. Also, $q_{i,j}(r+1)$ and $q_{i,j}(r)$ represents the updated and recent output weight values, respectively. Now, the explicit technique of Eq. (11) is illustrated as

$$
\begin{bmatrix} q_{0,0} \\ \vdots \\ q_{n,m} \end{bmatrix}_{r+1} = \begin{bmatrix} q_{0,0} \\ \vdots \\ q_{n,m} \end{bmatrix}_r - \frac{\left(\nabla E_{i,j}(q)_r\right)^T \nabla E_{i,j}(q)_r}{\left(\nabla E_{i,j}(q)_r\right)^T Q_r \nabla E_{i,j}(q)_r} \nabla E_{i,j}(q)_r + \gamma \begin{bmatrix} \Delta q_{0,0} \\ \vdots \\ \Delta q_{n,m} \end{bmatrix}_{r-1} \tag{12}
$$

where

$$
\nabla E_{i,j}(q) = \left( \frac{\partial E_{i,j}}{\partial q_{0,0}}, \ldots, \frac{\partial E_{i,j}}{\partial q_{n,m}} \right)^T
$$

and

$$
Q = \begin{bmatrix} \frac{\partial^2 E_{i,j}}{\partial q_{0,0}^2} & \frac{\partial^2 E_{i,j}}{\partial q_{0,0} \partial q_{1,1}} & \cdots & \frac{\partial^2 E_{i,j}}{\partial q_{0,0} \partial q_{n,m}} \\ \frac{\partial^2 E_{i,j}}{\partial q_{1,1} \partial q_{0,0}} & \frac{\partial^2 E_{i,j}}{\partial q_{1,1}^2} & \cdots & \frac{\partial^2 E_{i,j}}{\partial q_{1,1} \partial q_{n,m}} \\ \cdots & & \ddots & \\ \frac{\partial^2 E_{i,j}}{\partial q_{n,m} \partial q_{0,0}} & \frac{\partial^2 E_{i,j}}{\partial q_{n,m} \partial q_{1,1}} & \cdots & \frac{\partial^2 E_{i,j}}{\partial q_{n,m}^2} \end{bmatrix} \tag{13}
$$

are calculated at the current mesh points $(x_i, t_j)$ In this case, $Q$ is the Hessian matrix with components $\frac{\partial^2 E_{i,j}}{\partial q_{i,j} \partial q_{\tilde{i}\tilde{j}}}$ (for $i, \tilde{i} = 0, \ldots, n; j, \tilde{j} = 0, \ldots, m$). It is clear that the convergence speed is in direct relation at par with the learning rate parameter. For attaining the optimal learning rate in concerned with rapid convergence of our learning optimization rule, the inverse of Hessian matrix $Q$ of the error function $E_{i,j}$ is introduced at the current mesh points. The Newton technique mentioned above is very much capable for scaling the descent step in each step. Now, the partial derivative $\frac{\partial E_{i,j}}{\partial q_{i,j}}$ can be extracted at the current weight values.

## 3    Numerical Results and Discussion

All computations mentioned in the following tables are obtained by utilizing Matlab.

**Example 1.** The Buckley-Leverett differential equation is portrayed as [27]

$$
\frac{\partial v(x,t)}{\partial t} + \frac{\partial g(v(x,t))}{\partial x} = \frac{\partial^2 A(v(x,t))}{\partial x^2}
$$

Where

$$
g(v) = \frac{v^2}{v^2 + (1-v)^2}, a(v) = 4\varepsilon v(1-v),
$$

on the domain $(x, t) \in [0, 1] \times [0, 0.5]$ with initial condition

$$v_0(x) = \begin{cases} 0 & x < 0.1 \\ 1 & otherwise \end{cases}$$

and boundary conditions

$$v(0, t) = 1, \quad v(1, t) = 0$$

also, $\varepsilon = 0.01$

The following characteristics are taken into consideration

I. *Time step:* $k' = 0.98 \frac{k^2}{k\|g'\|_\infty + 2\varepsilon\|a\|_\infty}$

II. $L^1$-error $E_{mid} = \left( \sum_{i=1}^{n} \sum_{j=1}^{n'} |v(x_i, t_j)| \right)^{-1} \sum_{i=1}^{n} \sum_{j=1}^{n'} |v(x_i, t_j) - v_m(x_i, t_j)|$

where $v_m(x_i, t_j)$ as well as $v(x_i, t_j)$ are termed as the calculated solution and exact value of reference solution at grid point $(x_i, t_j)$, respectively.

This problem is solved by applying the methodology of neural network proposed in this paper. Comparisons between the suggested algorithm on different training steps and the discrete mollification scheme [27] with support parameter $\vartheta$ are displayed in Table 1.

**Table 1.** Approximation errors of neural network based algorithm and mollified method

| $1/k$ | Mollified method | | | r | NN method | | |
|---|---|---|---|---|---|---|---|
| | $\vartheta = 3$ | $\vartheta = 5$ | $\vartheta = 8$ | | $m = 7$ | $m = 11$ | $m = 17$ |
| 64 | 2.6105e−2 | 2.5327e−2 | 2.5055e−2 | 15 | 2.4021e−2 | 1.8035e−2 | 1.0012e−2 |
| 128 | 1.4932e−2 | 1.4287e−2 | 1.4133e−2 | 30 | 8.0207e−3 | 6.8025e−3 | 5.5215e−3 |
| 256 | 8.3709e−3 | 7.9698e−3 | 7.6833e−3 | 45 | 1.9548e−3 | 1.0008e−3 | 8.9025e−4 |
| 512 | 4.5075e−3 | 4.3271e−3 | 4.1141e−3 | 60 | 4.9925e−4 | 3.1875e−4 | 1.9011e−4 |
| 1024 | 1.9997e−3 | 1.9335e−3 | 1.8279e−3 | 75 | 7.3081e−5 | 5.8295e−5 | 4.6952e−5 |

**Example 2.** The following wave equation is taken into consideration that models the motion associated with the guitar string of length L

$$\frac{\partial^2 v(x, t)}{\partial t^2} = c^2 \frac{\partial^2 v(x, t)}{\partial x^2}$$

with the boundary conditions

$$v(0, t) = \sin(\pi t), \quad v(L, t) = 0$$

on the domain $(x,t) \in [0,L] \times [0,T]$ initial position and velocity

$$v(x,0) = 0, \quad v_t(x,0) = \pi \cos(\pi x)$$

In the proposed problem $c^2 = \frac{T_s}{\rho}$, where $T_s$ is taken to be the tension in the string, also $\rho$ is the density of the string. The specifications mentioned here are given by $L = 1, T = 4, T_s = 2\frac{N}{m}$ and $\rho = 2\frac{kg}{m^3}$. The exact solution related to the problem is $v(x,t) = \cos(\pi x)\sin(\pi t)$.

We use dynamic Bernstein neural network (DNN) to approximate the solution. To compare our results, we use the other two popular methods: 3-point explicit method and optimal explicit method [28]. The exact solution is displayed in Fig. 3. Corresponding approximated error plots are shown in Fig. 4.

**Example 3.** Two semi-infinite strings of different densities are joined as [29]

$$\frac{\partial^2 v(x,t)}{\partial t^2} = \left(c_1^2 + c_2^2\right)\frac{\partial^2 v(x,t)}{\partial x^2}$$

with the boundary conditions

$$v(0,t) = \cos(\pi t), \quad v(L,t) = 0$$

On the domain $(x,t) \in [0,T]$ initial position and velocity

$$v(x,0) = \cos(\pi x), \quad v_t(x,0) = 0$$

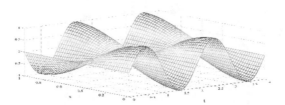

**Fig. 3.** Plot of the exact solution

See Fig. 5. In the proposed problem $c = \sqrt{\frac{E}{\rho}}$, where E is the Young's modulus, also $\rho$ is the density of the rod. The specifications mentioned here are given by $L = 1$, $T = 5, E_1 = 2\frac{kg}{m.s^2}, \rho_1 = 2.882\frac{kg}{m^3}, E_2 = 4.3\frac{kg}{m.s^2}$ and $\rho_1 = 15.136\frac{kg}{m^3}$. The exact solution related to the problem is $v(x,t) = \frac{1}{2}(\cos(\pi(x+t)) + \cos(\pi(x-t)))$. The exact solution is displayed in Fig. 6. Figure 7 shows the approximated error with DNN.

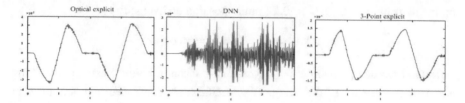

**Fig. 4.** Plot of the approximated error with 3-point explicit, optimal explicit, SNN and DNN

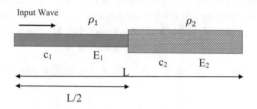

**Fig. 5.** Two semi-infinite strings of different densities

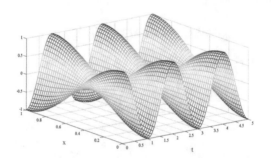

**Fig. 6.** Plot of the exact solution

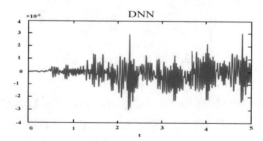

**Fig. 7.** Plot of the approximated error with DNN for r = 70

# 4   Conclusions

This paper proposes a method based on the neural networks for finding the solutions of the second-order nonlinear PDEs. It is controller design process. The trial solution related to the PDE is stated as an addition of two parts. The primary part suffices the initial and boundary conditions, also does not have adjustable parameters. The secondary part includes a neural network having adjustable parameters (weights and biases). Also, a type of Bernstein neural networks namely dynamic model is proposed in order to resolve the PDEs. For obtaining the superior estimated solution of PDEs, the adjustable parameters at par with the Bernstein neural network are adjusted in suitable manner by implementing quasi-Newton learning algorithm.

Numerical examples as well as comparison with solution obtained by the employing other numerical methodologies open up that the use of neural networks based on quasi-Newton learning rule provides solutions with superior generalization and major accuracy. The future work is the application of the mentioned methodologies for system of PDEs.

# References

1. Jin, L.: Homotopy perturbation method for solving partial differential equations with variable coefficients. Int. J. Contemp. Math. Sci. **3**, 1395–1407 (2008)
2. Ayaz, F.: On the two-dimensional differential transform method. Appl. Math. Comput. **143**, 361–374 (2003)
3. Vitanov, N.K., Dimitrova, Z.I., Kantz, H.: Modified method of simplest equation and its application to nonlinear PDEs. Appl. Math. Comput. **216**, 2587–2595 (2010)
4. Wazwaz, A.M.: The variational iteration method: A reliable analytic tool for solving linear and nonlinear wave equations. Comput. Math Appl. **54**, 926–932 (2007)
5. Kharab, A., Kharab, R.: Spreadsheet solution of hyperbolic partial differential equation. IEEE Trans. Educ. **40**, 103–110 (1997)
6. Kincaid, D., Cheney, W.: Numerical Analysis, Mathematics of Scientific computing. Brooks/Cole, Pacific Grove (1991)
7. Evje, S., Karlsen, K.H.: Monotone difference approximations of BV solutions to degenerate convection-diffusion equations. SIAM. J. Numer. Anal. **37**, 1838–1860 (2000)
8. Bulbul, B., Sezer, M.: Taylor polynomial solution of hyperbolic type partial differential equations with constant coefficients. Int. J. Comput. Math. **88**, 533–544 (2011)
9. Guo, S., Mei, L., Zhou, Y.: The compound G G' - expansion method and double non-traveling wave solutions of (2 + 1)-dimensional nonlinear partial differential equations. Comput. Math Appl. **69**, 804–816 (2015)
10. Falletta, S., Monegato, G., Scuderi, L.: A space-time BIE method for nonhomogeneous exterior wave equation problems. The Dirichlet case. IMA J. Numer. Anal. **32**, 202–226 (2012)
11. Martinez, P.: A new method to obtain decay rate estimates for dissipative systems. ESAIM Control Optim. Calc. Var. **4**, 419–444 (1999)
12. Gibson, J.S.: An analysis of optimal modal regulation: convergence and stability. SIAM J. Control Optim. **19**, 686–707 (1981)
13. Kroner, A., Kunisch, K.: A minimum effort optimal control problem for the wave equation. Comput. Optim. Appl. **57**, 241–270 (2014)

14. Cybenko, G.: Approximation by superposition of a sigmoidal function. Math. Control Sig. Syst. **2**, 303–314 (1989)
15. Jafari, R., Yu, W.: Uncertainty nonlinear systems modeling with fuzzy equations. In: Proceedings of the 16th IEEE International Conference on Information Reuse and Integration, pp. 182–188, San Francisco, Calif, USA, August (2015)
16. Jafari, R., Yu, W.: Uncertainty nonlinear systems control with fuzzy equations. In: IEEE International Conference on Systems, Man, and Cybernetics, pp. 2885–2890 (2015)
17. Jafari, R., Yu, W.: Uncertainty nonlinear systems modeling with fuzzy equations. Math. Probl. Eng. **2017**, 10 (2017)
18. Jafari, R., Yu, W., Li, X.: Numerical solution of fuzzy equations with Z-numbers using neural networks. Intell. Autom. Soft Comput. **23**, 1–7 (2017)
19. Jafari, R., Yu, W., Li, X., Razvarz, S.: Numerical solution of fuzzy differential equations with Z-numbers using Bernstein neural networks. Int. J. Comput. Intell. Syst. **10**, 1226–1237 (2017)
20. Razvarz, S., Jafari, R., Granmo, O.C., Gegov, A.: Solution of dual fuzzy equations using a new iterative method. In: Asian Conference on Intelligent Information and Database Systems, pp. 245–255 (2018)
21. Dissanayake, M.W.M.G., Phan-Thien, N.: Neural-network based approximations for solving partial differential equations. Commun. Numer. Meth. Eng. **10**, 195–201 (2000)
22. He, S., Reif, K., Unbehauen, R.: Multilayer neural networks for solving a class of partial differential equations. Neural Netw. **13**, 385–396 (2000)
23. Montelora, C., Saloma, C.: Solving the nonlinear Schrodinger equation with an unsupervised neural network: estimation of error in solution. Opt. Commun. **222**, 331–339 (2003)
24. Alli, H., Ucar, A., Demir, Y.: The solutions of vibration control problems using artificial neural networks. J. Franklin Inst. **340**, 307–325 (2003)
25. Sukavanam, N., Panwar, V.: Computation of boundary control of controlled heat equation using artificial neural networks. Int. Commun. Heat Mass Transf. **30**, 1137–1146 (2003)
26. Curtis, S., Ghosh, S.: A variable selection approach to monotonic regression with Bernstein polynomials. J. Appl. Stat. **38**, 961–976 (2011)
27. Acosta, C.D., Burger, R., Mejia, C.E.: Monotone difference schemes stabilized by discrete mollification for strongly degenerate parabolic equations. Numer. Meth. Part Differ. Equ. **28**, 38–62 (2012)
28. Dehghan, M.: On the solution of an initial-boundary value problem that combines neumann and integral condition for the wave equation. Numer. Meth. Partial Differ. Equ. **21**, 24–40 (2005)
29. Tongue, B.H.: Principles of Vibration. Oxford University Press, New York (2001)

# Daily Energy Price Forecasting Using a Polynomial NARMAX Model

Catherine McHugh[1(✉)], Sonya Coleman[1], Dermot Kerr[1], and Daniel McGlynn[2]

[1] Intelligent Systems Research Centre (ISRC), School of Computing, Engineering and Intelligent Systems, Ulster University, Londonderry, UK
{mchugh-c24, sa.coleman, d.kerr}@ulster.ac.uk
[2] Click Energy, Londonderry, UK
daniel.mcglynn@clickenergyni.com

**Abstract.** Energy prices are not easy to forecast due to nonlinearity from seasonal trends. In this paper a Nonlinear AutoRegressive Moving Average model with eXogenous input (NARMAX model) is created using nonlinear energy price data. To investigate if a short-term forecasting model is capable of predicting energy prices a model was developed using daily data from 2017 over a period of five weeks: observing 1 input lag prediction up to 12 input lag prediction for low-order polynomials (linear, quadratic, and cubic). Various input factors were explored (energy demand and previous price) with different combinations to observe which factors, if any, had an impact on the current price prediction. The results show that the generated NARMAX model is good at describing the input-output relationship of energy prices. The model works best with a low-order input regression parameter and linear polynomial degree. It was also noted that including energy demand as an input factor slightly improves the model validation results suggesting that there is a relationship between demand and energy prices.

**Keywords:** NARMAX modelling · Energy price forecasting · Polynomial Machine learning

## 1 Introduction

Many computational intelligence methods have been used for forecasting within the energy sector. Forecasting energy price is quite a difficult task due to the complex behavior of the data, for instance displaying nonlinearity [1]; therefore it is best to use machine learning algorithms trained with realistic data to generate a model that can predict future outputs. Machine learning algorithms are designed to make predictions by learning from data without relying on rules based programming [2]. The NARMAX system identification technique is used to obtain a model based on measures of the system inputs and outputs and when the data shows nonlinear traits since information regarding past error can be incorporated into this model to help enhance future prediction [3].

© Springer Nature Switzerland AG 2019
A. Lotfi et al. (Eds.): UKCI 2018, AISC 840, pp. 71–82, 2019.
https://doi.org/10.1007/978-3-319-97982-3_6

When predicting electricity prices for energy trading there are other factors (for example, demand, fuel costs, weather [4]), as well as historical electricity price, that need to be considered as input parameters in the forecasting model. Along with input factors, another important feature is the length of the prediction window. Short-term energy forecasting (days or weeks) tends to be more favorable since, due to the volatile nature of energy, there is a smaller timeframe to balance demand and supply [5].

This paper examines price forecasting through utilizing energy data with a trained NARMAX model over different periods of time. The methodology applied is appropriate to deal with non-linear energy data and is a good technique to model input-output relationships. The proposed model aims to predict day-ahead energy prices and this knowledge would be beneficial to the energy industry to know when is best to buy or sell in the market.

The remainder of the paper is divided into the following sections: literature relating to energy price forecasting and non-linear models is discussed in the remainder of Sect. 1; the approaching to computational modelling is detailed in Sect. 2 and results are presented and summarized in Sect. 3, with concluding remarks discussed in Sect. 4.

## 1.1   Energy Price Forecasting

Energy market participants who use algorithms that provide accurate price predictions can increase their profits over time in buying and selling of electricity [6]. Mosbah and El-Hawary [1] looked at short-term (next month) forecasting by applying a multilayer neural network to train with previous hourly data (load, gas, and temperature) and observed that averaging the output errors (parallel topology) enhanced performance compared to collecting errors at each stage of the training (cascade topology).

The balance between supply and demand needs to be stable to conquer volatility. Energy time-series data display spikes and nonstationary behavior hence energy price forecasting is vital to help tackle volatility [5]. Vijayalakshmi and Girish [6] investigated if Artificial Neural Network (ANN) models were the answer for short-term forecasting and found that time-series models did provide better predictions.

Price forecasting varies in length from short (day or weeks), medium (weeks or months), or long (months or years). Gao et al. [7] analyzed the output from two models (ANN and AutoRegressive Integrated Moving Average [ARIMA]) for price forecasting with a training period of eight weeks. From their results it was noted that the ARIMA root mean squared error value was better than the ANN value. However, as the forecasting window increased, the two models became less precise. Therefore they concluded that short-term forecasting provides a stronger relationship between past and estimated values [7].

From this it is clear that forecasting algorithms help to improve energy costs and can spot when spikes will occur which makes them a valuable tool. Nonetheless since the nature of energy prices is unstable, short-term forecasting proves to be better at making accurate predictions. For this reason the NARMAX model developed in this work analyzed daily day-ahead price.

## 1.2    Factors Influencing Energy Trading

Time-series price forecasting is influenced by several market data factors (system marginal price, load, generators, etc.). Pandey and Upadhyay [4] outlined that price fluctuation is very normal as a result of economic and technical elements. They examined the key factors and noted that demand was the main contributor since the price fluctuates when demand varies.

Many different factors can cause uncertainty to the market clearing price (price when demand and supply curves meet [8]) over a long period, thus price forecasting helps make energy trading successful [5]. Therefore, for a time-series model to predict accurately, appropriate selection of input parameters should be considered. Performing a regression analysis on lagged explanatory variables over time will show if any relationship between the variables exist [9]. Li et al. [10] mentioned how correlation of peaks in energy load data appear at 24-h lags implying strong correlation between same hour loads and hence are suitable as input parameters.

Energy suppliers, who buy electricity from the energy market and sell to customers, will become more adaptable to forecasting if they can train models using historical data [11]. Severiano et al. [12] looked at short-term solar forecasting applying a fuzzy time-series model to check how desirable the information provided was to solar energy. Their results emphasized considerable improvement in predicting accuracy. For that reason, short-term forecasts can be beneficial to improve future outcomes within solar energy trading.

Since the selection of initial parameters is important for the NARMAX modelling technique the current approach includes energy demand as an input factor to determine if it is has any impact on the predicted price. The model was tested using various lags to compare regression orders and to check whether any particular lag improves model prediction accuracy.

## 2    NARMAX Methodology

A NARMAX model (Leontaritis and Billings 1985 [13]) is outlined as follows:

$$y(t) = F^{\ell}\big[y(t-1), \ldots, y(t-N_y), u(t), \ldots, u(t-N_u), \varepsilon(t-1), \ldots \varepsilon(t-N_\varepsilon)\big] + \varepsilon(t)$$

$$(1)$$

where $F^{\ell}$ is a nonlinear function, $u(t)$ is the input time-series, $y(t)$ is the output time-series, $\varepsilon(t)$ is the prediction error, $N_u$ is the regression order of the input, $N_y$ is the regression order of the output, and $N_\varepsilon$ is the regression order of the prediction error [13]. The model attempts to find $F^{\ell}$ as well as significant model terms through different stages: (i) finding the structure, (ii) estimating the parameters, (iii) validating the model, (iv) prediction, and (v) analysis [14].

A polynomial NARMAX model estimates parameters from simple algorithms and has the benefit of simple performance with the process of inputting and outputting variables [15]. Zito and Landau [16] considered a polynomial NARMAX because of the straightforward process of input and structure selection when choosing a model to compare the relationship between air pressure and turbine command in diesel engines; therefore polynomial models can be an option when applying NARMAX to industry data. Nepomuceno and Martins [17] emphasize that even though a polynomial NARMAX model aims to forecast a random number of steps into the future and is validated by free-run simulation, the required precision from the run should be checked against the lower bound error to ensure the model's reliability.

The NARMAX model has advantages as it prevents linear regression methods (stepwise regression) from occurring in the model identification stage by limiting the function $F^\ell$ to multivariable polynomials [18]. The difficulty in a polynomial NARMAX model is working out what polynomial degree and interaction terms are needed, but typically this is discovered through trial and error of all likely combinations of degrees and inputs [19].

A polynomial NARMAX model was considered here for energy prediction as it can handle nonlinear data and is good at predicting unknown parameters. The method applied includes algorithms from the procedure proposed by Korenberg et al. [13] and the model validation follows the processes outlined by Billings and Voon [20].

A basic polynomial model is described by Pearson [18], who highlights that the model's behavior depends on the input factors, as follows:

$$y(t) = ay(t-1) + bu(t-1) + cu(t-1)y(t-1) + dy^2(t-1) \qquad (2)$$

where $a$, $b$, $c$, and $d$ are arbitrary parameters. A polynomial NARMAX is beneficial to estimate $F^\ell$ by creating one large polynomial from each of the input factor polynomials and removing any unnecessary terms [19].

The first stage of the method uses an orthogonal estimation algorithm which firstly estimates parameters independently of each other, not including the $\varepsilon(t)$ term; secondly the prediction errors are estimated; and lastly the $\varepsilon(t)$ term. This process is repeated until convergence is met and the model coefficients can be estimated [13]. Polynomial models have an advantage as the parameters are linear which allows the parameter values to be obtained using simple methods [19].

The final stage of identification is the model validation and involves unsupervised learning where the model tries to predict price +1day. This phase looks at the residuals to identify unmodeled nonlinear relationships [21]. Billing and Voon's [20] method developed tests to function in the worst possible combinations since the validation process has little influence over input or residuals.

The most important part of the method is the input stage as all results are dependent on what inputs have been selected [21]. The model is tested using various lag periods to establish if increasing the input regression has any effect on the predicted day ahead price. In this paper we test the lag periods from 1 h to 12 h.

## 3   Results

The hourly price data used in the models were retrieved from the N2EX market published report of yearly prices available on the Nordpool website [22] and the hourly demand data were taken from the Great Britain (GB) Balancing Mechanism

(BM) demand report, recorded in five-minute intervals, available on the BM website [23]. Historical data (previous price and demand) from 01 May 2017 until 04 June 2017 were tested as the input and historical data (price) from 02 May 2017 until 05 June 2017 were used as the target data (day-ahead prediction). In total 35 days of data (with 24 hourly prices each day) were analyzed resulting in 840 data records.

In Fig. 1 we plot the distribution of energy prices illustrating that energy prices do not follow a normal distribution since the tail of the distribution of variables over the five-week period tends to skew to the right (positively skewed) and the plot on normal distribution displays a slight non-linear pattern. This can be due to the seasonality of the data resulting in peaks or spikes over time and emphasizes that a NARMAX model is the best method to use as it can handle non-linearity.

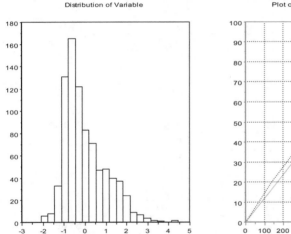
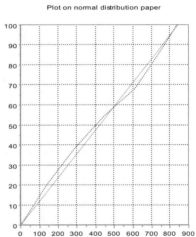

**Fig. 1.** Normal distribution plot of energy prices from 02 May 2017 up until 05 June 2017.

The model is split into an estimation stage (the first 420 records are used for model estimation) and the validation stage (for testing purposes the remaining 420 records are used to test the model accuracy on unseen data). The first NARMAX model developed used only previous energy price as input and the results are outlined in Table 1.

Table 1 summarizes the percentage error at both the estimation and validation stages: the closer the error value is to zero, the better the precision. From these results we can see that as the lag increased the percentage error decreased for estimation and appears to improve when the polynomial degree is also increased.

During the model validation stage a linear polynomial provided a better percentage error overall. In particular, an input regression order of 2 (lag 2 h) with 5 terms remaining at the validation stage had the lowest error of 54.80%. The detailed validated NARMAX model with 5 terms outputted as:

**Table 1.** Previous price percentage error for model estimation and model validation.

| Input parameters | Lag input | Polynomial degree | Model estimation percentage error | Model validation percentage error |
|---|---|---|---|---|
| Previous price | 1 | 1 | 22.95 | 55.18 |
| | | 2 | 22.01 | 56.70 |
| | | 3 | 21.76 | 64.39 |
| | 2 | 1 | 22.35 | **54.80** |
| | | 2 | 21.38 | 58.23 |
| | | 3 | 20.85 | 64.29 |
| | 3 | 1 | 21.82 | 57.10 |
| | | 2 | 20.70 | 65.52 |
| | | 3 | 19.74 | 68.76 |
| | 4 | 1 | 21.73 | 59.22 |
| | | 2 | 20.04 | 89.96 |
| | | 3 | 19.27 | 56.27 |
| | 5 | 1 | 21.75 | 60.05 |
| | | 2 | 19.81 | 122.23 |
| | | 3 | 19.29 | 63.77 |
| | 6 | 1 | 21.77 | 59.43 |
| | | 2 | 19.60 | 129.95 |
| | | 3 | 20.06 | 96.16 |
| | 7 | 1 | 21.33 | 58.67 |
| | | 2 | 18.83 | 85.90 |
| | | 3 | 18.26 | 105.04 |
| | 8 | 1 | 21.16 | 57.68 |
| | | 2 | 18.44 | 65.69 |
| | | 3 | 16.43 | 60.12 |
| | 9 | 1 | 21.03 | 57.40 |
| | | 2 | 17.88 | 66.63 |
| | | 3 | 18.14 | 91.15 |
| | 10 | 1 | 21.11 | 57.56 |
| | | 2 | 17.84 | 68.63 |
| | 11 | 1 | 20.77 | 58.73 |
| | | 2 | 16.96 | 65.65 |
| | | 3 | **16.21** | 311.49 |
| | 12 | 1 | 20.65 | 59.43 |

$$y(t) = +5.76511\ldots$$
$$+ 0.57867 * u(n)\ldots$$
$$- 0.24956 * u(n-1)\ldots \tag{3}$$
$$- 0.14951 * u(n-2)\ldots$$
$$+ 0.70593 * y(n-1)\ldots$$

Figures 2 and 3 display the models with lowest percentage error for both, estimation $\{N_u = 11, N_y = 1, N_\varepsilon = 0, N_p = 3\}$ and, validation $\{N_u = 2, N_y = 1, N_\varepsilon = 0, N_p = 1\}$ respectively, where $N_u$ is the input regression order, $N_y$ is the output regression order, $N_\varepsilon$ is the prediction regression order, and $N_p$ is the polynomial degree.

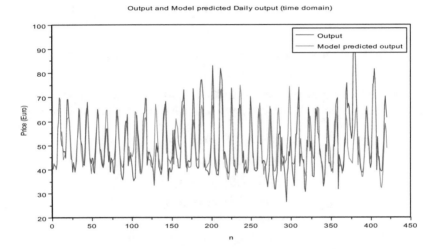

**Fig. 2.** Best model estimation with previous price as input {lag 11, cubic}

The cubic polynomial with lag 11 had the lowest percentage error overall for model estimation (Fig. 2), however the high percentage error (311.49%) during model validation with unseen data shows that this model is over-fitted. Therefore a low-order linear polynomial is the best-fitting model. Figure 3 illustrates that this predicted output is close to the actual output, but it also has difficulty reaching the high peaks (outliers).

The NARMAX model was next modified to include both previous price and demand as input parameters in order to see if demand improved the parameter estimates and model accuracy. Table 2 presents percentage error results using two input parameters.

From this set of results, the percentage error decreased during the estimation stage as the lag difference increased, which is similar to the behavior noted when the previous price parameter was used on its own. Again in the model validation the percentage error on average was lower with a linear polynomial degree, with an input regression order of 2 (lag of 2 h) with 7 terms remaining giving the lowest percentage error. In this instance the percentage error was 52.20%.

Output and Model predicted Daily output (time domain)

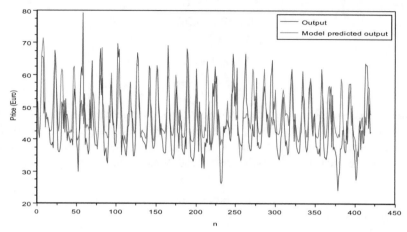

**Fig. 3.** Best model validation with previous price as input {lag 2, linear}

The top performing validated NARMAX model for previous price (displayed as 1 in the model equation) and demand (displayed as 2 in the model equation) with 7 terms is:

$$
\begin{aligned}
y(t) = {} & +3.10297\ldots \\
& + 0.51544 * u(n, 1)\ldots \\
& - 0.23868 * u(n-1, 1)\ldots \\
& - 0.09377 * u(n-2, 1)\ldots \\
& + 0.00004 * u(n, 2)\ldots \\
& - 0.00003 * u(n-2, 2)\ldots \\
& + 0.70725 * y(n-1)\ldots
\end{aligned} \tag{4}
$$

Figure 4 (best performing model estimation, $\{N_u = 5, N_y = 1, N_\varepsilon = 0, N_p = 3\}$) and Fig. 5 (best performing model validation, $\{N_u = 2, N_y = 1, N_\varepsilon = 0, N_p = 1\}$) illustrate the models with lowest percentage error.

The presented figures show that the predicted output is close to the actual output and is better at predicting spikes than the previous NARMAX model that only uses previous price as input. Like the previous price only model the cubic polynomial is slightly over-fitted with a percentage error of 93.66%, so the best-fitting model for previous price and demand is linear.

To summarize the best polynomial degree for both previous price model and previous price with energy demand model is a low-order linear polynomial. Since the model validation did not always fit for a quadratic or cubic polynomial, it would be best to use a linear polynomial for a reliable NARMAX model that will not over-fit and works for all input regression orders.

**Table 2.** Previous price & demand percentage error for model estimation and model validation.

| Input parameters | Lag input | Polynomial degree | Model estimation percentage error | Model validation percentage error |
|---|---|---|---|---|
| Previous price & demand | 1 | 1 | 22.68 | 54.43 |
| | | 2 | 18.07 | 62.06 |
| | | 3 | 17.44 | 101.81 |
| | 2 | 1 | 21.76 | **52.20** |
| | | 2 | 17.91 | 67.57 |
| | | 3 | 16.69 | 74.54 |
| | 3 | 1 | 21.38 | 53.21 |
| | | 2 | 16.69 | 67.88 |
| | 4 | 1 | 21.21 | 54.69 |
| | | 2 | 16.35 | 74.53 |
| | 5 | 1 | 21.29 | 54.29 |
| | | 3 | **14.60** | 93.66 |
| | 6 | 1 | 21.30 | 53.62 |
| | | 2 | 15.31 | 84.72 |
| | | 3 | 16.97 | 81.65 |
| | 7 | 1 | 20.93 | 53.71 |
| | | 2 | 15.70 | 75.52 |
| | | 3 | 15.81 | 113.86 |
| | 8 | 1 | 20.79 | 53.05 |
| | 9 | 1 | 20.68 | 53.20 |
| | | 2 | 15.93 | 62.71 |
| | | 3 | 14.87 | 90.99 |
| | 10 | 1 | 20.75 | 53.09 |
| | | 2 | 15.15 | 72.01 |
| | 11 | 1 | 20.43 | 53.68 |
| | | 2 | 14.88 | 61.88 |
| | 12 | 1 | 20.27 | 54.66 |
| | | 2 | 15.14 | 75.31 |

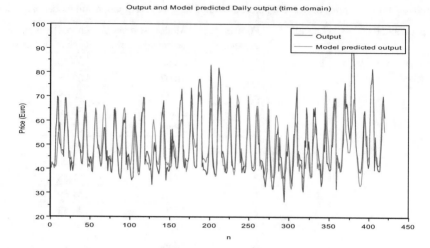

**Fig. 4.** Best model estimation with previous price & demand as input {lag 5, cubic}

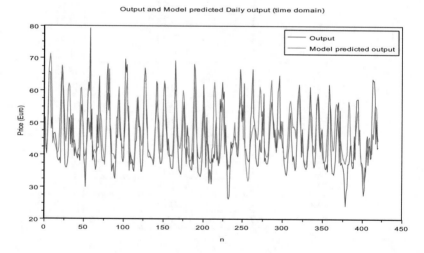

**Fig. 5.** Best model validation with previous price & demand as input {lag 2, linear}

## 4   Conclusion

This paper has examined a polynomial NARMAX model for forecasting future day-ahead energy prices. It can be noted that a linear NARMAX with a low regression input order is best at predicting the input-output relationship. The developed polynomial models had less than ten terms remaining, and were shown to be adequate for predicting energy prices using a small number of terms in a NARMAX model [13]. The results showed that the model estimation stage had lower percentage errors than during model validation, outlining that the predicted output was better throughout estimation.

However even if the model algorithm identifies and estimates parameters, the validation stage needs to provide a suitable fit with unseen data before the model can be approved [13]. Therefore the best model was selected from the lowest percentage error outputted in the validation stage, rather than during model estimation.

The findings established that a NARMAX model used to predict energy prices gives a more precise prediction when both previous price and demand are included as input parameters. Current prices show a relationship to historical price and demand through dynamic regression [8], so the conclusion of a relationship between price and demand is logical. Future work could examine additional inputs, for example temperature or wind, to see if other factors influence energy price and if they also need to be considered in short-term energy price forecasting models.

**Acknowledgment.** This work was funded via DfE CAST scholarship in collaboration with Click Energy.

# References

1. Mosbah, H., El-Hawary, M.: Hourly electricity price forecasting for the next month using multilayer neural network. Can. J. Electr. Comput. Eng. **39**, 283–291 (2016). https://doi.org/10.1109/CJECE.2016.2586939
2. Gupta, S., Mohanta, S., Chakraborty, M., Ghosh, S.: Quantum machine learning-using quantum computation in artificial intelligence and deep neural networks quantum, pp. 268–274 (2017)
3. Acuna, G., Ramirez, C., Curilem, M.: Comparing NARX and NARMAX models using ANN and SVM for cash demand forecasting for ATM. In: Proceedings of International Joint Conference on Neural Networks, pp. 10–15 (2012). https://doi.org/10.1109/ijcnn.2012.6252476
4. Pandey, N., Upadhyay, K.G.: Different price forecasting techniques and their application in deregulated electricity market : a comprehensive study. In: International Conference on Emerging Trends in Electrical Electronics & Sustainable Energy Systems (ICETEESES), pp. 1–4 (2016). https://doi.org/10.1109/iceteeses.2016.7581342
5. Amjady, N., Hemmati, M.: Energy price forecasting: problems and proposals for such predictions (2006)
6. Vijayalakshmi, S., Girish, G.P.: Artificial neural networks for spot electricity price forecasting: a review. Int. J. Energy Econ. Policy. **5**, 1092–1097 (2015)
7. Gao, G., Lo, K., Fan, F.: Comparison of ARIMA and ANN models used in electricity price forecasting for power market. Energy Power Eng. **9**, 120–126 (2017). https://doi.org/10.4236/epe.2017.94B015
8. Georgilakis, P.S.: Artificial intelligence solution to electricity price forecasting problem. Appl. Artif. Intell. **21**, 707–727 (2007). https://doi.org/10.1080/08839510701526533
9. Ghalehkhondabi, I., Ardjmand, E., Weckman, G.R., Young, W.A.: An overview of energy demand forecasting methods published in 2005–2015. Energy Syst. **8**, 411–447 (2017). https://doi.org/10.1007/s12667-016-0203-y
10. Li, P., Arci, F., Reilly, J., Curran, K., Belatreche, A., Shynkevich, Y.: Predicting short-term wholesale prices on the Irish single electricity market with artificial neural networks. In: 2017 28th Irish Signals System Conference, ISSC 2017 (2017). https://doi.org/10.1109/issc.2017.7983623

11. Green, A.: Machine learning in energy - part two. http://adgefficiency.com/machine-learning-in-energy-part-two. Accessed 21 Dec 2017
12. Severiano, C.A., Silva, P.C.L., Sadaei, H.J., Guimaraes, F.G.: Very short-term solar forecasting using fuzzy time series. In: 2017 IEEE International Conference on Fuzzy System, pp. 1–6 (2017). https://doi.org/10.1109/fuzz-ieee.2017.8015732
13. Korenberg, M., Billings, S.A., Liu, Y.P.: An orthogonal parameter estimation algorithm for nonlinear stochastic systems. Acse report 307 (1987)
14. Nehmzow, U.: Robot Behaviour: Design, Description, Analysis and Modelling, pp. 169–171. Springer (2009)
15. Pagano, D.J., Filho, V.D., Plucenio, A.: Identification of polinomial narmax models for an oil well operating by continuous gas-lift. IFAC Proc. **39**, 1113–1118 (2006). https://doi.org/10.3182/20060402-4-BR-2902.01113
16. Zito, G., Landau, I.D.: A methodology for identification of narmax models applied to diesel engines. IFAC Proc. **16**, 374–379 (2005). https://doi.org/10.3182/20050703-6-CZ-1902.00063
17. Nepomuceno, E.G., Martins, S.A.M.: A lower bound error for free-run simulation of the polynomial NARMAX. Syst. Sci. Control Eng. **4**, 50–58 (2016). https://doi.org/10.1080/21642583.2016.1163296
18. Pearson, R.K.: Nonlinear input/output modelling. J. Process Control **5**, 197–211 (1995). https://doi.org/10.1016/0959-1524(95)00014-H
19. Warnes, M.R., Glasseyfl, J., Montague, G.A., Kara, B.: On Data-Based Modelling Techniques for Fermentation Processes. Process Biochem. **31**, 147–155 (1996)
20. Billing, S.A., Voon, W.S.F.: Correlation based model validity tests for nonlinear models. Acse report 285 (1985)
21. Billings, S.A., Fadzil, M.B.: The practical identification of systems with nonlinearities. IFAC Proc. **18**, 155–160 (1985). https://doi.org/10.1016/S1474-6670(17)60551-2
22. Nordpool: N2EX Market Prices. https://www.nordpoolgroup.com/historical-market-data. Accessed 09 Mar 2018
23. BM:    Demand.    https://bmreports.com/bmrs/?q=demand/rollingsystemdemand/historic. Accessed 09 Mar 2018

# Model Selection in Online Learning for Times Series Forecasting

Waqas Jamil$^{(\boxtimes)}$ and Abdelhamid Bouchachia

Machine Intelligence Group, Bournemouth University, Poole, UK
{wjamil,abouchachia}@bournemouth.ac.uk

**Abstract.** This paper discusses the problem of selecting model parameters in time series forecasting using aggregation. It proposes a new algorithm that relies on the paradigm of prediction with expert advice, where online and offline autoregressive models are regarded as experts. The desired goal of the proposed aggregation-based algorithm is to perform not worse than the best expert in the hindsight. The theoretical analysis shows that the algorithm has a guarantee that holds for any data sequence. Moreover, the empirical evaluation shows that the algorithm outperforms other popular model selection criteria such as Akaike and Bayesian information criteria on cyclic behaving time series.

**Keywords:** Model selection · Online learning
Aggregation algorithm · Time series

## 1 Introduction

Model selection is about choosing a model from a set of fitted models that performs better on a given data. In statistics Akaike information criterion (AIC) [1] and Schwarz criterion [17] are popular model selection techniques. These criteria are not competing rules since they are useful for different scenarios. For instance, AIC achieves asymptotic efficiency [18], while BIC originates from Bayesian hypothesis testing of the regular exponential family using an asymptotic approximation to identify the correct model when the sample size increases [14]. These criteria are developed by considering batch learning, and sometimes they achieve a slower rate of convergence as shown in [5] where the presented algorithm has a similar flavour to the fixed-share algorithm [7].

To fit a model one needs to input the coefficients and the parameters. In batch learning this can be done by for example using cross validation. In contrast, in online learning due to the sequential arrival of the data methods like cross validation can't be used. Over the past, numerous attempts have been made to address the problem of online model selection of time series. For instance,

The European Commission has supported Waqas Jamil and Abdelhamid Bouchachia under the Horizon 2020 Grant 687691 related to the project: *PROTEUS: Scalable Online Machine Learning for Predictive Analytic and Real-Time Interactive Visualisation*.

Noshad et al. [12] proposed a dynamic algorithm that operates sequentially to select a suitable model. In [10] an adaptive algorithm is used for automatically selecting the best model but is not applicable generally. It only allows to reduce data communication in wireless sensor networks. The approach discussed by Prado and Lopes [13] addresses the issue of parameter selection in the state space representation of time series. Sato [16] uses variational Bayes to provide complete online model selection mechanism with a guarantee, by averaging over an ensemble of models. Our work differs in the sense that guarantee is held for sure, not on average or high probability as discussed in the past.

In this study, we investigate model selection in the context of online time series forecasting. There have been some recent but very scarce attempts to address the problem of time series prediction using online learning. For instance, Anava et al. [2] proposed an online version of Auto-Regressive-Moving-Average (ARMA), while Liu et al. [11] investigated online Auto-Regressive-Integrated-Moving-Average (ARIMA). The fundamental idea behind the online version of ARMA and ARIMA models is that ARMA is a subset of AR. The online ARIMA model is an extension of the online ARMA model. These papers present a reasonable approach to handle the uni-variate time series prediction problem.

This paper presents an approach that uses aggregation, similarly to the approach presented by Romanenko [15], but focuses on mixing online ARMA models leading to a novel utterly online framework. The proposed approach does not require the use of any information criterion and has the guarantee of not being too far from the best model. Furthermore, it has the possibility of beating the best model for time series prediction. The bounds of the competitive online algorithms, such as the strong Aggregation Algorithm (AA) by Vovk [21], are guaranteed to hold. That is, the error bounds of competitive online statistics algorithms do not contain only with high probability or on average as one can encounter with many forecast combination algorithms, but such bounds do hold with certainty.

In our work we include a formal proof of the bound of the ARMA-OGD along with derivations of AA's substitution functions associated with square loss prediction games. The novelty of our work lies in the modification of AA and ARMA Gradient Descent (ARMA-OGD) algorithms. More precisely:

1. We plug ARMA-OGD into AA to perform online model selection.
2. We explicitly show that for our suggested approach, the following type of bound holds regardless of the data generating mechanism:

$$L_T \leq L_T^* + \frac{1}{\eta} \log n$$

where $L_T$ is the cumulative loss incurred by of the learning algorithm up until time $T$, $L_T^*$ is the cumulative loss of the best learning strategy in the hindsight. The learning rate is denoted by $\eta$ and $n$ is a finite integer denoting the number of experts.

Following is the organisation. In next two sections we provide context to our work by briefly discussing the essential features of Aggregation Algorithm (AA) and ARMA-OGD that we later are used to combine the two algorithms.

In Sect. 3 we provide an explicit algorithm that combines AA and ARMA-OGD by modifying them. Section 4 illustrates the guarantee of AA+ARMA-OGD on two real world datasets. Section 5 concludes our work.

# 2 Background

## 2.1 Aggregating Algorithm

Let $\Gamma$ be the prediction space and $\Omega = [Y_1, Y_2]$ be the outcome space, such that $Y_2 > Y_1$, the $n$ number of experts $\theta_k$ for $k = 1, 2, ..., n < \infty$, makes predictions $\gamma_t^{\theta_k} \in \Gamma$ on each trial $t \in \mathbb{Z}$; the learner makes a prediction by aggregating experts predictions; nature chooses an outcome; each expert $Loss_{expert} = \sum_{t=1}^{T} \lambda(\gamma_t^{\theta_k}, \omega_t)$ and learner loss $Loss_{learner} = \sum_{t=1}^{T} \lambda(\gamma_t, \omega_t)$ is calculated using square loss. It is not assumed that there is a model generating the outcomes and the nature is considered as an oblivious adversary. The initialisation of the experts weights is done uniformly (each expert is assigned the same weight initially). AA [21,22] works under the protocol of Prediction With Experts Advice (PWEA), which is as follows:

---

**Protocol 1.** Prediction With Expert Advise

---
1: **for** $t = 1, 2, ...$ **do**
2:     Experts $\theta_k \in \Theta$ predicts $\gamma_t^{\theta_k} \in \Gamma$, $k = 1, 2, ..., n$
3:     Learner output $\gamma_t \in \Gamma$
4:     Nature output $\omega_t \in \Omega$
5:     Learner suffers loss $\lambda(\gamma_t, \omega_t)$
6:     Experts $\theta \in \Theta$ suffers loss $\lambda(\gamma_t^{\theta_k}, \omega_t)$
7: **end for**

---

AA generalises the weighted majority algorithm providing an exponentially weighted average that has bounds in the case of mixable game. For $\eta > 0$, a loss function is called $\eta-$mixable if there exists a substitution function (more on it later) for it such that [19,21]:

$$\lambda(\omega, \gamma) \leq g(\omega) = \log_\beta \int \beta^{(\omega-p)^2} P(dp) \qquad (1)$$

where $\forall \gamma \in \mathbb{R}$, $\lambda(Y_1, \gamma) \leq g(Y_1)$ & $\lambda(Y_2, \gamma) \leq g(Y_2)$ such that $\beta = e^{-\eta}$ for $\eta > 0$, and $g$ represents the generalised prediction corresponding to the probability distribution $P \in \mathbb{R}$, such that $\omega, p \in [Y_1, Y_2]$ for $Y_2 > Y_1$.

The loss of AA cannot be much larger than that of the best expert, for a mixable finite experts game by equally initialising the weights of the experts.

$$Loss(AA) \leq Loss_{best}(\theta) + \frac{\log n}{\eta} \qquad (2)$$

where $\theta \in \Theta$, $\eta$ is the learning rate, and $n$ is the number of experts. This bound (Eq. 2) is shown in [20] to be optimal i.e. it cannot be improved by any other prediction algorithm. It is for this reason, that we have chosen AA in this paper to apply for time series prediction.

AA takes two parameters, the learning rate $\eta > 0$ and a prior probability which indicates the initial weights of the experts. At every step $t$, we update the weights. So intuitively, if an expert makes a mistake, we would reduce its weight. AA uses a *substitution function* which maps the generalised prediction $g(\omega)$ into $\Gamma$,

---

**Algorithm 1.** Aggregation Algorithm

---

1: Initialise weights $w_0^\theta, \theta = 1, 2, ...n$
2: **for** $t = 1, 2, ..$ **do**
3:     Notice experts prediction $\gamma_t^\theta$
4:     Normalise experts weight $w_t^\theta = \frac{w_{t-1}^\theta}{\sum_{i=1}^{N} w_{t-1}^i}$
5:     Use substitution function to obtain $\gamma_t$
6:     Notice actual outcomes $\omega_t$
7:     Update the experts weights $w_t^\theta = w_t^\theta e^{-\eta\lambda(\gamma_t^\theta, \omega_t)}$
8: **end for**

---

Next we explain line 5 of Algorithm 1 by focusing on work done in [21]. Consider $\Omega = \{-1, 1\}$ and $\gamma \in [-1, 1]$. AA's prediction without using the substitution function is $(g(-1), g(1))$ (a point on a plane), which does not lie on the losses curve $((-1 - \gamma)^2, (1 - \gamma)^2)$. AA's prediction $(g(-1), g(1))$ is transformed to the point $(e^{-\eta(-1-\gamma)^2}, e^{-\eta(1-\gamma)^2})$ by the use of substitution function and the set of permitted predictions becomes $(e^{-\eta(-1-\gamma)^2}, e^{-\eta(1-\gamma)^2})$ (for more details see [19]). To find the learning rate $\eta$, for which the curve $(e^{-\eta(-1-\gamma)^2}, e^{-\eta(1-\gamma)^2})$ is convex is equivalent to the problem of finding the values of second derivative for which $(e^{-\eta(-1-\gamma)^2}, e^{-\eta(1-\gamma)^2})$ is less or equal to 0 for all values of $\gamma \in [-1, 1]$. Therefore:

$$(u, v) = (e^{-\eta(-1-\gamma)^2}, e^{-\eta(1-\gamma)^2})$$

$$\frac{\partial u}{\partial \gamma} = -2\eta(1 + \gamma)e^{-\eta(1+\gamma)^2}$$

$$\frac{\partial v}{\partial \gamma} = 2\eta(1 - \gamma)e^{-\eta(1-\gamma)^2}$$

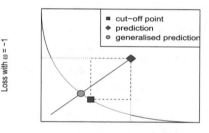

**Fig. 1.** $((-1 - \gamma)^2, (1 - \gamma)^2)$ curve where $\gamma \in [-1, 1]$.

The inf-sup of the ratio $\frac{g(\omega)}{\lambda(\omega,\gamma)}$ is obtained where the line $((0,0), (g(1), g(-1)))$ intersects with the losses curve. From Fig. 1 the intersection between losses (the line intersecting the red curve) and the between the (red) curve is at $((\gamma - 1)^2, (\gamma + 1)^2)$ (cut-of point in green see Fig. 1), thus the inf-sup of the ratio is $\frac{(\gamma-1)^2}{(\gamma+1)^2} = \frac{g(1)}{g(-1)}$, which is non-linear (difficult to use in practice), so instead we use a different point (prediction point in blue, see Fig. 1) on the curve (prediction and generalised prediction are mapped by a square in Fig. 1).

**Lemma 1.** *The restricted square loss game is $\eta$-mixable if and only if $\eta \leq \frac{1}{2}$.*

*Proof.* Applying the chain rule we obtain:

$$\frac{\partial v}{\partial u} = \frac{2\eta(1-\gamma)e^{-\eta(1-\gamma)^2}}{-2\eta(1+\gamma)e^{-\eta(1+\gamma)^2}} = -\frac{1-\gamma}{1+\gamma}e^{4\eta\gamma} = \frac{(\gamma-1)e^{4\eta\gamma}}{\gamma+1}$$

To achieve the minimum or inflection point, the second derivative must be negative or null, we proceed as follows:

$$\frac{\partial^2 v}{\partial u^2} = \frac{\partial v}{\partial u \partial \gamma}\frac{\partial \gamma}{\partial u} = \frac{e^{4\eta\gamma}}{(1+\gamma)^2}(4\eta(1-\gamma)(1+\gamma)-2)2\eta(1+\gamma)e^{-\eta(1+\gamma)^2}$$

$$= \frac{e^{4\eta\gamma}}{(1+\gamma)^2}\left(\frac{4\eta(1-\gamma^2)-2}{2\eta(1+\gamma)e^{-\eta(1+\gamma)^2}}\right) \qquad (3)$$

The term in Eq. (3) will be negative or zero if and only if: $4\eta(1-\gamma^2)-2 \leq 0$ which implies $\eta \leq 0.5$, since $\gamma^2 \in [0,1]$. $\qquad \square$

**Proposition 1.** *For a game of square loss with $\Omega = \{-1,1\}$, then $\gamma = \frac{g(-1)-g(1)}{4}$ is a substitution function.*

*Proof.* The curve $((\gamma-1)^2, (\gamma+1)^2)$ for $\gamma \in [-1,1]$ contains all possible values of $\gamma$. The point $(g(1), g(-1))$ represents generalised prediction. The substitution function maps generalised prediction to actual predictions, thus $(\gamma+1)^2 - g(-1) = (\gamma-1)^2 - g(1)$. By doing simple algebraic manipulation, we get: $\gamma = \frac{g(-1)-g(1)}{4}$. $\qquad \square$

**Lemma 2.** *The square loss game $\Omega = [-Y, Y]$ where $Y \in \mathbb{R}$ is $\eta$-mixable if and only if $\eta \leq \frac{1}{2Y^2}$.*

*Proof.* We find the values of $\eta$ for which the game is mixable by exponentiation of the generalised prediction. We have $e^{(-Y-\gamma)^2}$ and $e^{(Y-\gamma)^2}$ where $\gamma \in [-Y, Y]$. If we instead use $e^{\frac{\hat{\eta}}{Y^2}(-Y-\gamma)^2}$ and $e^{\frac{\hat{\eta}}{Y^2}(Y-\gamma)^2}$ our game becomes restricted square loss game for which as seen in Lemma 1 the game is mixable if and only if, $\eta \leq 0.5$. By writing $\eta = \frac{\hat{\eta}}{Y^2}$, we have $\eta Y^2 \leq 0.5$ which implies that $\eta \leq \frac{1}{2Y^2}$. $\qquad \square$

**Proposition 2.** *For a square loss game with $\Omega = [-Y, Y]$, where $Y \in \mathbb{R}$ then:*

$$\gamma = \frac{g(-Y)-g(Y)}{4Y}$$

*is a substitution function.*

*Proof.* By solving $(\gamma+Y)^2 - g(-Y) = (\gamma-Y)^2 - g(Y)$, we obtain our desired result. $\qquad \square$

**Lemma 3.** *The square loss game* $\Omega = [Y_1, Y_2]$ *where* $Y_1, Y_2 \in \mathbb{R}$ *and* $Y_1 < Y_2$ *is* $\eta$−*mixable if and only if* $\eta \leq \frac{2}{(Y_2-Y_1)^2}$.

*Proof.* We need to prove for the curve $(u, v) = (e^{-\eta(\gamma-Y_1)^2}, e^{-\eta(\gamma-Y_2)^2})$ that:

$$\frac{\partial^2 v}{\partial u^2} = \frac{\frac{\partial^2 v}{\partial \gamma \partial u}}{\frac{\partial u}{\partial \gamma}} \leq 0$$

By performing above differentiation for the curve, we get $\frac{1}{Y_1-\gamma} + 2\eta(Y_2 - \gamma) \leq 0 \Rightarrow \eta \leq \frac{1}{2(Y_2-\gamma)(\gamma-Y_1)}$. We notice that $\max_{\gamma \in [Y_1,Y_2]}(Y_2-\gamma)(\gamma-Y_1) = \frac{1}{4}(Y_2-Y_1)^2$ and the curve is concave $\forall \gamma$ provided that $\eta \leq \frac{2}{(Y_2-Y_1)^2}$.  □

**Proposition 3.** *For a square loss game with* $\Omega = [Y_1, Y_2]$, *where* $Y_1, Y_2 \in \mathbb{R}$ *and* $Y_1 < Y_2$ *then:*

$$\gamma = \frac{Y_2 + Y_1}{2} - \frac{g(Y_2) - g(Y_1)}{2(Y_2 - Y_1)}$$

*is a substitution function.*

*Proof.* To find $\gamma$, consider $(Y_1 - \gamma)^2 + g(Y_1) = (Y_2 - \gamma)^2 + g(Y_2)$. By using the fact $(Y_1^2 - Y_2^2) = (Y_1 + Y_2)(Y_1 - Y_2)$ and re-arranging, $2\gamma(Y_2 - Y_1) = g(Y_2) - g(Y_1) - (Y_1 + Y_2)(Y_1 - Y_2)$, we get the substitution function.  □

## 2.2 ARMA-OGD

---
**Algorithm 2.** ARMA-OGD$(p, q)$
---
1: ARMA order $p, q$, Learning rate $\eta$, $m = q.\log_{1-\beta}\left((TLM_{max})^{-1}\right)$
2: **for** $t = 1, 2, ...(T-1)$ **do**
3:   Predict $\hat{X}_t(\gamma_t) = \sum_{i=1}^{m+k} \gamma_t^i X_{t-i}$
4:   Observe $X_t$ and suffer loss $\lambda^m(\gamma_t, \omega_t)$

5:   Let $\nabla_t = \nabla \lambda^m(\gamma_t, \omega_t)$
6:   Set $\gamma_{t+1} \leftarrow \prod_{\mathcal{K}}\left(\gamma_t - \frac{1}{\eta}\nabla_t\right)$
7: **end for**
---

ARMA-OGD$(p, q)$ was introduced by Anava et al. [2]. The pseudo-code of the algorithm is presented in Algorithm 2. We proceed by defining some notation. The prediction set $\mathcal{K}$ contains $m + p$−dimensional coefficient vectors and is defined as $\mathcal{K} = \{\gamma \in \mathbb{R}^{m+p}, |\gamma_j| \leq c, j = 1, ..., m\}$. We denote the diameter of $\mathcal{K}$ by $D$ and bound $D = 2c\sqrt{(m+p)}$. The upper bound of convex loss $|| \nabla \lambda(\gamma, \omega)||$ for all $t$ $\gamma \in \mathcal{K}$ on sequence $|X_t| \leq X_{max}$, is denoted by $G = D(X_{max})^2$. We say $M_{max}$ is the upper bound on $|W_t|$ for all $t = 1, 2, ..., T$ if we assume that noise is adversarial and when noise are i.i.d then $\mathbb{E}(|\beta_t|) < M_{max} < \infty$ and $L$ denotes Lipshitz constant which is assumed to be greater than zero. The coefficients $|\alpha_i|$ are less then some constant $c \in \mathbb{R}$ and $\sum_{i=1}^{q} |\theta_i| < 1 - \beta$ where $\beta > 0$. We next present the proof of Theorem 5 mentioned in [2] but not shown due to the similarity to Theorem 1 of their paper.

**Theorem 1.** *For any data sequence $\{X_t\}_{t=1}^T$ such that $p, q \geq 1$, and set $\eta = \frac{1}{X_{max}^2 \sqrt{T}}$, Algorithm 2 predicts using a convex loss function, with the following guarantee:*

$$\sum_{t=1}^{T} \lambda(\gamma_t, \omega_t) - min_{\alpha, \beta} \sum_{t=1}^{T} \mathbb{E}[f_t(\alpha, \beta)] = O(4c(m+p)X_{max}^2 \sqrt{T})$$

*Proof.* Let $(\alpha^*, \beta^*) = \text{argmin}_{\alpha, \beta} \sum_{t=1}^{T} \mathbb{E}[f_t(\alpha, \beta)]$. We know for any convex loss function we have [23]:

$$\sum_{t=1}^{T} \lambda^m(\gamma_t, \omega_t) - \min \sum_{t=1}^{T} \lambda^m \left( X_t, \left( \sum_{i=1}^{m+k} \gamma_t^i X_{t-i} \right) \right) = O(4c(m+p)X_{max}^2 \sqrt{T})$$

Now by using the fact that $\text{ARMA}(p, q)$ can be represented by $\text{AR}(\infty)$ [6], by using entire past, we can recursively write:

$$X_t^\infty(\alpha, \beta) = \sum_{i=1}^{p} \alpha_i X_{t-i} + \sum_{i=1}^{q} \beta_i (X_{t-i} - X_{t-i}^\infty(\alpha, \beta))$$

plugging in initial condition $X_t^\infty = X_1$, we get the loss suffered as:

$$f_t^\infty(\alpha, \beta) = \lambda(X_t, X_t^\infty(\alpha, \beta))$$

which is not convex. The loss function here considers entire data. We need to replace $f_t^\infty$ by $f_t$, which can be done by considering some weight $w_i(\alpha, \beta)$ function and write our loss function as follows:

$$f_t^\infty(\alpha, \beta) = \lambda(X_t, \sum_{i=1}^{t} w_i(\alpha, \beta)X_{t-i})$$

This allows the loss function to update prediction by using only the last outcome in contrast to using the entire history. By setting $m \in \mathbb{N}$, the prediction can be rewritten as:

$$X_t^m(\alpha, \beta) = \sum_{i=1}^{p} \alpha_i X_{t-i} + \sum_{i=1}^{q} \beta_i (X_{t-i} - X_{t-i}^{m-i})$$

Plugging in the initial condition $X_t^m(\alpha, \beta) = X_t$ for all $t, m \leq 0$, the loss suffered by the prediction at time $t$ becomes:

$$f_t^m(\alpha, \beta) = \lambda(X_t, X_t^m(\alpha, \beta))$$

By considering last $(m + k)$ observations and since $\min_\gamma \lambda^m(\gamma_t, \omega_t) \leq f_t^m(\alpha^*\beta^*)$ (Lemma 2 in [2]), we have:

$$\sum_{t=1}^{T} \lambda^m(\gamma_t, \omega_t) - \sum_{t=1}^{T} f_t^m(\alpha^*\beta^*) = O(4c^2(m+p)X_{max}^2\sqrt{T})$$

From Lemma 3 in [2] we know that the following holds:

$$\left|\sum_{t=1}^{T} \mathbb{E}[f_t^\infty(\alpha, \beta) - \sum_{t=1}^{T} \mathbb{E}[f_t^m(\alpha, \beta)]\right| = \mathcal{O}(1) \implies$$

$$\sum_{t=1}^{T} \lambda^{q\log_{1-\epsilon}\left((TLM_{max})^{-1}\right)}(\gamma_t, \omega_t) - \sum_{t=1}^{T} f_t^{q\log_{1-\epsilon}\left((TLM_{max})^{-1}\right)}(\alpha^*, \beta^*)$$

$$= \sum_{t=1}^{T} \lambda(\gamma_t, \omega_t) - \min_{\alpha,\beta} \sum_{t=1}^{T} \mathbb{E}[f_t(\alpha, \beta)]$$

From Lemma 4 in [2] we know that the following holds:

$$\left|\sum_{t=1}^{T} \mathbb{E}[f_t^\infty(\alpha, \beta) - \sum_{t=1}^{T} \mathbb{E}[f_t(\alpha, \beta)]\right| = \mathcal{O}(1) \implies$$

$$\sum_{t=1}^{T} \lambda^{q\log_{1-\epsilon}\left((TLM_{max})^{-1}\right)}(\gamma_t, \omega_t) - \sum_{t=1}^{T} f_t(\alpha^*, \beta^*)$$

$$= O\left(4c\left(q\log_{1-\epsilon}(TLM_{max})^{-1} + p\right)X_{max}^2\sqrt{T}\right)$$

which was to be proven. □

## 3  AA+ARMA-OGD

In this section, we provide an explicit algorithm for AA+ARMA-OGD$(p, q)$ (Algorithm 3). Each of our expert is an ARMA-OGD model with different values of parameters $p, q$. To obtain a competitive guarantee we combine the ARMA-OGD models using AA. Algorithms 2 uses Online Gradient Decent (OGD). The analysis done in [23] shows that the OGD attains the following regret when the learning rate is defined to be $\frac{1}{\sqrt{t}}$:

$$L_T - L_T^* = \mathcal{O}\left(\sqrt{T}\right) \tag{4}$$

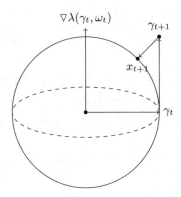

**Fig. 2.** Online gradient descent

where $L_T$ denotes the cumulative loss of the algorithm and $L_T^*$ denotes the cumulative loss of the best strategy in the hindsight. We now explain the details of Algorithm 2 projection step with the aid of Fig. 2. In OGD we have some prediction $\gamma$ which is a point in a convex set. For a given convex loss function we move in the direction of the first derivative (gradient) of the loss incurred. By moving in the direction of the gradient we might go outside the convex set as there is no restriction that will stop us from going out the convex set (notice $\lambda(\gamma_t, \omega_t)$ is slightly outside the sphere in Fig. 2). To keep the prediction inside the convex set, we do a projection by finding the closest point in the convex set. We predict $\gamma_{t+1} = \prod_{\mathcal{K}} \left( \gamma_t - \frac{1}{\eta} \nabla \lambda(\gamma_t, \omega_t) \right)$, where $\nabla \lambda(\gamma_t, \omega_t)$ denotes the gradient of the current loss and $\prod_{\mathcal{K}}$ represents Euclidean projection onto set $\mathcal{K}$ i.e. $\prod_{\mathcal{K}}(\gamma) = \text{argmin} ||\gamma - x||_2$.

Algorithm 3 has the following guarantee that holds for all $T$, regardless of the data generating mechanism:

$$Loss_{\text{Best ARMA-OGD}} - Loss_{\text{AA+ARMA-OGD}} \geq -\frac{\log n}{\eta} \qquad (5)$$

where $n$ denotes the number of experts and for the details of the learning rate please see Lemma 3.

---

**Algorithm 3.** AA+ARMA-OGD $(p, q)$

---

1: Input for each expert parameters $p^{\theta_1,\dots,n}, q^{\theta_1,\dots,n}, \eta > 0$. Initialise experts weight $w_0^{\theta_k} = \mathbf{1}$
2: **for** $k = 1, 2, \dots, n$ **do**
3:     **for** $t = (max(p^{\theta_k}, q^{\theta_k}) + 1), \dots$ **do**
4:         Read experts predictions $\gamma_t^{\theta_k} = \hat{X}_t^{\theta_k}(\hat{\gamma}_t^{\theta_k}) = \sum_{i=1}^{p^{\theta_k}} \alpha_t^{\theta_k} X_{t-i}^{\theta_k} + \sum_{i=1}^{q^{\theta_k}} \beta_t^{\theta_k} \epsilon_{t-i}^{\theta_k}$
5:         Normalise experts weights $w_t^{\theta} = \frac{w_{t-1}^{\theta}}{\sum_{i=1}^{N} w_{t-1}^i}$
6:         Predict $\gamma_t = \frac{Y_2 + Y_1}{2} + \frac{g(Y_2) - g(Y_1)}{2(Y_2 - Y_1)}$ # This is AA+ARMA-OGD prediction
        using proposition 3
7:         Notice actual outcomes $\omega_t \in \mathbb{R}$
8:         Calculate error $\epsilon_t^{\theta_k} = \gamma_t^{\theta_k} - \omega_t$ # notice $\omega_t$ is a value, so $\omega_t$ is
        subtracted from each experts $(k = 1, 2, \dots n)$ prediction $\gamma_t^{\theta_k}$.
9:         Average $\epsilon_t^{\theta_k} = \frac{\sum_i^t \epsilon_i^{\theta_k}}{t - max(p^{\theta_k}, q^{\theta_k})}$
10:        Apply Gradient Decent on $\alpha^{\theta_k}$ and $\beta^{\theta_k}$

$$\alpha_{OGD}^{\theta_k} = -2\epsilon_t^{\theta_k} \sum_{i=1}^{p^{\theta_k}} X_{t-i}^{\theta_k} \quad , \quad \beta_{OGD}^{\theta_k} = -2\epsilon_t^{\theta_k} \sum_{i=1}^{q^{\theta_k}} \epsilon_{t-i}^{\theta_k}$$

11:        Calculate $\alpha^{\theta_k}$ and $\beta^{\theta_k}$:

$$\alpha_t^{\theta_k} = \alpha_{t-1}^{\theta_k} - \frac{\alpha^{\theta_k}}{\sqrt{t}} \alpha_{OGD}^{\theta_k} \quad , \quad \beta_t^{\theta_k} = \beta_{t-1}^{\theta_k} - \frac{\beta^{\theta_k}}{\sqrt{t}} \beta_{OGD}^{\theta_k}$$

12:        Project $\alpha^{\theta_k}$ and $\beta^{\theta_k}$ to simplex:

$$\alpha_t^{\theta_k} = \frac{\alpha_{t-1}^{\theta_k}}{max\left(1, \sum_{i=1}^t \sqrt{(\alpha_i^{\theta_k})^2}\right)} \quad , \quad \beta_t^{\theta_k} = \frac{\beta_{t-1}^{\theta_k}}{max\left(1, \sum_{i=1}^t \sqrt{(\beta_i^{\theta_k})^2}\right)}$$

13:        Update the experts weights $w_t^{\theta_k} = w_t^{\theta_k} e^{-\eta(\epsilon_t^{\theta_k})^2}$
14:     **end for**
15: **end for**

---

## 4   Empirical Evaluation

Figure 3 shows the behaviour of the two-time series, [3,4]. The two time-series refers to 3650 days and exhibits cyclic (stationary) behaviour. Minimum temperature time series lies in the range $[-0.8, 26.3]$ and maximum temperature time series lies in the range $[7, 43.3]$. By using Lemma 3, we calculate $\eta \approx 0.0027$ and $\eta \approx 0.0015$.

We set five ARMA-OGD$(p, q)$, $p = 1, 2, 3, 4, 5$ and $q = 0$ as our experts. We call the ARMA-OGD with the least loss as Best Online ARMA-OGD (BOARMA-OGD). Notice in Fig. 4 it is shown that the guarantee (5) given by Algorithm 3 holds. For minimum and maximum temperature time series the right side of the inequality (5) is $-\frac{\log 5}{\eta} \approx -591$ and $-\frac{\log 5}{\eta} \approx -1060$ respectively.

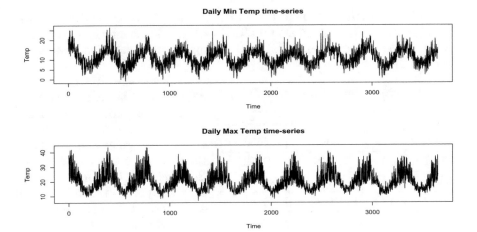

**Fig. 3.** Minimum and maximum temperature time-series

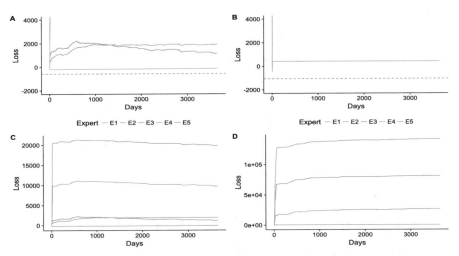

**Fig. 4.** Theoretical guarantee AA+ARMA-OGD. Plot A is zoomed plot C and both refer to the minimum temperature time series. Similarly, B is zoomed D and refers to the maximum temperature time series. The dotted red lines in plot A and B refer to AA+ARMA-OGD guarantee.

For the sake of comparison, we fit statistical ARMA model with fourier series [8]:

$$Y_t = \sum_{m=1}^{K} \left[ \alpha_m \sin\left(\frac{2\pi mt}{L}\right) + \beta_m \cos\left(\frac{2\pi mt}{L}\right) \right] + X_t \qquad (6)$$

where $X_t$ is stationary ARMA/ARIMA$(p, q)$, $\alpha \in \mathbb{R}^p$, $\beta \in \mathbb{R}^q$, and $Y_t$ is periodic on interval $[-L, L]$. We choose parameters of (6) using AIC and call it the Best

Batch ARMA (BBARMA) model. We then fit a set of ARMA models and perform aggregation using AA (AA+BARMA) for the details on the set of batch ARMA models used please see [9]. Table 1 reports the cumulative losses of all the fitted models.

**Table 1.** Cumulative losses.

| Model | Min temp | Max temp |
|---|---|---|
| BBARMA | 28097 | 105632 |
| AA+BARMA | 28049 | 102188 |
| BOARMA-OGD | 27768 | 86550 |
| AA+ARMA-OGD | 27634 | 86131 |

Our suggested Algorithm AA+ ARMA-OGD is the best performing model on both time series, but this is not what the model guarantees. The guarantee is that in the worst case the model will be close to BOARMA-OGD. We may say usually AA+ARMA-OGD will outperform the best performing model when there are several models performing close to each other. The prediction quality of AA+ARMA-OGD depends on the quality of the underlying experts.

## 5  Conclusion

In this paper, we introduced a way to tackle the problem of model selection in online learning for time series forecasting. Unlike statistical ARMA models our algorithm AA+ARMA-OGD is not restricted to the stationary time-series.

It has a guarantee for experts and their aggregation − experimental evaluation show how this guarantee holds.

In the future, we will investigate the spectral analysis of ARMA-OGD.

## References

1. Akaike, H.: Information theory and an extension of the maximum likelihood principle. In: Selected Papers of Hirotugu Akaike, pp. 199–213. Springer (1998)
2. Anava, O., Hazan, E., Mannor, S., Shamir, O.: Online learning for time series prediction. In: COLT, pp. 172–184 (2013)
3. Daily maximum temperatures in Melbourne, Australia. Australian Bureau of Meteorology (2012). https://datamarket.com/data/set/2323/daily-maximum-temperatures-in-melbourne-australia-1981-1990#!ds=2323&display=line
4. Daily minimum temperatures in Melbourne, Australia. Australian Bureau of Meteorology (2012). https://datamarket.com/data/set/2324/daily-minimum-temperatures-in-melbourne-australia-1981-1990#!ds=2324&display=line
5. Erven, T.V., Rooij, S.D., Grünwald, P.: Catching up faster in Bayesian model selection and model averaging. In: Advances in Neural Information Processing Systems, pp. 417–424 (2007)
6. Hamilton, J.D.: Time Series Analysis, vol. 2. Princeton University Press, Princeton (1994)
7. Herbster, M., Warmuth, M.K.: Tracking the best expert. Mach. Learn. **32**(2), 151–178 (1998)
8. Hyndman, R.J., Athanasopoulos, G.: Forecasting: Principles and Practice. OTexts (2014)

9. Jamil, W., Kalnishkan, Y., Bouchachia, A.: Aggregation algorithm vs. average for time series prediction. In: Proceedings of the ECMLPKDD 2016 Workshop on Large-Scale Learning from Data Streams in Evolving Environments STREAMEVOLV-2016, September 2016 (2016)
10. Le Borgne, Y.-A., Santini, S., Bontempi, G.: Adaptive model selection for time series prediction in wireless sensor networks. Sig. Process. **87**(12), 3010–3020 (2007)
11. Liu, C., Hoi, S.C., Zhao, P., Sun, J.: Online ARIMA algorithms for time series prediction. In: Thirtieth AAAI Conference on Artificial Intelligence (2016)
12. Noshad, M., Ding, J., Tarokh, V.: Sequential learning of multi-state autoregressive time series. In: Proceedings of the 2015 Conference on Research in Adaptive and Convergent Systems, pp. 44–51. ACM (2015)
13. Prado, R., Lopes, H.F.: Sequential parameter learning and filtering in structured autoregressive state-space models. Stat. Comput. **23**(1), 43–57 (2013)
14. Robert, C.: The Bayesian Choice: From Decision-Theoretic Foundations to Computational Implementation. Springer, New York (2007)
15. Romanenko, A.: Aggregation of adaptive forecasting algorithms under asymmetric loss function. In: International Symposium on Statistical Learning and Data Sciences, pp. 137–146. Springer (2015)
16. Sato, M.-A.: Online model selection based on the variational Bayes. Neural Comput. **13**(7), 1649–1681 (2001)
17. Schwarz, G., et al.: Estimating the dimension of a model. Ann. Stat. **6**(2), 461–464 (1978)
18. Shibata, R.: Selection of the order of an autoregressive model by akaike's information criterion. Biometrika **63**(1), 117–126 (1976)
19. Vovk, V.: Aggregating strategies. In: Proceedings of Third Workshop on Computational Learning Theory, pp. 371–383. Morgan Kaufmann (1990)
20. Vovk, V.: A game of prediction with expert advice. In: Proceedings of the Eighth Annual Conference on Computational Learning Theory, pp. 51–60. ACM (1995)
21. Vovk, V.: Competitive on-line statistics. In: International Statistical Review/Revue Internationale de Statistique, pp. 213–248 (2001)
22. Vovk, V., Zhdanov, F.: Prediction with expert advice for the brier game. J. Mach. Learn. Res. **10**, 2445–2471 (2009)
23. Zinkevich, M.: Online convex programming and generalized infinitesimal gradient ascent. Technical report CMU-CS-03-110, School of Computer Science, Carnegie Mellon University (2003)

# Fuzzy Modeling for Uncertain Nonlinear Systems Using Fuzzy Equations and Z-Numbers

Raheleh Jafari[1]([✉]) [ID], Sina Razvarz[2], Alexander Gegov[3],
and Satyam Paul[4]

[1] Centre for Artificial Intelligence Research (CAIR), University of Agder,
4879 Grimstad, Norway
jafari3339@yahoo.com
[2] Departamento de Control Automático, CINVESTAV-IPN
(National Polytechnic Institute), 07360 Mexico City, Mexico
[3] School of Computing, University of Portsmouth, Buckingham Building,
Portsmouth PO13HE, UK
[4] School of Engineering and Sciences, Tecnológico de Monterrey,
64849 Monterrey, Nuevo Leon, Mexico

**Abstract.** In this paper, the uncertainty property is represented by Z-number as the coefficients and variables of the fuzzy equation. This modification for the fuzzy equation is suitable for nonlinear system modeling with uncertain parameters. Here, we use fuzzy equations as the models for the uncertain nonlinear systems. The modeling of the uncertain nonlinear systems is to find the coefficients of the fuzzy equation. However, it is very difficult to obtain Z-number coefficients of the fuzzy equations.

Taking into consideration the modeling case at par with uncertain nonlinear systems, the implementation of neural network technique is contributed in the complex way of dealing the appropriate coefficients of the fuzzy equations. We use the neural network method to approximate Z-number coefficients of the fuzzy equations.

**Keywords:** Fuzzy modeling · Z-number · Uncertain nonlinear system

## 1 Introduction

An exceptional case of uncertain system modeling at par with fuzzy equation is fuzzy polynomial interpolation. Polynomials have been used with fuzzy coefficients in order to interpolate uncertain data that are expressed using fuzzy numbers [1]. Interpolation methodology has been broadly utilized for function approximation as well as system identification [2]. In [3], the fuzzy polynomial interpolation is applied for system modeling. The theory problem associated with polynomial interpolation is researched in [4]. It elaborates that the interpolation of the function includes time complexity at par with data points. In [5], two-dimensional polynomial interpolation is demonstrated. Smooth function approximation has been broadly implemented currently [6]. It yields a model by utilizing Lagrange interpolating polynomials at the points of product grids [1, 7].

© Springer Nature Switzerland AG 2019
A. Lotfi et al. (Eds.): UKCI 2018, AISC 840, pp. 96–107, 2019.
https://doi.org/10.1007/978-3-319-97982-3_8

However if it involves uncertainties in the interpolation points, the above suggested techniques will not work appropriately.

The fuzzy equation can be regarded as a generalized form of the fuzzy polynomial. Compared with the normal fuzzy systems, the fuzzy equations are more easy to be applied. There are several approaches to construct the fuzzy equations. [8] used the fuzzy number on parametric shapes and replaced the original fuzzy equations with crisp linear systems. [9] proposed the homotypic analysis technique. [10] used the Newton methodology. In [11] the solution associated to the fuzzy equations is studied by the fixed point technique. The numerical solution associated to the fuzzy equations can be extracted by iterative technique [12], interpolation technique [13] and the Runge-Kutta technique [14]. The neural networks may also be used to solve fuzzy equations. In [15], the simple fuzzy quadratic equation is resolved by the neural network method. [16] extended the result of [15] to fuzzy polynomial equations. In [17], the solution of dual fuzzy equation is obtained by neural networks. A matrix pattern associated with the neural learning has been quoted in [18]. The predictor-corrector approach is applied in [19].

The decisions are carried out based on knowledge. In order to make the decision fruitful, the knowledge acquired must be credible. Z-numbers connect to the reliability of knowledge [20]. Many fields related to the analysis of the decisions use the ideas of Z-numbers. Z-numbers are much less complex to calculate when compared to nonlinear system modeling methods. Z-number is abundantly adequate number than the fuzzy number. Although Z-numbers are implemented in many literatures, from theoretical point of view this approach is not certified completely. There are few structure based on the theoretical concept of Z-numbers [21]. [22] gave an inception which results in the extension of Z-numbers. [23] proposed a theorem to transfer Z-numbers to the usual fuzzy sets. In [20] a novel approach was followed for the conversion of Z-number into fuzzy number.

In this paper, we use fuzzy equations to model the uncertain nonlinear systems, where the coefficients and variables are Z-numbers. Z-number is a novel idea that is subjected to a higher potential in order to illustrate the information of the human being as well as to use in information processing [20]. Z-numbers can be regarded as to answer questions and carry out the decisions [24]. This paper is one of the first attempts in finding the coefficients of fuzzy equations based on Z-numbers. We use the neural network method to approximate the coefficients of the fuzzy equations. The standard backpropagation method is modified, such that Z-numbers in the fuzzy equations can be trained.

## 2 Nonlinear System Modeling with Fuzzy Equations and Z-Numbers

A general discrete-time nonlinear system can be described as

$$\bar{x}_{k+1} = \bar{p}[\bar{x}_k, w_k], s_k = \bar{q}[\bar{x}_k] \tag{1}$$

Here we consider $W_k \in \Re^u$ as the input vector, $\bar{x}_k \in \Re^l$ is regarded as an internal state vector and $S_k \in \Re^m$ is the output vector. $\bar{p}$ and $\bar{q}$ are noted as generalized nonlinear smooth functions $\bar{p}, \bar{q} \in C^\infty$. Define $S_k = \left[ s_{k+1}^T, s_k^T, \ldots \right]^T$ and $W_k = \left[ W_{k+1}^T, W_k^T, \ldots \right]^T$. Suppose $\frac{\partial S}{\partial \bar{x}}$ is non-singular at the instance $\bar{x}_k = 0$, $W_k = 0$, this will create a path towards the following model

$$s_k = \Omega[s_{k-1}^T, s_{k-2}^T, \cdots w_k^T, w_{k-1}^T, \cdots] \tag{2}$$

Where $\Omega(\cdot)$ is an nonlinear difference equation exhibiting the plant dynamics, $W_k$ and $S_k$ are computable scalar input and output respectively. The nonlinear system which is represented by (2) is implied as a NARMA model. The input of the system with incorporated nonlinearity is considered to be as

$$x_k = [s_{k-1}^T, s_{k-2}^T, \cdots w_k^T, w_{k-1}^T, \cdots]^T$$

Taking into consideration the nonlinear systems as mentioned in (plant), it can be simplified as the following linear-in-parameter model

$$s_k = \sum_{i=1}^{n} \sum_{j=1}^{m} b_{ij} p_i(x_k) q_j(y_k) \tag{3}$$

here $b_{ij}$ is considered to be the linear parameter, $p_i(x_k)$ and $q_j(y_k)$ are nonlinear functions. The variables related to these functions are quantifying input and output.

The modeling of uncertain nonlinear systems can be achieved by utilizing the linear-in-parameter models linked to fuzzy parameters. We assume the model of the nonlinear systems (3) has uncertainties in the $b_{ij}$, $x_k$ and $y_k$. These uncertainties are in the sense of Z-numbers [25].

**Definition 1.** If $v$ is: (1) normal, there exists $\vartheta_0 \in \Re$ in such a manner $v(\vartheta_0) = 1$, (2) convex, $v(\gamma\vartheta + (1-\gamma)\vartheta) \geq \min\{v(\vartheta, v(\theta))\}, \forall \vartheta, \theta \in \Re, \forall \gamma \in [0,1]$, (3) upper semi-continuous on $\Re, v(\vartheta) \leq v(\vartheta_0) + \varepsilon, \ \forall \vartheta \in N(\vartheta_0), \forall \vartheta_0 \in \Re, \forall \varepsilon > 0, N(\vartheta_0)$ is a neighborhood, (4) $v^+ = \{\vartheta \in \Re, v(\vartheta) > 0\}$ is compact, then $v$ is a fuzzy variable, $v \in E : R \to [0,1]$.

The fuzzy variable $v$ can be also represented as

$$v = (\underline{v}, \bar{v}) \tag{4}$$

Where $\underline{v}$ is the lower-bound variable and $\bar{v}$ is the upper-bound variable.

**Definition 2.** A $Z$ - number has two components $Z = [v(\vartheta), p]$. The primary component $v(\vartheta)$ is termed as a restriction on a real-valued uncertain variable $\vartheta$. The secondary component $p$ is a measure of reliability of $v$. $p$ can be reliability, strength of belief, probability or possibility. When $v(\vartheta)$ is a fuzzy number and $p$ is the probability distribution of, $Z$ - number is defined as $Z^+$ - number. When both $v(\vartheta)$ and $p$ are fuzzy numbers, $Z$ - number is defined as $Z^-$ - number.

$Z^+$ - number carries more information than $Z^-$ - number. In this paper, we use the definition of $Z^+$- number, i.e., $Z = [v,p]$, $v$ is a fuzzy number and $p$ is a probability distribution.

We use so called membership functions to express the fuzzy number. One of the most popular membership function is the triangular function

$$\mu_v = G(a, b, c) = \begin{cases} \frac{\vartheta-a}{b-a} & a \leq \vartheta \leq b \\ \frac{c-\vartheta}{c-b} & b \leq \vartheta \leq c \end{cases} \; otherwise \; \mu_v = 0 \tag{5}$$

and trapezoidal function

$$\mu_v = G(a, b, c, d) = \begin{cases} \frac{\vartheta-a}{b-a} & a \leq \vartheta \leq b \\ \frac{d-\vartheta}{d-c} & c \leq \vartheta \leq d \; otherwise \; \mu_v = 0 \\ 1 & b \leq \vartheta \leq c \end{cases} \tag{6}$$

The probability measure is expressed as

$$p = \int_R \mu_v(\vartheta)p(\vartheta)d\vartheta \tag{7}$$

where $p$ is the probability density of $\vartheta$ and R is the restriction on $p$. For discrete $Z$ - numbers

$$p(v) = \sum_{i=1}^{n} \mu_v(\vartheta_i)p(\vartheta_i) \tag{8}$$

**Definition 3.** The fuzzy number $v$ in association to the $\alpha$ - level is illustrated as

$$[v]^\alpha = \{a \in \Re : v(a) \geq a)\} \tag{9}$$

Where $0 < \alpha \leq 1, v \in E$.

Therefore $[v]^0 = v^+ = \{\vartheta \in \Re, v(\vartheta) > 0\}$ Since $\alpha \in [0, 1]$, $[v]^\alpha$ is a bounded mentioned as $\underline{v}^\alpha \leq [v]^\alpha \leq \bar{v}^\alpha$ The $\alpha$ - level of $v$ in midst of $\underline{v}^\alpha$ and $\bar{v}^\alpha$ is explained as

$$[v]^\alpha = (\underline{v}^\alpha, \bar{v}^\alpha)$$

$\underline{v}^\alpha$ and $\bar{v}^\alpha$ signify the function of $\alpha$. We state $\underline{v}^\alpha = d_A(\alpha), \bar{v}^\alpha d_B(\alpha), \alpha \in [0, 1]$.

**Definition 4.** The $\alpha$ - level of $Z$ - number $Z = (v, P)$ is demonstrated as

$$[Z]^\alpha = ([V]^\alpha, [p]^\alpha) \tag{10}$$

where $0 < \alpha \le 1$. $[p]^{\alpha}$ is calculated by the Nguyen's theorem

$$[p]^{\alpha} = p([v]^{\alpha}) = p([\underline{v}^{\alpha}, \overline{v}^{\alpha}]) = [\underline{P}^{\alpha}, \overline{P}^{\alpha}]$$

where $p([v]^{\alpha}) = \{p(\vartheta) | \vartheta \in [v]^{\alpha}\}$. So $[Z]^{\alpha}$ can be expressed as the form $\alpha$ - level of a fuzzy number

$$[Z]^{\alpha} = (\underline{Z}^{\alpha}, \overline{Z}^{\alpha}) = ((\underline{v}^{a}, \underline{P}^{a}), (\overline{v}^{a}, \overline{P}^{a})) \tag{11}$$

where $\underline{p}^{\alpha} = \underline{v}^{\alpha} p(\underline{\vartheta_i}^{\alpha}), \overline{p}^{\alpha} = \overline{v}^{\alpha} p(\overline{\vartheta_i}^{\alpha}), [\vartheta_i]^{\alpha} = (\underline{\vartheta_i}^{\alpha}, \overline{\vartheta_i}^{\alpha})$.

Similar with the fuzzy numbers [26–29], $Z$ - numbers are also incorporated with three primary operations: $\oplus, \ominus$ and $\odot$. These operations are exhibited by: sum subtract multiply and division. The operations in this paper are different definitions with [20]. The $\alpha$ - level of $Z$ - numbers is applied to simplify the operations.

Let us consider $Z_1 = (v_1, p_1)$ and $Z_2 = (v_2, p_2)$ be two discrete $Z$ - numbers illustrating the uncertain variables $\vartheta_1$ and $\vartheta_2$, $\sum_{k=1}^{n} p_1(\vartheta_{1k}) = 1$, $\sum_{k=1}^{n} p_2(\vartheta_{2k}) = 1$. The operations are defined

$$Z_{12} = Z_1 * Z_2 = (v_1 * v_2, p_1 * p_2)$$

where $* \in \{\oplus, \ominus, \odot\}$.

The operations for the fuzzy numbers are defined as [26]

$$\begin{aligned} [v_1 \oplus v_2]^{\alpha} &= [\underline{v_1}^{\alpha} + \underline{v_2}^{\alpha}, \overline{v_1}^{\alpha} + \overline{v_2}^{\alpha}] \\ [v_1 \ominus v_2]^{\alpha} &= [\underline{v_1}^{\alpha} - \underline{v_2}^{\alpha}, \overline{v_1}^{\alpha} - \overline{v_2}^{\alpha}] \\ [v_1 \odot v_2]^{\alpha} &= \left( \begin{array}{c} \min\{\underline{v_1}^{\alpha}\underline{v_2}^{\alpha}, \underline{v_1}^{\alpha}\overline{v_2}^{\alpha}, \overline{v_1}^{\alpha}\underline{v_2}^{\alpha}, \overline{v_1}^{\alpha}\overline{v_2}^{\alpha}\} \\ \max\{\underline{v_1}^{\alpha}\underline{v_2}^{\alpha}, \underline{v_1}^{\alpha}\overline{v_2}^{\alpha}, \overline{v_1}^{\alpha}\underline{v_2}^{\alpha}, \overline{v_1}^{\alpha}\overline{v_2}^{\alpha}\} \end{array} \right) \end{aligned} \tag{13}$$

For all $p_1 * p_2$ operations, we use convolutions for the discrete probability distributions

$$p_1 * p_2 = \sum_i p_1(\vartheta_{1,i}) p_2(\vartheta_{2,(n-i)}) = p_{12}(\vartheta)$$

The above definitions satisfy the Hukuhara difference [30–32],

$$Z_1 \ominus_H Z_2 = Z_{12}$$
$$Z_1 = Z_2 \oplus Z_{12}$$

Here if $Z_1 \ominus_H Z_2$ prevails, the $\alpha$ - level is

$$[Z_1 \ominus_H Z_2]^{\alpha} = [\underline{Z}_1^{\alpha} - \underline{Z}_2^{\alpha}, \overline{Z}_1^{\alpha} - \overline{Z}_2^{\alpha}]$$

Obviously, $Z_1 \ominus_H Z_1 = 0$, $Z_1 \ominus_H Z_1 \ne 0$.

Also the above definitions satisfy the generalized Hukuhara difference [33]

$$Z_1 \ominus_{gH} Z_2 = Z_{12} \Leftrightarrow \begin{cases} 1)\ Z_1 = Z_2 \oplus Z_{12} \\ 2)\ Z_2 = Z_1 \oplus (-1)Z_{12} \end{cases} \tag{14}$$

It is convenient to display that 1) and 2) in combination are genuine if and only if $Z_{12}$ is a crisp number. With respect to $\alpha$ - level what we got are $\left[Z_1 \ominus_{gH} Z_2\right]^\alpha = \left[\min\{\underline{Z}_1^\alpha - \underline{Z}_2^\alpha, \bar{Z}_1^\alpha - \bar{Z}_2^\alpha\}, \max\{\underline{Z}_1^\alpha - \underline{Z}_2^\alpha, \bar{Z}_1^\alpha - \bar{Z}_2^\alpha\}\right]$ and If $Z_1 \ominus_{gH} Z_2$ and $Z_1 \ominus_H Z_2$ subsist, $Z_1 \ominus_H Z_2 = Z_1 \ominus_{gH} Z_2$. The circumstances for the inerrancy of $Z_{12} = Z_1 \ominus_{gH} Z_2 \in E$ are

$$(1) \begin{cases} \underline{Z}_{12}^\alpha = \underline{Z}_1^\alpha - \underline{Z}_2^\alpha \ and\ \bar{Z}_{12}^\alpha = \bar{Z}_1^\alpha - \bar{Z}_2^\alpha \\ with\ \underline{Z}_{12}^\alpha\ increasing\ \bar{Z}_{12}^\alpha\ decreasing\ \underline{Z}_{12}^\alpha \leq \bar{Z}_{12}^\alpha \end{cases}$$
$$(2) \begin{cases} \underline{Z}_{12}^\alpha = \bar{Z}_1^\alpha - \bar{Z}_2^\alpha \ and\ \bar{Z}_{12}^\alpha = \underline{Z}_1^\alpha - \underline{Z}_2^\alpha \\ with\ \underline{Z}_{12}^\alpha\ increasing\ \bar{Z}_{12}^\alpha\ decreasing,\ \underline{Z}_{12}^\alpha \leq \bar{Z}_{12}^\alpha \end{cases} \tag{15}$$

where $\forall \alpha \in [0,1]$

If $v$ is a triangular function, the absolute value of Z –number $Z = (v,p)$ is

$$|Z(\vartheta)| = (|a_1| + |b_1| + |c_1|,\ p(|a_2| + |b_2| + |c_2|)) \tag{16}$$

If $v_1$ and $v_2$ are triangular functions, the supremum metric for Z - numbers $Z_1 = (v_1, p_1)$ and $Z_2 = (v_2, p_2)$ is given as

$$D(Z_1, Z_2) = d(v_1, v_2) + d(p_1, p_2)$$

in this case $d(\cdot, \cdot)$ is the supremum metrics considering fuzzy sets [26]. $D(Z_1, Z_2)$ is incorporated with the following possessions:

$$D(Z_1 + Z,\ Z_2 + Z) = D(Z_1,\ Z_2)$$
$$D(Z_2,\ Z_1) = D(Z_1,\ Z_2)$$
$$D(\zeta Z_1,\ k Z_2) = |\zeta| D(Z_1,\ Z_2)$$
$$D(Z_1,\ Z_2) \leq D(Z_1,\ Z) + D(Z,\ Z_2)$$

where $\in \Re$, $Z = (v,p)$ is Z - number and $v$ is triangle function.

**Definition 5.** Let $\tilde{Z}$ denotes the space of Z - numbers. The $\alpha-$ level of Z - number valued function $G : [0, a] \rightarrow \tilde{Z}$ is

$$G(v, \alpha) = [\underline{G}(v, \alpha), \bar{G}(v, \alpha)]$$

where $\in \tilde{Z}$, for each $\alpha \in [0, 1]$.

With the definition of Generalized Hukuhara difference, the gH-derivative of $G$ at $v_0$ is expressed as

$$\frac{d}{dt} G(v_0) = \lim_{h \to 0} \frac{1}{h} \left[ G(v_0 + h) \ominus_{gH} G(v_0) \right] \tag{17}$$

In (17), $G(v_0 + h)$ and $G(v_0)$ exhibits similar style with $Z_1$ and $Z_2$ respectively included in (14).

Now we utilize the fuzzy Eq. (3) to model the uncertain nonlinear system (1). Modeling with fuzzy equation (or fuzzy polynomial) can be regarded as fuzzy interpolation. In this paper, we utilize the fuzzy Eq. (1) to model the uncertain nonlinear system (1), in such a manner that the output related to the plant $s_k$ can approach to the desired output $s_k^*$,

$$\min_{W_k} \| s_k - s_k^* \| \tag{18}$$

This modeling object can be regarded as to detect $b_{i,j}$ for the following fuzzy equation

$$s_k^* = \sum_{i=1}^{n} \sum_{j=1}^{m} b_{ij} p_i(x) \, q_j(y) \tag{19}$$

where $x_r = \left[ s_{k-1}^T, s_{k-2}^T, \ldots, w_k^T, w_{k-1}^T, \ldots \right]^T$.

## 3   Z-Number Parameter Estimation with Neural Networks

We design a neural network to represent the fuzzy Eq. (3), see Fig. 1. The inputs to the neural network are $x_k$ and $y_k$, the output is Z-number $Z_k$. The main idea is to detect appropriate weight of neural network $b_{i,j}$ in such a manner that the output of the neural network $Z_k$ converges to the desired output $s_k^*$.

The input Z-numbers $x_k$ and $y_k$ are first applied to $\alpha$ - level as in (11)

$$\begin{aligned} [x_k]^\alpha &= \left( \underline{x}_k^\alpha, \bar{x}_k^\alpha \right) \\ [y_k]^\alpha &= \left( \underline{y}_k^\alpha, \bar{y}_k^\alpha \right) \end{aligned} \tag{20}$$

Then in the first hidden units we have

$$\begin{aligned} [\Phi_i]^\alpha &= \left( p_i\left(\underline{x}_k^\alpha\right), p_i\left(\bar{x}_k^\alpha\right) \right) \quad i = 1, \ldots, n \\ [\Phi_j]^\alpha &= \left( q_j\left(\underline{y}_k^\alpha\right), q_j\left(\bar{y}_k^\alpha\right) \right) \quad j = 1, \ldots, m \end{aligned} \tag{21}$$

and in the second hidden units we have

$$[\Phi_{ij}]^\alpha = \Big\{ \sum_{i,j \in N} \underline{\Phi}_i^\alpha \underline{\Phi}_j^\alpha + \sum_{i,j \in O} \overline{\Phi}_i^\alpha \overline{\Phi}_j^\alpha + \sum_{i,j \in Q} \underline{\Phi}_i^\alpha \overline{\Phi}_j^\alpha, \\ \sum_{i,j \in N'} \overline{\Phi}_i^\alpha \overline{\Phi}_j^\alpha + \sum_{i,j \in O'} \underline{\Phi}_i^\alpha \underline{\Phi}_j^\alpha + \sum_{i,j \in Q'} \overline{\Phi}_i^\alpha \underline{\Phi}_j^\alpha \Big\} \tag{22}$$

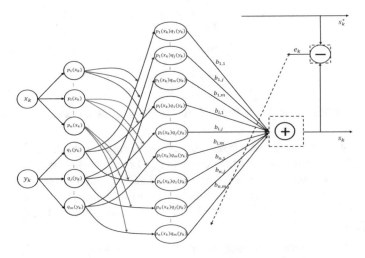

**Fig. 1.** Fuzzy equation in the form of neural network.

where $N = \left\{ i,j \middle| \underline{\Phi_i}^\alpha \geq 0, \ \underline{\Phi_j}^\alpha \geq 0 \right\}, O = \{i,j| \overline{\Phi_i}^\alpha < 0, \overline{\Phi_j}^\alpha < 0\}, \ Q = \{i,j| \underline{\Phi_i}^\alpha <$
$0, \ \overline{\Phi_j}^\alpha \geq 0\}, N' = i,j| \overline{\Phi_i}^\alpha \geq 0, \ \overline{\Phi_j}^\alpha \geq 0\},$
$O' = i,j \middle| \underline{\Phi_i}^\alpha < 0, \ \underline{\Phi_j}^\alpha < 0\}, \ Q' = i,j \middle| \overline{\Phi_i}^\alpha < 0, \ \underline{\Phi_j}^\alpha < 0\}.$

The neural network output is

$$[s_k]^\alpha = \{ \textstyle\sum_{i,j \in N} \underline{\Phi_{i,j}}^\alpha \underline{b_{i,j}}^\alpha + \sum_{i,j \in O} \overline{\Phi_{i,j}}^\alpha \overline{b_{i,j}}^\alpha + \sum_{i,j \in Q} \underline{\Phi_{i,j}}^\alpha \overline{b_{i,j}}^\alpha , $$
$$\textstyle\sum_{i,j \in N'} \overline{\Phi_{i,j}}^\alpha \overline{b_{i,j}}^\alpha + \sum_{i,j \in O'} \underline{\Phi_{i,j}}^\alpha \underline{b_{i,j}}^\alpha + \sum_{i,j \in Q'} \overline{\Phi_{i,j}}^\alpha \underline{b_{i,j}}^\alpha \}$$

(23)

where $N = \left\{ i,j \middle| \underline{\Phi_{i,j}}^\alpha \geq 0, \underline{b_{i,j}}^\alpha \geq 0 \right\}, O = \{i,j| \overline{\Phi_{i,j}}^\alpha < 0, \overline{b_{i,j}}^\alpha < 0\}, Q =$
$\{i,j \middle| \underline{\Phi_{i,j}}^\alpha < 0, \overline{b_{i,j}}^\alpha \geq 0\}, N' = \{i,j| \overline{\Phi_{i,j}}^\alpha \geq 0, \overline{b_{i,j}}^\alpha \geq 0\}, O' = \left\{ i,j \middle| \underline{\Phi_{i,j}}^\alpha < 0, \underline{b_{i,j}}^\alpha < 0 \right\},$
$Q' = \left\{ i,j \middle| \overline{\Phi_{i,j}}^\alpha < 0, \underline{b_{i,j}}^\alpha < 0 \right\}.$

In order to train the weights, we need to define a cost function for the fuzzy numbers. The error of the training is

$$e_k = s_k^* - s_k$$

where $\left[ s_k^* \right]^\alpha = \left( \underline{s_k^{*\alpha}}, \overline{s_k^{*\alpha}} \right), [s_k]^\alpha = \left( \underline{s_k^\alpha}, \overline{s_k^\alpha} \right), [e_k]^\alpha = \left( \underline{e_k}^\alpha, \overline{e_k}^\alpha \right).$ The cost function is defined as

$$\Upsilon_k = \underline{\Upsilon}^\alpha + \overline{\Upsilon}^\alpha$$
$$\underline{\Upsilon}^\alpha = \tfrac{1}{2}\left(\underline{s_k^{*\alpha}} - \underline{\overline{s}_k^{*\alpha}}\right)^2$$
$$\overline{\Upsilon}^\alpha = \tfrac{1}{2}\left(\overline{s_k^{*\alpha}} - \underline{s_k^{*\alpha}}\right)^2 \tag{24}$$

$\Upsilon_k$ is considered to be a scalar function. It is quite obvious, $\Upsilon_k \to 0$ means $[s_k]^\alpha \to [s_k^*]^\alpha$.

The vital positiveness lies within the least mean square (24) is that it has a self-correcting feature that makes it suitable to function for arbitrarily vast duration without shifting from its constraints. The mentioned gradient algorithm is subjected to cumulative series of errors and is convenient for long runs in absence of an additional error rectification procedure. It is more robust in statistics, identification and signal processing [34].

Now we use gradient method to train Z-number weight $b_{i,j} = \left(\underline{b_{i,j}}, \overline{b_{i,j}}\right)$. We compute $\frac{\partial \Upsilon_k}{\partial \underline{b_{i,j}}}$ and $\frac{\partial \Upsilon_k}{\partial \overline{b_{i,j}}}$ as

$$\frac{\partial \Upsilon_k}{\partial \underline{b_{i,j}}} = \frac{\partial \Upsilon_k^\alpha}{\partial \underline{s_k^\alpha}} \frac{\partial \underline{s_k^\alpha}}{\partial \underline{b_{ij}^\alpha}} \frac{\partial \underline{b_{ij}^\alpha}}{\partial \underline{b_{ij}^\alpha}} + \frac{\partial \overline{\Upsilon}_k^\alpha}{\partial \overline{s_k^\alpha}} \frac{\partial \overline{s_k^\alpha}}{\partial \underline{b_{ij}^\alpha}} \frac{\partial \underline{b_{ij}^\alpha}}{\partial \underline{b_{ij}^\alpha}}$$
$$= -\left(\underline{s_k^{*\alpha}} - \underline{s_k^\alpha}\right)\sum_{i,j\in N} \underline{\Phi_{i,j}}^\alpha \Gamma - \left(\overline{s_k^{*\alpha}} - \overline{s_k^\alpha}\right)\left(\sum_{i,j\in O'}\underline{\Phi_{i,j}}^\alpha + \sum_{i,j\in O'}\overline{\Phi_{i,j}}^\alpha\right)\Gamma$$

where
and

$$\frac{\partial \Upsilon_k}{\partial \overline{b_{i,j}}} = \frac{\partial \Upsilon_k^\alpha}{\partial \underline{s_k^\alpha}} \frac{\partial \underline{s_k^\alpha}}{\partial \overline{b_{ij}}^\alpha} \frac{\partial \overline{b_{ij}}^\alpha}{\partial \overline{b_{ij}}^\alpha} + \frac{\partial \overline{\Upsilon}_k^\alpha}{\partial \overline{s_k^\alpha}} \frac{\partial \overline{s_k^\alpha}}{\partial \overline{b_{ij}}^\alpha} \frac{\partial \overline{b_{ij}}^\alpha}{\partial \overline{b_{ij}}^\alpha}$$
$$= -\left(\underline{s_k^{*\alpha}} - \underline{s_k^\alpha}\right)\left(\sum_{i,j\in O}\overline{\Phi_{i,j}}^\alpha + \sum_{i,j\in Q}\underline{\Phi_{i,j}}^\alpha\right)\Gamma_1 - \left(\overline{s_k^{*\alpha}} - \overline{s_k^\alpha}\right)\sum_{i,j\in N'}\overline{\Phi_{i,j}}^\alpha \Gamma_1$$

where
The coefficient $b_{i,j}$ is updated as

$$\underline{b_{i,j}}(k+1) = \underline{b_{i,j}}(k) - \eta \frac{\partial \Upsilon_k}{\partial \underline{b_{i,j}}}$$
$$\overline{b_{i,j}}(k+1) = \overline{b_{i,j}}(k) - \eta \frac{\partial \Upsilon_k}{\partial \overline{b_{i,j}}} \tag{25}$$

where $\eta$ is the training rate $\eta > 0$. For the requirement of increasing the training process, the adding of the momentum term is mentioned as

$$\underline{b_{i,j}}(k+1) = \underline{b_{i,j}}(k) - \eta \frac{\partial \Upsilon_k}{\partial \underline{b_{i,j}}} + \Upsilon\left[\underline{b_{i,j}}(k) - \underline{b_{i,j}}(k-1)\right]$$
$$\overline{b_{i,j}}(k+1) = \overline{b_{i,j}}(k) - \eta \frac{\partial \Upsilon_k}{\partial \overline{b_{i,j}}} + \Upsilon\left[\overline{b_{i,j}}(k) - \overline{b_{i,j}}(k-1)\right] \tag{26}$$

where $\gamma > 0$.

**Learning Algorithm**

(1) Step 1: Choose the training rates $\eta > 0$, $\gamma > 0$ and the stop criterion $\overline{\Upsilon} > 0$. The initial Z-number vector $B = (b_{1,1}, \ldots, b_{n,m})$ is selected randomly. The initial learning iteration is $k = 1$ the initial learning error $\Upsilon = 0$.

(2) Repeat the following steps for $\alpha = \alpha_1, \ldots, \alpha_m$, until all training data are applied

    (a) Forward calculation: Calculate the $\alpha$-level of Z-number output $s_k$ with the $\alpha$-level of Z-number input vectors $(x_k, y_k)$, and Z-number connection weight $B$.

    (b) Back-propagation: Adjust Z-number parameters $b_{i,j}$, $i = 1, \ldots, n$, $j = 1, \ldots, m$, by using the cost function for the $\alpha$-level of Z-number output $s_k$, and Z-number target output $s_k^*$.

    (c) Stop criterion: calculate the cycle error $\Upsilon_k$, $\Upsilon = \Upsilon + \Upsilon_k . k = k + 1$ If $\Upsilon > \overline{\Upsilon}$ let $\Upsilon = 0$, a new training cycle is initiated. Go to (a).

# 4  Conclusion

In this paper, the classical fuzzy equation is modified such that its coefficients and variables are Z-numbers. However, the parameters of the fuzzy equations cannot be obtained directly. We use the neural network method to approximate Z-number coefficients of the fuzzy equations. The neural model is constructed with the structure of fuzzy equations. With modified backpropagation method, the neural network is trained. Further work is to study the stability of training algorithms.

# References

1. Barthelmann, V., Novak, E., Ritter, K.: High dimensional polynomial interpolation on sparse grids. Adv. Comput. Math. **12**, 273–288 (2000)
2. Jafarian, A., Jafari, R., Mohamed Al Qurashi, M., Baleanud, D.: A novel computational approach to approximate fuzzy interpolation polynomials, SpringerPlus **5**, 1428 (2016). https://doi.org/10.1186/s40064-016-3077-5
3. Neidinger, R.D.: Multi variable interpolating polynomials in newton forms. In: Proceedings of the Joint Mathematics Meetings, Washington, DC, USA, pp. 5–8 (2009)
4. Schroeder, H., Murthy, V.K., Krishnamurthy, E.V.: Systolic algorithm for polynomial interpolation and related problems. Parallel Comput. **17**, 493–503 (1991)
5. Zolic, A.: Numerical Mathematics. Faculty of mathematics, Belgrade, pp. 91–97 (2008)
6. Szabados, J., Vertesi, P.: Interpolation of Functions. World Scientific Publishing Co., Singapore (1990)
7. Xiu, D., Hesthaven, J.S.: High-order collocation methods for differential equations with random inputs. SIAM J. Sci. Comput. **27**, 1118–1139 (2005)
8. Friedman, M., Ming, M., Kandel, A.: Fuzzy linear systems. Fuzzy Sets Syst. **96**, 201–209 (1998)
9. Abbasbandy, S.: The application of homotopy analysis method to nonlinear equations arising in heat transfer. Phys. Lett. A **360**, 109–113 (2006)

10. Abbasbandy, S., Ezzati, R.: Newton's method for solving a system of fuzzy nonlinear equations. Appl. Math. Comput. **175**, 1189–1199 (2006)
11. Allahviranloo, T., Ahmadi, N., Ahmadi, E.: Numerical solution of fuzzy differential equations by predictor-corrector method. Inform. Sci. **177**, 1633–1647 (2007)
12. Kajani, M., Asady, B., Vencheh, A.: An iterative method for solving dual fuzzy nonlinear equations. Appl. Math. Comput. **167**, 316–323 (2005)
13. Waziri, M., Majid, Z.: A new approach for solving dual fuzzy nonlinear equations using Broyden's and Newton's methods. Adv. Fuzzy Syst. (2012). Article 682087, 5 pages
14. Pederson, S., Sambandham, M.: The Runge-Kutta method for hybrid fuzzy differential equation. Nonlinear Anal. Hybrid Syst. **2**, 626–634 (2008)
15. Buckley, J., Eslami, E.: Neural net solutions to fuzzy problems: the quadratic equation. Fuzzy Sets Syst. **86**, 289–298 (1997)
16. Jafarian, A., Jafari, R., Khalili, A., Baleanud, D.: Solving fully fuzzy polynomials using feed-back neural networks. Int. J. Comput. Math. **92**(4), 742–755 (2015)
17. Jafarian, A., Jafari, R.: Approximate solutions of dual fuzzy polynomials by feed-back neural networks. J. Soft Comput. Appl. (2012). https://doi.org/10.5899/2012/jsca-00005
18. Mosleh, M.: Evaluation of fully fuzzy matrix equations by fuzzy neural network. Appl. Math. Model. **37**, 6364–6376 (2013)
19. Allahviranloo, T., Otadi, M., Mosleh, M.: Iterative method for fuzzy equations. Soft. Comput. **12**, 935–939 (2007)
20. Zadeh, L.A.: Toward a generalized theory of uncertainty (GTU) an outline. Inform. Sci. **172**, 1–40 (2005)
21. Gardashova, L.A.: Application of operational approaches to solving decision making problem using Z-Numbers. J. Appl. Math. **5**, 1323–1334 (2014)
22. Aliev, R.A., Alizadeh, A.V., Huseynov, O.H.: The arithmetic of discrete Z-numbers. Inform. Sci. **290**, 134–155 (2015)
23. Kang, B., Wei, D., Li, Y., Deng, Y.: Decision making using Z-Numbers under uncertain environment. J. Comput. Inf. Syst. **8**, 2807–2814 (2012)
24. Kang, B., Wei, D., Li, Y., Deng, Y.: A method of converting Z-number to classical fuzzy number. J. Inf. Comput. Sci. **9**, 703–709 (2012)
25. Zadeh, L.A.: A note on Z-numbers. Inf. Sci. **181**, 2923–2932 (2011)
26. Jafari, R., Yu, W.: Fuzzy control for uncertainty nonlinear systems with dual fuzzy equations. J. Intell. Fuzzy Syst. **29**, 1229–1240 (2015)
27. Jafari, R., Yu, W.: Uncertainty nonlinear systems modeling with fuzzy equations. In: Proceedings of the 16th IEEE International Conference on Information Reuse and Integration, San Francisco, Calif, USA, pp. 182–188, August 2015
28. Jafari, R., Yu, W.: Uncertainty nonlinear systems control with fuzzy equations. In: IEEE International Conference on Systems, Man, and Cybernetics, pp. 2885–2890 (2015)
29. Razvarz, S., Jafari, R., Granmo, O.Ch., Gegov, A.: Solution of dual fuzzy equations using a new iterative method. In: Asian Conference on Intelligent Information and Database Systems, pp. 245–255 (2018)
30. Aliev, R.A., Pedryczb, W., Kreinovich, V., Huseynov, O.H.: The general theory of decisions. Inform. Sci. **327**, 125–148 (2016)
31. Jafari, R., Yu, W., Li, X.: Solving fuzzy differential equation with Bernstein neural networks. In: IEEE International Conference on Systems, Man, and Cybernetics, Budapest, Hungary, pp. 1245–1250 (2016)

32. Jafari, R., Yu, W., Li, X., Razvarz, S.: Numerical solution of fuzzy differential equations with Z-numbers using Bernstein neural networks. Int. J. Comput. Intell. Syst. **10**, 1226–1237 (2017)
33. Suykens, J.A.K., Brabanter, JDe, Lukas, L., Vandewalle, J.: Weighted least squares support vector machines: robustness and sparse approximation. Neurocomputing **48**, 85–105 (2002)
34. Bede, B., Stefanini, L.: Generalized differentiability of fuzzy-valued functions. Fuzzy Sets Syst. **230**, 119–141 (2013)

# Medical Expert Systems – A Study of Trust and Acceptance by Healthcare Stakeholders

Ioannis Vourgidis[1](✉), Shadreck Joseph Mafuma[1], Paul Wilson[1](✉), Jenny Carter[2](✉), and Georgina Cosma[3]

[1] De Montfort University, The Gateway, Leicester LE1 9BH, UK
ioannis.vourgidis@my365.dmu.ac.uk,
p08256218@alumni365.dmu.ac.uk, pdcwilson@dmu.ac.uk
[2] Huddersfield University, Queensgate, Huddersfield HD1 3DH, UK
j.carter@hud.ac.uk
[3] Nottingham Trent University, Clifton Campus, Nottingham NG11 8NS, UK
georgina.cosma@ntu.ac.uk

**Abstract.** The increasing prevalence of complex technology in the form of medical expert systems in the healthcare sector is presenting challenging opportunities to clinicians in their quest to improve patients' health outcomes. Medical expert systems have brought measurable improvements to the healthcare outcomes for some patients. This paper highlights the importance of trust and acceptance in the healthcare industry amongst receivers of the care as well as other stakeholders and between large healthcare organizations. Studies show that current conceptual trust models, which are being used to measure the degree of trust relationships in different healthcare settings, cannot be easily evaluated because of the resistance of organizational and social changes which are to be implemented. Research findings also suggest that the use of medical expert systems do not automatically guarantee improved patient healthcare outcomes. Furthermore, during the building of predictive and diagnostic expert medical systems, studies recommend the use of algorithms which can deal with noisy and imprecise data which is typical in healthcare data. Such algorithms include fuzzy rule based systems.

**Keywords:** Healthcare NHS · Medical expert systems · Trust
Acceptance · Artificial intelligence · Systematic literature review

## 1 Introduction

The systems specialized for medical diagnosis and treatment using Artificial Intelligence (AI) are designated as Medical Expert Systems (MES). Expert Systems are of the most frequent applications of artificial intelligence. Expert Systems model human knowledge into software applications and utilize it to solve problems within a specialized domain which necessitates human expertise. An expert system is also called Knowledge Based System [36]. The increasing interest in trust and acceptance of the use of MES has been highlighted in a number articles. Improving patient trust is key to improving patient access to, experience of and outcomes from healthcare [2, 3]. To understand what has been done in this research area, thoroughly, in an unbiased and

© Springer Nature Switzerland AG 2019
A. Lotfi et al. (Eds.): UKCI 2018, AISC 840, pp. 108–119, 2019.
https://doi.org/10.1007/978-3-319-97982-3_9

professional manner. This paper aims to: (1) provide a review of existing research relating to trust and acceptance of medical expert systems; (2) to identify any research that has been done on Medical Expert Systems (MES); and (3) to discuss gaps in the existing research in order to suggest any areas for further research studies.

The paper analyses how medical expert systems (MES) are trusted and accepted in the healthcare field, and how these systems are now playing a pivotal role in the provision of healthcare in this medical field. This paper is structured as follows. Section 2 provides a comprehensive literature review on medical expert systems; Sect. 3 provides the results of the review; and Sect. 4 provides a conclusion.

## 2 Literature Review

### 2.1 Overview

Expert systems are artificial intelligence programs that can achieve expert level competence ability in solving problems in a specific task area using the knowledge base about that specific field. Expert systems are computer systems that have artificial intelligence capabilities that use knowledge from a given domain represented in a form of decision production rules. According to [4], expert systems use human knowledge in a computer to solve problems that are difficult enough to ordinarily require human expertise. According to [8], a medical expert system is a computer system that performs diagnosis and makes therapy recommendations. According to [8], there is evidence to suggest that advice from a computer will be convincing if supported with explanation facilities. It is also suggested that evidence shows that trials with CAPSULE, an expert system which gives advice to general practitioners about prescribing drugs indicate "finding the most effective way of presenting the explanation is an important goal for future development of computer support for prescribing drugs" [7]. During the early years of development of expert systems in medicine, a lot of research systems were developed to assist clinicians in the process of diagnosis classically with the intention of being used in a clinical encounter with a patient. Although most of them were not developed further, others have continued to be improved and have been transformed into educational systems and are leading the way in which clinicians make knowledge-based decisions. Expert systems have quickly evolved from an academic notion into a proven and highly marketable product [3, 8]. During the early 2000s, medical expert system were routinely expected to be used in drug prescribing or in reminding clinicians to engage in preventive interventions through immunizations. In 1998 a systematic review was published of clinical trials on clinical expert systems, which found clear evidence for the effectiveness of such systems with 43 out of 65 trials showing an improvement of clinician performance [37]. The trials which involved a variety of expert system tasks that included diagnostic aid, preventive care, and reminder systems. However, only 20% of the diagnostic aid trials in this study were effective, as compared with 74% of trials based on preventive care and reminder systems. Studied conducted by [12, 13] show that clinical expert systems improved practitioner performance in 62 (64%) of the 97 studies assessing this outcome. Of these trials, 21 were based on reminder systems, with 16 of them (76%) deemed to be successful, and 10 of these trials were diagnostic based with 4 (40%) successful. There has been a dramatic improvement in the use of diagnostic expert systems which reflects an increasing trend in recent years as will be shown later in this article. There has been an increase in other medical expert systems which are being used in both the hospitals and patient homes in aiding them in activities of daily living. Application of fuzzy logic to a healthcare diagnostic system was described in which a set of sensors monitors body temperature, blood pressure and heart rate. Many other systems are known that employ artificial intelligence techniques in the medical field and healthcare diagnostics to mention a few [14].

Expert systems are routinely employed in medical settings as decision support and/or decision-making systems. Decision support systems are utilized to remind experienced and qualified decision makers of all options and issues to be considered when arriving at decisions. Decision making systems are adopted to aid experts and non-experts during decision making processes. For example, they can be utilized to support experts during decision making by providing a diagnosis or a prediction; and they can also be used by people to obtain advice on symptoms, who have no medical background.

## 2.2   Former and Current Implementations of Medical Expert Systems

Current expert systems are systems which can be adopted for aiding clinicians during the task of making a diagnosis and identifying best treatment options for patients. Components of expert system models include knowledge base, sometimes based on fuzzy rules, inference engine, and knowledge acquisition utilities, to edit knowledge base [33]. Below are some medical expert systems.

- PEIRS: Stands for Pathology Expert Interpretative Reporting System, which interpreted (during its period of operation until 1994) about 80–100 reports a day with a diagnostic analytical accuracy of about 95% [38].
- MYCIN: This is possibly the most widely known first of all medical expert systems thus far developed [5]. It was developed at Stanford University exclusively as a research effort to provide support to physicians in their diagnosis and treatment of meningitis and bacteremia infections it has never been put into actual practice.
- Apache III: Was designed to predict an individual's risk of dying in the hospital. The medical profile of each individual is compared against 18,000 cases in its memory before reaching a prognosis that is (on average) 95% accurate. This expert system (APACHE III) is very much in use 16 hospitals in the USA. There are more than 40 hospitals worldwide where the APACHE III expert system is used to generate reports which compare their actual average ICU outcomes with those predicted by the APACHE III methodology [15].
- Iliad: Has been under development for several years, primarily for diagnosis in internal medicine. It is said to cover about 1500 diagnoses in this domain based on several thousand findings. It is currently being used as a teaching tool for medical students overall the world [34].
- Jeremiah: This is an expert system designed to provide dentists with orthodontic treatment plans for cases suitable for treatment by general dental practitioners with

knowledge of removable orthodontic techniques. This system became commercially available in 1992 [35].

- SNOMED CT: There is a long-term goal in the medical field of converting natural language in clinical documents into structured machine language which help in the clinical decision support, medical error detection, generating medical statistics, bio surveillance, and clinical queries. An important step in automating this problem is being tackled by medical expert system called SNOMED CT. SNODMED CT stands for Systematized Nomenclature of Medicine-Clinical Terms, [16, 17], is regarded as a comprehensive terminology which is now widely used in electronic healthcare record systems for documentation purposes and reporting [17]. SNOMED CT is a uniform representation of clinical concepts whose richness and lucidity makes it suitable for precisely encoding clinical phrases. A concept in SNOMED CT is defined in terms of its relationships with other concepts and it currently includes around four hundred thousand pre-defined clinical notions [16, 17]. SNOMED CT was developed and maintained by College of American Pathologists. It is a comprehensive clinical reference terminology which contains more than 360,000 concepts and over 1 million relationships.
- PathoSys: This is a software reporting system which was designed in partnership with consultant Pathologists which incorporates the Royal College of Pathologists. PathoSys provides standardized reporting and structured data capture facilitating rapid real-time audit and the export well-structured pathology data to authorized third parties of cancer diagnostics. Pathosys became the first medical software system recognized by the Royal College of Pathologists responsible for Cancer reporting universal system in 2001 [39].

## 2.3    The Stakeholders

According to [19], current medical Decision Support Systems (DSS) are developed and classified into four different groups of decision makers while two of whom have a medical background (clinicians and medical researchers), the other two (administrators and patients) who may not have enough medical background. General research has shown that expert physicians do make use of explanation facilities, and their requirements are very different to that of other users. Experts tend to use feedback rule-trace explanations and are more likely—than non-experts—to use explanations for resolving anomalies, such as disagreements with system advice, exploring alternative diagnoses, and verification of assumptions [20]. However, non-experts such as trainee physicians are more likely to use a range of explanations types for short- and long-term learning. Authors in [20] argued that non-experts tend to use both feedback and feedforward justification explanations as well as terminological feedforward explanations. Although administrators do not have direct clinical interaction with patients, they are responsible for offering the management of healthcare options and facilities. Patients are the largest group of decision-makers with the least amount of medical knowledge or training. However, chronic patients inadvertently often become expert patients because they gain understanding over a long period of chronic illness. Chronic patients are likely to acquire skills to help them manage their illness better in their residential environments [21].

## 2.4     What Is Trust in a Medical Context?

Trust can be characterized as a multi-layered concept consisting of a cognitive element and an effective dimension, grounded on relationships and effective bonds generated through interaction, empathy and identification with others [22, 23]. According to [22], Trust tends to be necessary when there is a risk of uncertainty, whether high medium or low. It is usually derived from the individual's uncertainty regarding the intentions, motives and future actions of the other party whom the individual is dependent. Trust may differ in quality depending on the relationship between the individuals concerned and in the professional or social context. According to several studies in the healthcare sector there is evidence to suggest that the concept seems to embrace competence (skill and knowledge) as well as whether the trustee is working the best interests of the trustor [24, 25]. The expectation tends to cover honesty, confidentiality, caring, and showing respect whereas the trustee may also show social, technical and communication skills. According to [29], the vulnerability associated with being sick may specifically lead trust in medical settings to have a stronger emotional instinctive component.

### 2.4.1     Trust by Patients

For the medical expert systems to be accepted, trust must exist among the stakeholders. This means that there must be trust from the patients about the services they get from clinicians. The clinicians, healthcare professionals and administrators must trust the medical expert systems they rely on to provide services to the patients. According to [40], many studies investigating patients' experience of healthcare provision, trust appeared instinctively as a quality indicator with a lot of patients which suggested that high-quality doctor-patient communications are characterized by high levels of trust. Although trust is highly interconnected with patient experience, it is a distinct concept about forward-looking, reflecting and healthy altitude to a new or old relationship whereas satisfaction tends to be based on past experience and is usually associated to the assessment by the patients of the service providers' performance [26]. According to [29] several research studies stated that there is evidence to support that trust appears to mediate therapeutic processes and has an indirect influence on the patients' health outcomes through its impact on patient satisfaction. The use of "Patient Choice" highlights the importance of trust among the stakeholders, patients, clinicians and administrators in using medical expert systems. Patient choice which is at the heart of the UK NHS healthcare system reform programme is very well demonstrated on the public health (Department of Health 2004b) and subsequently adopted by the adult Social care (Department of Health 2005).

### 2.4.2     Trust by Clinicians

Trust by clinicians in the context of using medical expert systems is when they are confident of the validity of information and advice supplied by expert systems in order to make informed decisions and make a difference to the healthcare outcomes of the patients. The ability of a system to provide the clinicians the credit and satisfaction ultimately leads to other healthcare providers to follow suit in pursuance of collaboration because of the system reliability. According to [30], the new policy framework of clinical governance was intended to shift focus from trust relationship between people to confidence in abstract systems such rules and regulations which can also be used as in expert systems. With limited research exists on the topic of trust in the medical sector within the NHS healthcare establishments, the studies seem to suggest that while there are high levels of trust between patients and clinicians ('BE YOUR OWN DOCTOR') there were lower levels of trust within the healthcare institution ('doctors in general') [27–29]. One example where trust has led to organizational collaboration is the invention of Intelligent Pathology System. The cooperation of clinical experts to create expert reporting not only gives a standard benchmark but also work in partnership with other professionals in healthcare delivery.

## 3   Results and Analysis

In this section the authors analyzed the number of articles and then categorized them in accordance with their relevance to the research topic. This preliminary search resulted in a total of 1144 hits of articles including research papers. These results were obtained using a combination of three different sets of keywords. Articles which clearly did not indicate any relevance to the search study were automatically excluded and managed to identify 118 as relevant to the research study. The keywords that were used in the search were "Medical expert systems" and "trust on medical expert systems" and "Acceptance on Medical expert system" used the inclusion and exclusion procedure such as peer reviewed, fully text, and duplicate articles were used due to the high number that was found. Data collected from the 58 research papers on Trust and acceptance on Medical Expert Systems were collected with the help of systems review procedures. The 58 research studies identified covered the thesis using case studies, empirical studies, surveys, academic studies etc. Each research study was reviewed, analyzed using the background of the study, research questions, and observed confirmation of the result. The studies were put into the following categories to easily classify the studies and analyze the results properly.

- Medical Expert Systems
- Trust on Medical Expert Systems
- Acceptance of Medical Expert Systems
- Medical Expert Systems in Emerging Economies

A total of seven (7) studies from the 58 relevant studies did not fit into the categories mentioned above as the content was not very relevant to the research study. Most of the research papers found were on this research study were based on empirical studies. About five (5) studies had some relevance towards trust on medical expert systems. Although most research papers from this systematic literature review presented some relevance to the research topic, they were more generic in nature.

### 3.1   Methodology

This section describes the characteristics of the studies including the summary of the studies and description in the categories. The type of research studies found were diverse from empirical, survey, case study, academic, and a combination of case study, empirical, and academic although empirical study had the highest number of research papers as shown in Table 1.

**Table 1.**  Type of studies reviewed

| Type of study | Number of papers | Percentage |
|---|---|---|
| Empirical study | 35 | 60.34% |
| Case study | 8 | 13.79% |
| Academic | 7 | 12.08% |
| Comparative | 8 | 13.79% |
| **Total** | **58** | **100%** |

There highest number of reviewed studies were classified as Empirical 35 (60.34%), Case study 8 (13.79%) Academic 7 (12.08%), Comparative 8 (13.79%).

From the 58 studies, 14 were published journals and 44 were conference papers

## 3.2    Trust in Medical Expert Systems

Most of studies that are included in this study are found in the category of *Trust on Medical Expert Systems*. There is a clear disparity between the importance of trust for the operation of a healthcare system and the priority given to research on trust. The studies vary from trust of medical expert systems by patients, by clinicians, and between healthcare providers. These studies also looked at the factors that influence the acceptance of these systems in the provision of healthcare services. A total of fifteen (15) studies out of 58 can be classified in this category. From these studies, the one in [26] addressed the issue of trust relations in the NHS through theoretical method-ological challenges. The study carried out in [27] also covered the factors that are associated with patients' trust in their GPs. A theoretical framework is also used to highlight the changing dynamics in motivation for trust from affect based trust to cognitive based trust as patients, clinicians and managers are being encouraged to become proactive partners in trust relations. Other research articles of the study covered how much trust is now dynamic. The study presented in [25] states that the shift of trust among GPs and hospital doctors and in relation with other practitioners has created the need to be primary care led and other care professionals have become responsible for delivery of services, which has created the need for the trust to be earned.

## 3.3    Acceptance of Medical Expert Systems

From the search results 18 studies were found and evaluated. The proportion of cor-rectly diagnosed cases was between 43.1 and 99.9%. The acceptance of these medical expert systems relies on the performance on the sensitivity and specificity which were in the range of 62 to 100 and 88 to 98%, obtained respectively from the study presented in [7], the validation process was in general underappreciated. This review identifies optimal characteristics to increase the survival rate of expert systems due to acceptance by the medical fraternity and may serve as valuable information for future developments in the field. The study in [31], suggested that the way a CPOE system is configured has a bearing on user acceptance because of the ease of system use task behavior of clinicians in ordering drugs and medical error detection of the system. From this study, it was concluded that the system design can enhance physicians' acceptance of these systems. Hence, there is evidence to suggest that Medical expert systems that are properly designed to be more friendly and safer are easily accepted by the clinicians which will improve patients' healthcare outcomes.

## 4  Conclusions

### 4.1  Characteristics of an Expert System

An expert system is characterized by three main elements: consistency, availability, and comprehensiveness. These are described below.

- Consistency: Once an expert system is programmed to ask for the required inputs, it cannot be prone to poor memory which is opposite and different from a human being. As a machine, if a line of reasoning is acceptable, it will remain so indifferent consultations. This makes it unique in that it is different to even the best experts who can make mistakes or may forget important points.
- Availability: Expertise is acquired depending on the number of years (time) someone has been working in that given area. They are to be trained and then practiced. It generally takes over five years for someone to acquire expertise in an area. In contrast to the human, an expert system has all the expertise in-built, it never gets tired or dies.
- Comprehensiveness: An expert can only draw upon his knowledge and experience in solving complex situations. In other domain, an expert system could summarize the knowledge of more than one expert and subsequently offer several alternatives.

Trust is vital in the provision of healthcare as perceived risk is high for both patient and clinician [1]. Cases presented in [25, 26], were also used to reinforce the factors to appreciate trust when using medical expert system. Studies in [7, 31] highlighted the need for better system design as it makes the user friendly and efficient expert system. This makes it easier for user acceptance a precondition for the successful implementation of Hospital Information Systems (HISs).

### 4.2  Limitations in Medical Expert Systems

There is a need to check that all different types of knowledge-based systems that are used and appear in clinical settings remain up to date. Since technology is always evolving, there is need to keep all medical experts to continually educate both patients and clinical professionals and also keep the medical expert systems abreast with any changes as education for clinicians is an on-going process. The dynamics of trust on medical expert systems need more research as there is no rigid approach when dealing with all stakeholders especially the patient who has become more knowledgeable and cannot be taken for granted. Therefore, a possible limitation is that medical expert system rules have to be kept up-to-date as today's technology is always evolving at short notice, hence there is need to ensure there are no bureaucratic barriers which can hinder the approval of changes within the given medical system. Another limitation of medical expert systems is that they were created to solve certain situations and problems hence the information stored and processed was bounded to each problem's area. This means that when knowledge is captured (tangible), the intangible knowledge is not captured hence additional knowledge or data is not available as only what is programmed is what is used. Many medical expert systems have been invented which integrate with others providing a more feasible definition and understanding. However, there is still a lot of collaboration needed amongst the many healthcare providers to engage and create trust amongst themselves. It is suggested from the studies that the provision of information and patient involvement in their care through shared decision – making in the GP- patient or doctor-patient relationships has brought greater trust between patients and clinicians.

### 4.3    Limitations in This Research

There are number of limitations with the research which became noticeable and could possibly affect the validity of the research study. These limitations are explained below.

- The review study included only the articles and journals that were searched in the recorded in the digital databases in the literature review procedure.
- The review did not contain any chapters from the books which dealt with Medical expert systems or artificial intelligence systems.
- Only research articles and journals were used for this research study.
- The review included articles that were only available in full text.
- The articles that were reviewed were predominantly from the medical sector.

## 5    Future Work

Judging by the rate at which technology has been evolving, developers and healthcare settings must keep up with personal and organizational advancement with technology. There is need for more research to be undertaken to understand how human understanding can contribute to change within the healthcare institutions through self-development actualization. Digital revolution necessitates organizations of all sizes, across all different sectors especially medicine, to adapt to changing expectations regarding speed, scale and operational flexibility. The conceptual complexity of rigid factors in trust and the lack on many empirical research studies deserve pursuance by the researchers to go in this direction for possible research studies. According to [32], there exists a 'Vaccine confidence gap' where public assurance in vaccines is linked with low levels of public trust in the broader health care system. They argue that public trust in inoculations is exceptionally variable and the building of trust among members of the public hinge on factors such as the supposed risk of the vaccine being detrimental rather than benefit, political and religious beliefs and socioeconomic status. There is need for research not to only focus on the safety and effectiveness of a vaccine but also on the psychosomatic, emotional, social and political factors affecting the public's trust in it.

# References

1. Alaszewski, A.: Risk, trust and health. Health Risk Soc. **5**(3), 235–239 (2003). https://doi.org/10.1080/13698570310001606941
2. Jones, J., Barry, M.M.: Developing a scale to measure trust in health promotion partnerships. Health Promot. Int. **26**(4), 484–491 (2011)
3. Ward, P.R., et al.: A qualitative study of patient (dis)trust in public and private hospitals: the importance of choice and pragmatic acceptance for trust considerations in South Australia. BMC Health Serv. Res. **15**(1), 297 (2015). https://doi.org/10.1186/s12913-015-0967-0
4. Turban, E., Aronson, J.E.: Decision Support Systems and Intelligent Systems. Prentice Hall, Upper Saddle River (2001)
5. Musen, M.A., Shahar, Y., Shortliffe, E.H.: Clinical decision-support systems. In: Biomedical Informatics Computer Applications in Health Care and Biomedicine, 3rd edn., pp. 698–736. Springer, USA (2006)
6. Musen, M.A., Middleton, B., Greenes, R.A.: Clinical decision-support systems. In: Shortliffe, E.H., Cimino, J.J. (eds.) Biomedical Informatics: Computer Applications in Health Care and Biomedicine, pp. 643–674. Springer, London (2014)
7. Alder, H., et al.: Computer-based diagnostic expert systems in rheumatology: where do we stand in 2014? Int. J. Rheumatol. (2014)
8. Walton, R.: An evaluation of CAPSULE, a computer system giving advice to general practitioners about prescribing drugs. J. Innov. Health Inform. [S.l.], 2–7 (1996). ISSN 2058-4563
9. Darlington, K.W.: Designing for explanation in health care applications of expert systems. SAGE Open **1**(1) (2011). 2158244011408618
10. Metaxiotis, K., Psarras, J.: Expert systems in business: applications and future directions for the operations researcher. Ind. Manag. Data Syst. **103**(5), 361–368 (2003). https://doi.org/10.1108/02635570310477
11. Grol, R., Grimshaw, J.: From best evidence to best practice: effective implementation of change in patients' care. Lancet **362**(9391), 1225–1230 (2003)
12. Ax, G., et al.: Effects of computerized clinical decision support systems on practitioner performance and patient outcomes: a systematic review. JAMA **293**(10), 1223–1238 (2005)
13. Castaneda, C., et al.: Clinical decision support systems for improving diagnostic accuracy and achieving precision medicine. J. Clin. Bioinform. **5**(1), 4 (2015)
14. Madkour, M.A., Roushdy, M.: Methodology for medical diagnosis based on fuzzy logic. In: Proceedings of Fifth International Conference on Soft Computing, vol. 2, pp. 1–14 (2016)
15. Stone, D.J., Csete, M.: Actuating critical care therapeutics. J. Crit. Care **35**, 90–95 (2016)
16. https://www.snomed.org/
17. Benson, T.: SNOMED CT. In: Benson, T. (ed.) Principles of Health Interoperability HL7 and SNOMED, pp. 189–215 (2010). https://doi.org/10.1007/978-1-84882-803-2
18. Appleby, J., Harrison, A., Devlin, N.: What Is the Real Cost of More Patient Choice?. King's Fund, London (2003)

19. Leroy, G., Chen, H.: Introduction to the special issue on decision support in medicine. Decis. Support Syst. **43**(4), 1203–1206 (2007)
20. Arnold, V., Clark, N., Collier, P.A., Leech, S.A., Sutton, S.G.: The differential use and effect of knowledge-based system explanations in novice and expert judgement decisions. MIS Q. **30**(1), 79–97 (2006)
21. Berger, J.: Writing is an offshoot of something deeper (2014). https://www.theguardian.com/books/2014/dec/12/john-berger-writing-is-an-off-shoot-of-something-deeper
22. Mayer, R.C., Davis, J.H., Schoorman, F.D.: An integrative model of organizational trust. Acad. Manag. Rev. **20**, 709–734 (1995). Mayer, R.C., Gavin, M.B.: Trust (2005)
23. Lewicki, R.J., Bunker, B.B.: Developing and maintaining trust in work relationships. Trust Organ. Front. Theory Res. 114–139 (1996)
24. Hall, G., Longman, J.: The Postgraduate Companion. Sage Publications, London (2008). Chapters 4–7 Eds
25. Mechanic, D., Meyer, S.: Concepts of trust among patients with serious illness. Soc. Sci. Med. **51**(5), 657–668 (2000)
26. Calnan, M., Rowe, R.: Researching trust relations in health care: conceptual and methodological challenges – an introduction. J. Health Organ. Manag. **20**(5), 349–358 (2006)
27. Tarrant, C., Stokes, T., Baker, R.: Factors associated with patients' trust in their general practitioner: a cross-sectional survey. Br. J. Gen. Pract. **53**(495), 798–800 (2003)
28. Mainous III, A.G., Baker, R., Love, M.M., Pereira Gray, D., Gill, J.M.: Continuity of care and trust in one's physician: evidence from primary care in the United States and the United Kingdom. Fam. Med. **33**, 22–27 (2001)
29. Calnan, M.W., Sanford, E.: Public trust in health care: the system or the doctor? BMJ Qual. Safety **13**(2), 92–97 (2004). http://qualitysafety.bmj.com/content/13/2/92
30. Harrison, S., Smith, C.: Neo-bureaucracy and public management: the case of medicine in the national health service. Competition Change **7**(4), 243–254 (2003). https://doi.org/10.1080/1024529042000197077
31. Khajouei, R., Jaspers, M.W.M.: The impact of CPOE medication systems' design aspects on usability, workflow and medication orders a systematic review. Methods Inf. Med. **49**(1), 3–19 (2010)
32. Larson, H.J., et al.: Addressing the vaccine confidence gap. Lancet **378**(9790), 526–535 (2011). http://www.sciencedirect.com/science/article/pii/S0140673611606788
33. Systems, B.I.: A fuzzy expert system for response determining diagnosis and management movement impairments syndrome Fatemeh Mohammadi Amiri. Ameneh Khadivar Alireza Dolatkhah **24**(1), 31–50 (2017)
34. Lepage, E., et al.: ILIAD: an expert system for diagnostic assistance and teaching: implementation in France. In: Adlassnig, K.-P., et al. (eds.) Medical Informatics Europe 1991, pp. 629–633. Springer, Heidelberg (1991)
35. Mackin, N., Stephens, C.D.: Development and testing of a fuzzy expert system - an example in orthodontics. In: Proceedings of Fuzzy Logic: Applications and Future Directions, pp. 61–71. Unicom Seminars Ltd, Uxbridge, Middlesex (1997)
36. Nohria, R.: Medical expert system-A comprehensive review. Int. J. Comput. Appl. **130**(7), 975–8887 (2015)
37. Hunt, D.L., et al.: Effects of computer-based clinical decision support systems on physician performance and patient outcomes - a systematic review. JAMA J. Am. Med. Assoc. **280** (15), 1339 (1998)

38. Edwards, G., Compton, P., Malor, R., Srinivasan, A., Lazarus, L.: PEIRS: a pathologist maintained expert system for the interpretation of chemical pathology reports. Pathology **25**, 27–34 (1993)
39. http://www.aesgrp.com/medical/medical-products/pathosys
40. Safran, D.G., et al.: Linking primary care performance to outcomes of care. J. Family Pract. **47**, 213–220 (1998)

# Learning and Adaptation

# Agent Based Micro-simulation of a Passenger Rail System Using Customer Survey Data and an Activity Based Approach

Omololu Makinde$^{(\boxtimes)}$, Daniel Neagu$^{(\boxtimes)}$, and Marian Gheorghe$^{(\boxtimes)}$

University of Bradford, Bradford, UK
oamakind@student.bradford.ac.uk,
{D.Neagu,M.Gheorghe}@bradford.ac.uk

**Abstract.** Passenger rail overcrowding is fast becoming a problem in major cities worldwide. This problem therefore calls for efficient, cheap and prompt solutions and policies, which would in turn require accurate modelling tools to effectively forecast the impact of transit demand management policies. To do this, we developed an agent-based model of a particular passenger rail system using an activity based simulation approach to predict the impact of public transport demand management pricing strategies. Our agent population was created using a customer/passenger mobility survey dataset. We modelled the temporal flexibility of passengers, based on patterns observed in the departure and arrival behavior of real travelers. Our model was validated using real life passenger count data from the passenger rail transit company, after which we evaluated the use of peak demand management instruments such as ticketing fares strategies, to influence peak demand of a passenger rail transport system. Our results suggest that agent-based simulation is effective in predicting passenger behavior for a transportation system, and can be used in predicting the impact of demand management policies.

**Keywords:** Agent based simulation · Customer survey data
Passenger rail system

## 1 Introduction

Due to consistent increase in urbanization and societal changes in economic status, tremendous pressure has been mounted on transportation infrastructure, leading to the rise in research on travel demand models (i.e. models created to reflect the travel behavior of a population), as opposed to continuously building new infrastructure to meet growing demand.

The first generation of travel demand models were developed in the 1960's and are popularly called four step models [1]. These models fast became obsolete as their limitations became obvious. Firstly they consider individual trips separately, ignoring the fact that people make travel plans while considering entire trip chains. Also, sub-modules within the step sequences were inconsistent, and in the heart of the model itself lay aggregation based procedures [2]. These limitations, made these models inefficient in testing policies. This led to the development of two new approaches –disaggregate

© Springer Nature Switzerland AG 2019
A. Lotfi et al. (Eds.): UKCI 2018, AISC 840, pp. 123–137, 2019.
https://doi.org/10.1007/978-3-319-97982-3_10

choice and activity based travel demand models [3]. Although, disaggregate choice travel demand models were accepted for a while, they maintained a fundamental error of their predecessor which analyzed each trip independent of another trip made by the same individual, thereby limiting its policy testing ability. Activity based travel demand models have gradually gained popularity since the 1990's.

Activity based approach to travel demand views travel as a derived demand, derived from the need to pursue activities distributed in time and space [1]. Its approach allows the creation of travel models in which represented individuals from the complex system can be trace throughout the simulation. Therefore by implementing the activity based approach using agent based methods, micro simulations can be created with generated activities and trips for every person in the study area. This and in addition to the fact that computing capabilities have soured over the years have led to the increased development of large scale agent based traffic demand microsimulations for large metropolis such as Netherlands [4], Germany [5], Buffalo-Niagara [6], Zurich [7], Toronto [8], and south Africa [9].

One of the early activity based models created is the AMOS model [10] which was implemented in Washington Dc's traffic flow prediction, in which six policies were considered, including congestion pricing. The model input is a customized stated survey in which the respondents answered questions on how they would respond to the application of a control measure in context of their activity and travel behavior a day before. Fourie et al. [11] created a fictitious model of the transportation demand of the city of Sioux-fall, focusing on its public and private transportation with the aim of using the model as a general test bed for different transportation policies. They used real world household survey data in creating an agent population and land-use data in extracting spatial locations of activities of individuals travelling within the area of focus which was then mapped to a physical model of the relevant transportation infrastructure. The model also took into cognizance the socio demographic characteristics of each represented individual, such as age, sex and income. Many others such as Kickhofer et al. [12], and Rolando et al. [13], used similar data and methods in creating their models, with some of them validating their models using traffic signal data. These models were created as test beds for testing various policies and scenarios which revealed travel demand impact on environment [14], energy usage [15], evacuation and many more. Managing peak congestions have also become popular over the years. Lovric et al. [14] created an agent based traffic demand model using smart card dataset from a Dutch public transit company in which he analyzed passenger behavior while applying three different pricing strategies. This growing research in demand modeling has encouraged the development of different simulation toolkits such as: MATSim [16], DynaMIT [17], TRANSIM [18], and AIMSUN [19].

As much as progress has been made in this area, agent based demand models using the activity based approach have mostly used travel data from national census, of which are never available at spatial micro level. Not as much research has been done in creating agent based models from PTO's (public transport organization's) passenger survey dataset [20, 21]. Therefore, our contribution is as follows: we propose a truly agent based micro simulation model created using a PTO's passenger survey datasets. This model is then used as a test bed to simulate the travel demand impact of seven common strategies used by PTOs.

The unique daily travel behavior of each rail passenger is modelled in individual agents so as to form an agent population that would reflect the demand on the rail system. A model of the physical rail system consisting of the links and coordinates of the real life rail tracks, stops, characteristics of individual trains (such as speed, capacity and engine size), and the different schedules (train time table) of the various routes is used in simulating a game-like interplay between supply and demand such that agents try to find the best times to leave their destinations and the best route to apply in getting to their destination without being late. An evolutionary algorithm is applied to this problem such that day plans created for each agent are optimized on each simulation iteration till it gets to an equilibrium were no better change to its day plans can be made.

The remainder of this paper is as follows: Sect. 2 gives an overview of the simulation discussing its basic modules. Section 3 explains how the demand model is created using an agent population created from the customer survey dataset. Section 4 details how the score of an agent's day plan is calculated. Section 5 compares the simulated data with real life passenger count data, in an attempt to validate the model. Section 6 uses the model as a test bed to simulate common fare pricing policies to see their effect in influencing passenger peak demand and Sect. 7 concludes.

## 2  Simulation Overview

Our simulation is constructed around the notion of agents that make independent decisions about their actions. Each passenger of the rail system is modelled as an individual agent (NB the simulation has been scaled down to 10% of the actual system), traveling from one activity (e.g. "Home") to another (e.g. "Work"). There are three important pieces of this simulation:

(1) Each agent generates his initial day plan, encoded with information of his activities (i.e. duration, route and end times within a day). An example of an initial day plan for 20 agents can be found in [22].

(2) All agents execute a plan simultaneously in a virtual model of the physical rail system. This is known as the traffic flow or mobility simulation. (NB a model of the physical rail system consisting of the links and coordinates of the real life rail tracks, stops, the different schedules of the various routes, and characteristics of individual trains -such as speed, capacity and engine size.). Therefore agents can miss the train, be late for an activity or not be able to get in the vehicle if it is already filled to capacity.

(3) Agents learn from experience by performing an evolutionary process on their day plans. An iteration is completed by evaluating the agents' experiences with the selected day plans (scoring). The scoring of the plans is described in detail in Sect. 4. So for every iteration the agent could either chose the plan with the highest score, re-evaluate plans with bad scores (with an option of discarding them) or create new plans by modifying existing ones (modification can be done by changing time or route). The iterative process is repeated until the average population score stabilizes.

A basic assumption of this simulation is the assumption that customers have no alternative mode aside from this public transport train line.

The simulation is built using the MATSim (Multi Agent Transportation Simulation) java library, developed to provide a structure for implementing large-scale agent-based transport simulations. Currently, MATSim offers a framework for demand-modelling, agent-based mobility-simulation (traffic flow simulation), re-planning, a controller to iteratively run simulations as well as methods to visualize some outputs generated by the modules. A thorough description of MATSim can be found in Horni et al. [16].

## 2.1    The Four Core Modules

The simulation is organized into 4 core modules as shown in Fig. 1 below.

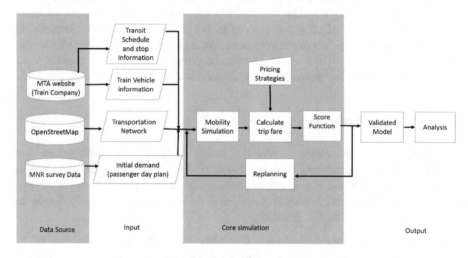

**Fig. 1.** Simulation overview

**Data Source.** To create the day plan for each agent we use a travel survey dataset made openly available by Metro-North Commuter Railroad (MNR) [20]. The survey was conducted on board of every inbound train operated on the Harlem, Hudson and new Haven routes in 2007, and customers filled forms detailing their trip boarding and alighting times, traveling purpose, and how often they use the rail system. The survey dataset contains 92,726 rows and 99 columns all stored in a Microsoft access database. Each column represents the survey questions asked in a passenger survey executed on board the trains in the Harlem, Hudson and new Haven routes of the MNR, while each row represents each passenger's data who filled the survey. After pre-processing, we were able to capture each passenger's travel purpose, station check-in, station check-out and activity durations and purpose. The stations geographical location and the rail network map was extracted from OpenStreetMap, while train schedules and capacities were gotten from the PTO's website.

**Input.** Input files used in this simulation refer to the post-processed data in machine readable format, from which the major simulation input parameters are extracted from. These files include the initial demand file which contains the individual day plan of each agent, an XML representation of the rail network of Metro Northern Commuter Rail, describing the physical layout of the network and the coordinates of its stations and links between them. Two GTFS files containing the transit system's latest daily schedule and physical characteristics of all train vehicles on chosen active transit lines were also used. Train's physical properties such as: capacity, was given a constant of 100 seats for each carriage (this was extracted from the GTFS files). Each link free-flow speed was set to 18 m/s, which allows a realistic time for the vehicle to reach the next stop according to the schedule. An early arrival vehicle will wait till the next departure time before leaving the station. The inputs has been deposited in a GitHub account [22].

**Core Simulation.** This module is where the mobility simulation is executed such that virtual transit vehicles are simulated according to their schedules. A queue model is used to simulate the traffic flow such that when the agents enter into the train the train travels through a link (i.e. the rail) in free flow speed and only comes out (gets to its next station) at the time required to travel through the link. Therefore agent transition is possible if capacity of the link and the train allows for agent to enter it. Agents board these train vehicles in order to get to their activities within their daily plans. When this is over, the Agent's daily plans are scored (explained in Sect. 4) by incorporating people's preference such as price sensitivity, route and late/early arrival to an activity. Passenger agent interaction within this environment can lead to consequences such as lateness. To avoid this, agents may modify copies of their plans after each simulation iteration, in order to avoid situations that lead to bad scores. Such changes could include: change in departure time or change in route. Not all plan copies are discarded after each run, each agent has a maximum of 5 plans within the working memory. This aspect of the simulation is handled by the re-planning submodule of the core simulation module. These ultimately allows the agents to mimic real life passenger's adaptation to changes made by PTOs, such as fare pricing changes.

**Output.** In this module, the validation and analysis are done. Our simulation was validated by comparing passenger arrival and departure volume to real life passenger count data. The validated model was then used in predicting passenger behavior when different pricing strategies were applied.

The next section discusses how we use the PTO's travel survey dataset to extract an O-D matric (origin destination matric) of each passenger activity, to create an overall agent plan for the day, centered on the passenger's day activities. This will serve as the initial demand that will be passed into the core simulation module.

## 3   Demand Modelling

Traffic demand is generated based on the concept of daily activity demand from which the need for transportation is derived. Such activity could be activities that require you to be at home (Home activity), work (Work activity), leisure activity or shopping activity.

We segmented customers based on their travel patterns:

*Trip based passengers.* Customers that were only taking a single journey but were not coming back through the train system.

*Tour-based passengers.* Customers that were departing and arriving using the same stations, and were either leaving from home or would finally arrive at home at the end of the day.

*Pattern-based passengers.* These are tour-based passengers that use the system at least twice a week.

### 3.1   Activity Profile

Based on the information available in the survey we can derive an activity profile for a passenger which would be represented as a tuple (l, b, e, $\Delta$b, $\Delta$e). The profile is then recreated in an xml format to represent a day plan.

- the closest station to the activity location is represented as l in x-y coordinates. This could either be extracted from the Begstationidr, endstationidr, ob1stationidr or ob2stationidr columns in the survey dataset.
- the scheduled start time of the activity is represented as b. Extracted from the depmam_r column of the survey dataset.
- the scheduled end time is represented as e. Extracted from outdepmam_r column.
- train boarding time on the day of survey at beginning of the activity is represented as $\Delta$b, extracted from depmam column.
- train boarding time at the end of the activity represented as $\Delta$e, extracted from outdepmam column.
- the actual duration $t_{dur,q}$ of preferred activity q is b – e (the difference of e and b).

It should be noted that an activity is assumed to start at the time the passenger boards the train. (NB during the evolutionary/learning process, the start time and the route to an activity can be altered, but the duration cannot. Therefore plans with activities whose duration ends up surpassing its preferred end time due to chosen routes or start times will be scored as bad plans and would be likely discarded as superior plans are generated).

## 3.2    Creating the Initial Demand for the Simulation

The initial demand for the simulation is the combined initial individual day plan of all agents. This is represented in an xml format.

For our simulation we decided to use only pattern based passengers in creating our agent population as this is the group that places the most consistent demand on the rail system, and the group from which plans can also be generated, that start and end at home (this is a crucial condition in activity based demand generation). This will give frequent journey patterns of: home....outing....home. Where outing represents the major journey purposes extracted from the customer survey data, which are work, recreation, school, personal business and other. For our simulation we only used passengers going to/from "work", "recreation" and "school", in an aim to extract more consistent passengers.

Using the tuples generated for each day activities of each agent, MATSim Java library is used in generating each plan file. We assume the agent's activity takes place at the final station from which the agent alight to indulge in the activity.

In the next section we discuss how an agent's day plan is scored.

# 4    Scoring an Agent's Day Plan

An agent's day plan is scored after an iteration, in an attempt to weed out bad plans so that only viable day plans evolve to further iteration steps in the simulation [16]. The value of score S of an agent's plan is represented in Eq. 1

$$S = \sum_{q=0}^{N-1} \left( s_{act,q} + s_{act.q}^{late} \right) + \sum_{q=1}^{N-1} S_{trav,Rail\_PT(q)} + U_{mon} \qquad (1)$$

Where: N is number of activities, which is 2 (e.g. home and work or home and school), $s_{act,q}$ is the utility for performing activity q, and $S_{trav,Rail\_PT(q)}$ is the (dis)utility (i.e. negative utility or penalty) for traveling to activity q from the location of an activity q − 1 using the rail public transport mode. This implies that routes with longer distance will be less attractive as this will reduce the overall score. $U_{mon}$ Is the (dis) utility based on the total monetary cost of the combined day trip (therefore higher transport cost will reduce the overall cost of the plan). The penalty for performing activity q late is given as:

$$s_{act.q}^{late} = \beta_{late} * t_{late} \qquad (2)$$

Where $t_{late}$ is the duration of lateness and $\beta_{late}$ is a negative slope (we use the MATSim default of −18\$/h). The utility earned from performing an activity is given by:

$$s_{act,q} = \beta_{dur} * t_{typ,q} * \ln(t_{dur,q}/t_{0,q}) \qquad (3)$$

Where: $t_{dur,q}$ and $t_{typ,q}$ are the actual and typical duration of activity q respectively. The actual duration for activity q was extracted from each passenger's survey and mapped to the corresponding agent while the typical duration was the highest occurring duration of the activity q among the surveyed passengers. $\beta_{dur}$ is the marginal utility of the activity duration or time as a resource (we use the MATSim default of 6$/h). $t_{0,q}$ is the minimal duration. The mode-specific utility from traveling is described by Eq. 4

$$S_{trav,mode(q)} = C_{mode(q)} + \beta_{trav,mode(q)} * t_{trav.q} \qquad (4)$$

Where: $C_{mode(q)}$ is the alternative specific constant represented as 1, $t_{trav.q}$ is the travel time, $\beta_{trav,mode(q)}$ is the direct marginal utility of time spent traveling. This value is 0 since we are only concerned about the rail mode.

After the plan has been scored for each agent, it is then stored in the agent's memory. Plans selected for the next iteration are either previous plans whose scores have been improved either by changing the agent's route or adjusting its activity departure time, or plans picked from the agent memory randomly.

The proof for the equations stated above is beyond the scope of this paper. It can be found in Horni et al. [16].

## 5   Running the Simulation

We ran our experiment on a desktop PC with an Intel i7 3.6 GHz processor and 8 GB RAM. We ran 200 iterations to bring the simulation to equilibrium. The simulation was scaled down to 10% of the original passenger daily volume for MNR (this was because, after pre-processing the survey data, only about 12,000 records were usable). In response to this, the physical system model was also scaled down so that one agent will represent 10 individuals within the system. This will have minimal impact on the simulation results [16]. Figure 2 gives a graphical view of the simulation as it approaches equilibrium, as all the agents optimizes their plans in each iteration.

### 5.1   Validation

In addition to the surveys distributed, MNR field workers created a count dataset from counting the total number of passengers getting on, and the number of passengers getting off. We compared the simulated volume percentages of agent's arriving and departing at peak hours bins to that of real passengers getting off and on the trains. The bar chart for this can be seen in Figs. 3 and 5. Also, in an attempt to measure the accuracy of the model we analyzed the correlation of the simulated data to the count data in Figs. 4 and 6. Both arrival and departure data from the simulation seem to have high associations with the count data equivalents, giving correlation co-efficiencies of 0.88 and 0.95 respectively.

The simulated data is not expected to correspond perfectly with the count dataset, due to the fact that not all the counted passengers volunteered to fill the survey form. But it can be deduced that the volume of simulated agents getting off the train (simulated arrival), peak in the evening at about 6 pm while the volume of simulated agents getting on the train (simulated departure) peak in the morning at about 8am. This corresponds to the real passenger count conducted.

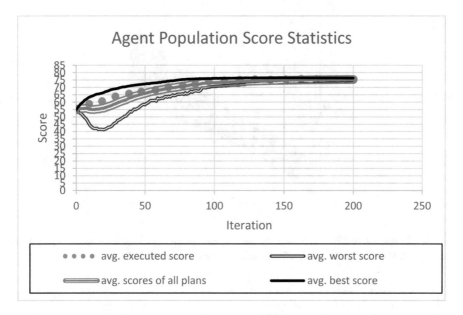

**Fig. 2.** The score statistics gives a graphical view of the scores of the agent population daily plans. This gives a picture of the evolution of the plans, as they get better and better till they reach an equilibrium phase where each agent seems to have reached its best plan which cannot be improved.

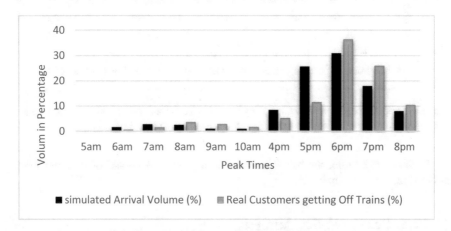

**Fig. 3.** Comparing agent volumes arriving at all virtual station links at peak time to the count data volumes of customers getting off the trains at the same time.

The highest level of difference can be seen in the evening peaks of Fig. 3 (especially at 5 pm). There seems to be a sharp arrival change between the 5 pm bin and the 6 pm bin in the passenger count dataset. This is not reflected in the simulation results as the simulation predicted such sharp change an hour too early i.e. between 4 pm and 5 pm.

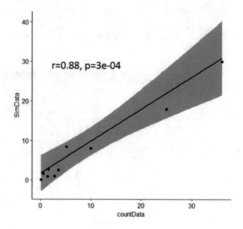

**Fig. 4.** Correlation test between the volume of agent arriving at virtual train stops per time and the number of passengers getting off the trains in the count data

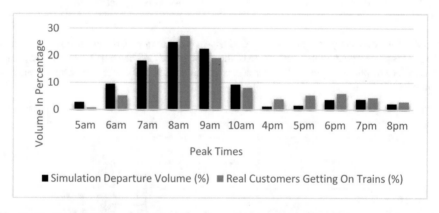

**Fig. 5.** Comparing agent volumes departing from all virtual train station links at peak time to the count data volumes of customers getting on the trains at the same time.

In the next section, we use this model to analyze the impact of different pricing policies on customer behavior.

## 6   Peak Pricing Policy Analysis

We intend to see how some basic pricing policies used by PTO's will influence customer behaviour. We will be considering seven different fare pricing policies shown in Fig. 7 below.

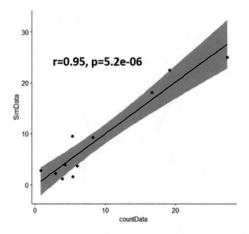

**Fig. 6.** Correlation test between the volume of agent departing from virtual train stop links per time and the number of passengers getting on the trains in the count data.

| Fare Policy | Morning | AM Peak | Mid Day | PM Peak | Evening |
|---|---|---|---|---|---|
| **Peak Surcharge** | | ▓ | | ▓ | |
| Non mid-day surcharge | ▓ | ▓ | | ▓ | ▓ |
| Mid-day discount | | | ▓ | | |
| **Off peak Discount** | ▓ | | ▓ | | ▓ |
| **Differential fare increase** | ▓ | ▓ | ▓ | ▓ | ▓ |
| **Differential fare reduction** | ▓ | ▓ | ▓ | ▓ | ▓ |
| Peak surcharge and off-peak discount | ▓ | ▓ | ▓ | ▓ | ▓ |

Initial Flat Rate ─────
Fare Change ▓▓▓▓

**Fig. 7.** Common daily fare policy used by many PTOs

*Peak Surcharge.* Fares at AM peaks and PM peak times.
*Non mid-day surcharge.* Charging base fares only at mid-day.
*Mid-day discount.* Fares are discounted at mid-day only.
*Off peak Discount.* Fare discount at early mornings, midday and evenings.
*Differential fare increase.* General increase in fares, especially during peak period.
*Differential fare reduction.* General reduction in fare, especially at off-peak times.
*Peak surcharge and off-peak discount.* Fare discount at off-peak times and fare increase at peak times.

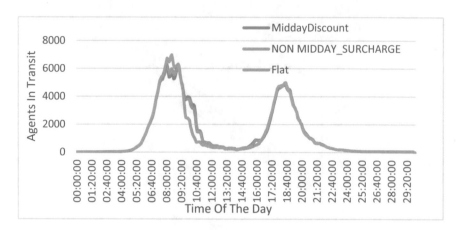

**Fig. 8.** Mid-day Discount vs Non Midday Surcharge vs Flat

Morning fares for the MNR rails starts from 12 am to 4:59, AM peak (morning peak) starts from 5 am to 9:59 am, mid-day starts is from 10:00 am to 3:59 pm, PM peak (evening peak) is from 4 pm to 8:59 pm while evenings starts from 9:00 pm to 11:59 pm. For the sake of analysis, we would assume that a flat price was in place at all hours of the day and that each customer bought a boarding ticket at point of entry.

By changing the fare price of each agent, different price strategies are simulated. For example, the peak surcharge policy is simulated by increasing the flat price by 40% during the AM and PM peak.

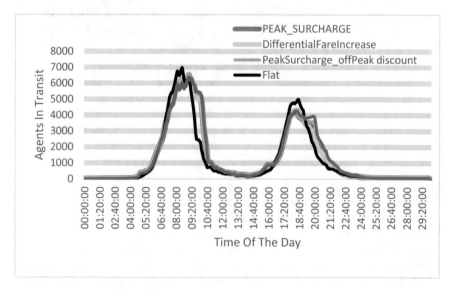

**Fig. 9.** Peak Surcharge vs Differential Fare Increase vs Peak surcharge_offpeak discount vs off_peak Discount vs Flat

**Reducing Demand Peak.** All the policies applied reduce the AM peak by at least 9%. The midday discount policy and the non-midday surcharge seem to produce the same effect, and reduce the am peak by up to 14%, with a slight widening of agents in the morning. This is common with other policies also. This could be the effect of agents with "recreation" and "school" travel purposes leaving a little later (or earlier) than normal to take advantage of price changes. The policies in Fig. 9 produce similar effects, reducing the evening peaks by about 16%.

**Shifting Peak Time.** The policies in Fig. 9 seem to have the ability to shift the peak times. The AM peak shifted by more than 30 min while the evening peak was widened. Therefore a PTO that seeks such effect in demand might consider any of these.

# 7    Conclusion and Future Works

This paper has majorly discussed simulating a commuter rail network using an agent based simulation. We used customer survey data to create agent profiles which was used in creating an agent population whose daily plans were optimized as they utilized a simulated commuter rail system. The purpose of all this is to create a model that can serve as a test bed for testing policies. This test bed was validated in Sect. 5.1, and policy testing was demonstrated by running a simulation to forecast passenger demand behavior on commuter rail system as pricing policies are applied in order to manage peak demand. We were able to determine policies that will likely generate similar behaviors (as shown in Figs. 8 and 9), and how different pricing policies have different effect on peak demand. All the data used in creating this simulation have come from open data sources easily accessible on-line, and our biggest challenge in creating this simulation has been in the accumulation of data.

In general, 70% of the commuters of MNR are going to or coming from "work". This explains why the flexibility of their behavior due to price change is not as pronounced, as it is difficult for such commuters to change their schedule or their route. By adding more municipality data, such as the spatial locations of businesses within the metropolitan area, we would be able to test the impact of adding other train lines, allowing the agents more choices in terms of route change. If we are also able to add the data of other modes such as buses, then mode shift impact analysis would also be a possibility, as accurate microscopic data of other optional transportation modes can be used in bridging the assumption made that customers will remain loyal after a price changes. With more data, a holistic view of the impact of a policy change such as price change, to the rail system can be estimated. Also the GTFS files used in the simulation is for recent years as compared to the customer survey data. This is more of a data challenge as we were unable to obtain GTFS files as far back as 2007. But there have been minimal changes to the schedule of the PTO over the years.

Finally, as at the time the survey dataset was created, the PTO already had a pricing strategy in motion, therefore in our future works we aim to improve the model such that a predicted flat price can be simulated which will be the basis for predicting and comparing the other pricing strategies.

# References

1. Rasouli, S., Timmermans, H.: Activity-based models of travel demand: promises, progress and prospects. Int. J. Urban Sci. **18**, 31–60 (2014)
2. Khademi, E., Timmermans, H.: Incorporating traveler response to pricing policies in comprehensive activity-based models of transport demand: literature review and conceptualisation. Procedia Soc. Behav. Sci. **20**, 594–603 (2011)
3. Jovicic, G.: Activity based travel demand modelling. Danmarks Transp. Skn (2001)
4. Melnikov, V.R., Krzhizhanovskaya, V.V., Lees, M.H., Boukhanovsky, A.V.: Data-driven travel demand modelling and agent-based traffic simulation in Amsterdam Urban Area. Procedia Comput. Sci. **80**, 2030–2041 (2016)
5. Hager, K., Rauh, J., Rid, W.: Agent-based modeling of traffic behavior in growing metropolitan areas. Transp. Res. Procedia **10**, 306–315 (2015)
6. Zhao, Y., Sadek, A.W.: Large-scale agent-based traffic micro-simulation: experiences with model refinement, calibration, validation and application. Procedia Comput. Sci. **10**, 815–820 (2012)
7. Balmer, M., Meister, K., Nagel, K.: Agent-based simulation of travel demand: Structure and computational performance of MATSim-T: ETH, Eidgenössische Technische Hochschule Zürich, IVT Institut für Verkehrsplanung und Transportsysteme (2008)
8. Chin, A., Lai, A., Chow, J.: Non-additive public transit fare pricing under congestion with policy lessons from Toronto case study. In: Transportation Research Board 95th Annual Meeting (2016)
9. Joubert, J.W.: Analyzing commercial through-traffic. Procedia Soc. Behav. Sci. **39**, 184–194 (2012)
10. Pendyala, R.M., Kitamura, R., Chen, C., Pas, E.I.: An activity-based microsimulation analysis of transportation control measures. Transp. Policy **4**, 183–192 (1997)
11. Chakirov, A., Fourie, P.: Enriched sioux falls scenario with dynamic and disaggregate demand. Arbeitsberichte Verkehrs-und Raumplanung, vol. 978 (2014)
12. Kickhofer, B., Hosse, D., Turner, K.: Creating an open MATSim scenario from open data: The case of Santiago de Chile. TU Berlin, Transport System Planning and Transport Telematics (2016). http://www.vsp.tu-berline.de/publication
13. Armas, R., Aguirre, H., Daolio, F., Tanaka, K.: An effective EA for short term evolution with small population for traffic signal optimization. In: 2016 IEEE Symposium Series on Computational Intelligence (SSCI), pp. 1–8 (2016)
14. Lovrić, M., Li, T., Vervest, P.: Sustainable revenue management: a smart card enabled agent-based modeling approach. Decis. Support Syst. **54**, 1587–1601 (2013)
15. Tomaschek, J., Kober, R., Fahl, U., Lozynskyy, Y.: Energy system modelling and GIS to build an Integrated Climate Protection Concept for Gauteng Province, South Africa. Energy Policy **88**, 445–455 (2016)
16. Horni, A., Nagel, K., Axhausen, K.W.: The multi-agent transport simulation MATSim (2016)
17. Ben-Akiva, M., Bierlaire, M., Koutsopoulos, H., Mishalani, R.: DynaMIT: a simulation-based system for traffic prediction. In: DACCORD Short Term Forecasting Workshop, pp. 1–12 (1998)
18. Lee, K.S., Eom, J.K., Moon, D.-S.: Applications of TRANSIMS in transportation: a literature review. Procedia Comput. Sci. **32**, 769–773 (2014)
19. Barceló, J., Casas, J.: Dynamic network simulation with AIMSUN. In: Simulation Approaches in Transportation Analysis, pp. 57–98 (2005)

20. MTA: Metro north origin-destination survey (2007, 15th March, 2018). http://web.mta.info/mta/planning/data.html#
21. M.T. Authority: MTA New York city travel survey, 2007', 24th August 2017. http://web.mta.info/mta/planning/data-nyc-travel.html
22. GitHub: Agent Based Micro-Simulation of a Passenger Rail System. https://github.com/lolumak. Accessed 08 Feb 2018

# Fintech Bitcoin Smart Investment Based on the Random Neural Network with a Genetic Algorithm

Will Serrano[✉]

Intelligent Systems and Networks Group, Electrical and Electronic Engineering
Imperial College London, London, UK
g.serrano11@imperial.ac.uk

**Abstract.** This paper presents the Random Neural Network in a Deep Learning Cluster structure with a new learning algorithm based on the genome model, where information is transmitted in the combination of genes rather than the genes themselves. The proposed genetic model transmits information to future generations in the network weights rather than the neurons. The innovative genetic algorithm is implanted in a complex deep learning structure that emulates the human brain: Reinforcement Learning takes fast and local decisions, Deep Learning Clusters provide identity and memory, Deep Learning Management Clusters take final strategic decisions and finally Genetic Learning transmits the learned information to future generations. This structure has been applied and validated in Fintech, a Bitcoin Smart Investment application based in an Intelligent Banker that performs Buy and Sell decisions on several Cryptocurrencies with an associated exchange and risk. Our results are promising; we have connected the human brain and genetics with Machine Learning based on the Random Neural Network model where Artificial Intelligence, similar as biology, is learning gradually and continuously while adapting to the environment.

**Keywords:** Genetic learning · Deep Learning Clusters
Reinforcement Learning · Random Neural Network · Smart Investment
Bitcoin · Fintech

## 1   Introduction

Biology is gradually and continuously learning while adapting to the environment using genetic changes to generate new complex structures in organisms that provides an increased reward to a goal function [1]. Random genetic changes have more probability to be successful in organisms that change in a systematic and modular manner where the new structures acquire the same set of sub goals in different combinations [2]. The adaptations learned from the living organisms affect and guide evolution even though the characteristics acquired are not transmitted to the genome [3], however, its gene functions are altered and transmitted to the new generation; this enables learning organisms to evolve much faster.

© Springer Nature Switzerland AG 2019
A. Lotfi et al. (Eds.): UKCI 2018, AISC 840, pp. 138–149, 2019.
https://doi.org/10.1007/978-3-319-97982-3_11

Successful Machine Learning and Artificial Intelligence models have been purely based on biology emulating the structures provided by nature during the learning, adaptation and evolution when interacting with the external environment. Neural networks and deep learning are based on the brain structure which is formed of dense local clusters of same neurons performing different functions which are connected between each other with numerous very short paths and few long distance connections [4]. The brain retrieves a large amount of data obtained from the senses; analyses the material and finally selects the relevant information [5] where the cluster of neurons specialization occurs due to their adaption when learning tasks.

This paper proposes a new genetic learning algorithm on Sect. 3 based on the genome and evolution; where the information transmitted to new generations is learned when interacting and adapting to the environment using reinforcement and deep learning respectively. Information in the proposed genetic algorithm is transmitted in the network weights through the different combinations of four different nodes (C, G, A, T) rather than the value of nodes themselves where the output layer of nodes replicates the input layer as the genome. This innovative genetic algorithm is inserted in a complex deep learning structure that emulates the human brain on Sect. 4: Reinforcement Learning takes fast local current decisions, Deep Learning clusters provide identity and memory, Deep Learning Management Clusters takes final strategic decisions and finally Genetic Learning transmits the information learned to future generations. This innovative model has been applied and validated in Fintech, a Bitcoin Smart Investment application on Sect. 5; an Intelligent Banker that performs Buy and Sell decisions on Cryptocurrencies with an associated Exchange and risk. The results shown on Sect. 6 are promising; the Intelligent Banker takes the right decisions, learns the variable asset price, makes profits on specific markets at minimum risk and finally it transmits the information learned to future generations.

## 2 Related Work

Artificial Neural Networks have been applied to make financial predictions. Leshno et al. [6] evaluate the bankruptcy prediction capability of several neural network models based on the firm's financial reports. Chen et al. [7] uses Artificial Neural Networks for a financial distress prediction model. Kara et al. [8] apply an Artificial Neural Network to predict the direction of Stock Market index movement. Guresen [9] evaluates the effectiveness of neural network models in stock market predictions. Zhang et al. [10] analyse Artificial Neural Networks in bankruptcy prediction. Kohara et al. [11] investigate different ways to use prior knowledge and neural networks to improve multivariate prediction ability. Sheta et al. [12] compares Regression, Artificial Neural Networks and Support Vector Machines for predicting the S&P 500 Stock Market Price Index. Tung et al. [13] includes Artificial Neural Networks and Fuzzy Logic for market predictions. Pakdaman et al. [14] use a feedforward multilayer perceptron and an Elman recurrent Network to predict a company's stock value. Iuhasz et al. [15] create a hybrid system based on a multi Agent Architecture to analyse Stock Market behaviour to improve the profitability in a short or medium time period investment. Nicholas et al. [16] examine the use of neural networks in stock performance modelling.

Machine learning has been applied to solve nonlinear models in continuous time in economics and finance by Duarte [17] and forecasting the volatility of asset prices by Stefani et al. [18]. Deep Learning has also recently incorporated in long short term memory Neural Networks for financial market predictions by Fischer et al. [19] and Hasan et al. [20]. Genetic Algorithms have been proposed as method to increase learning. Arifovic [21] analyses genetic algorithms in inflationary economies. Kim et al. [22] uses a genetic Algorithm to feature discretization in artificial neural networks for the prediction of stock market index. Ticona et al. [23] applies a hybrid model based on Genetic algorithm and Neural Networks to forecast Tax Collection. Hossain et al. [24] present a Genetic Algorithm based Deep Learning Method. Sremath [25] and David et al. [26] review of the latest deep learning structures and evolutionary algorithms that can be used to train them.

# 3   The Random Neural Network Genetic Deep Learning Model

## 3.1   The Random Neural Network

The RNN [27, 28] represents more closely how signals are transmitted in many biological neural networks where they travel as spikes or impulses, rather than as analogue signal. The RNN is a spiking recurrent stochastic model for neural networks. Its main analytical properties are the "product form" and the existence of the unique network steady state solution. The Random Neural network has been also applied in Genetics models [29, 30].

## 3.2   The Random Neural Network with Deep Learning Clusters

Deep Learning with Random Neural Networks is described by Gelenbe and Yin [31]. This model is based on the generalized queuing networks with triggered customer movement where customers are either "positive" or "negative" and customers can be moved from queues or leave the network.

## 3.3   Deep Learning Management Cluster

The Deep Learning management cluster was proposed by Serrano et al. [32]. It takes management decisions based on the inputs from different Deep Learning clusters.

## 3.4   Genetic Learning Algorithm Model

The proposed Genetic learning algorithm is based on the auto encoder presented by Gelenbe [31]; the auto encoder models the genome as it codes the replica of the organism that contains it. Network 1 is formed of U input neurons and C clusters and Network 2 has C input neurons and U clusters. The organism is represented as a set of data X which is a U vector $X \in [0, 1]^U$. The proposed Genetic learning algorithm fixes C to 4 neurons that represent the four different nucleoids G, C, A and T where $W_1$ is fixed to generate the four 4 different types of neurons rather than random values as shown on Fig 1.

Network 1 encodes the organism, it is defined as:

- $q_1 = \left(q_1^1, q_2^1, \ldots, q_u^1\right)$, a U-dimensional vector $q_1 \in [0,1]^U$ that represents the input state $q_u$ for neuron u;
- $W_1$ is the U x C matrix of weights $w_1^-(c,u)$ from the U input neurons to the neurons in each of the C clusters;
- $Q^1 = \left(Q_1^1, Q_2^1, \ldots, Q_c^1\right)$, a C-dimensional vector $Q^1 \in [0, 1]^C$ that represents state $q_c$ for the cluster c where $Q^1 = \zeta(W_1 X)$.

Network 2 decodes the genome, as the pseudo inverse of Network 1, it is defined as:

- $q_2 = \left(q_1^2, q_2^2, \ldots, q_c^2\right)$, a C-dimensional vector $q_2 \in [0, 1]^C$ that represents the input state $q_c$ for neuron c with the same value as $Q^1 = \left(Q_1^1, Q_2^1, \ldots, Q_c^1\right)$;
- $W_2$ is the C x U matrix of weights $w_2^-(c,u)$ from the C input neurons to the neurons in each of the U cells;
- $Q_2 = \left(q_1^2, q_2^2, \ldots, q_u^2\right)$, a U-dimensional vector $Q^2 \in [0, 1]^U$ that represents the state $q_u$ for the cell u where $Q^2 = \zeta(W_2 Q^1)$ or $Q^2 = \zeta(W_2 \zeta(X W_1))$.

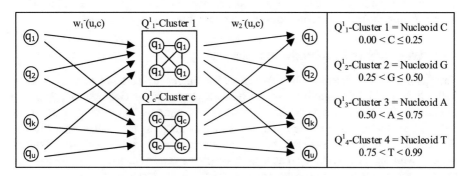

**Fig. 1.** Genetic learning algorithm

The learning algorithm is the adjustment of $W_1$ to code the organism X into the four different neurons or nucleoids and then calculate $W_2$ so that resulting decoded organism $Q^2$ is the same as the encoded organism X:

$$\min||X - \zeta(W_2\zeta(XW_1))|| \; s.t. W_1 \geq 0 \qquad (1)$$

Following the Extreme Learning Machine on [33]; W2 is calculated as:

$$W_2 = \text{pinv}(\zeta(XW_1))X \tag{2}$$

Where pinv is the Moore-Penrose pseudoinverse:

$$\text{pinv}(x) = (x^T x) x^T \tag{3}$$

## 4  Bitcoin Smart Investment Model

The Bitcoin Smart Investment model, called "GoldAI Sachs", combines there different learnings: Deep Learning and Genetic Learning presented in the previous section and Reinforcement Learning presented in the next Chapter. "GoldAI Sachs" is formed of clusters of Intelligent Bankers that take local fast local binary decisions "Buy or Sell" on a specific assets based on Reinforcement Learning through the interactions and adaptations with the environment where the Reward is the profit made. Each Asset Banker has an associated Deep Learning cluster that memorizes asset identity such as price and reward properties. Asset bankers are dynamically clustered to different properties such as investment reward, risk or market type and managed by a Market Banker Deep Learning Management Cluster that selects the best performing Asset Bankers. Finally, a CEO Banker Deep Learning Management Cluster manages the different Bankers and takes the final investment decisions based on the Market Reward and associated Risk prioritizing markets that generate more reward at a lower Risk as every banker would do. This approach enables decisions based on shared information where Intelligent Bankers work collaborative to achieve a bigger reward.

### 4.1  Asset Banker Reinforcement Learning

The Reinforcement Learning algorithm takes fast binary investment decisions "Buy or Sell", it is based on Cognitive Packet Network presented by Gelenbe [34, 35]. The Intelligent Banker is formed of two interconnected neurons "$q_0$ or Buy" and "$q_1$ or Sell" where the investment decision is taken according to the neuron that has the greater potential. The Reward R is based on the economic profit that the Asset Bankers achieve with the decisions they make, successive measured values of the R are denoted by $R_l$, $l = 1,2\ldots$ these are used to compute the Predicted Reward (PR):

$$PR_l = \alpha PR_{l-1} + (1 - \alpha) R_l \tag{4}$$

where $\alpha$ represents the investment reward memory.

If the observed measured Reward is greater than the associated Predicted Reward; Reinforcement Learning rewards the decision taken by increasing the network weight that point to it, otherwise; it penalises it.

## 4.2 Asset Banker Deep Learning Cluster

Deep Learning is used to memorize key investment values that generate asset identity. The Smart Investment model assigns a Deep Learning Cluster per Asset Banker. Each different Deep Learning Cluster learns the Asset Reward or Profit Prediction, the Asset price and the Asset Price Prediction. "GoldAI Sachs" groups Asset Bankers dynamically into market sectors according to their risk, profit or type. A set of x Asset Banker Deep Learning Clusters is defined as:

- $I_{Banker-x} = \left( i_1^{Banker-x}, i_2^{Banker-x}, \ldots, i_u^{Banker-x} \right)$ a U-dimensional vector where $i_1^{Banker-x}$, $i_2^{Banker-x}$, and $i_u^{Banker-x}$ are the same Banker number x;
- $w_{Banker-x}^{-}(u, c)$ is the U x C matrix of weights of the Deep Learning Cluster for Banker x;
- $Y_{Banker-x} = \left( y_1^{Banker-x}, y_2^{Banker-x}, \ldots, y_c^{Banker-x} \right)$ a C-dimensional vector where $y_1^{Banker-x}$ is the Reinforcement Learning Reward prediction, $y_2^{Banker-x}$ is the Dynamic Reward prediction, $y_3^{Banker-x}$ is the Transaction Price and $y_C^{Banker-x}$ is the Price Prediction for Banker Number x.

## 4.3 Market Banker Deep Learning Management Cluster

The Market Banker Deep Learning Management cluster analyses the predicted reward from its respective Asset Banker Deep Learning Clusters, prioritizes their values based on local market knowledge and finally reports to the CEO Banker Deep Learning Management Cluster the total predicted Profit that its Market can make. The model defines a set of x Market Banker Deep Learning Management Clusters as:

- $I_{MarketBanker-x}$, a C-dimensional vector $I_{MarkeBanker-x} \in [0, 1]^C$ with the values of the predicted Rewards from Asset Banker x;
- $w_{MarketBanker-x}^{-}(c)$ is the C-dimensional vector of weights that represents the priority of each Asset Banker x;
- $Y_{MarketBanker-x}$, a scalar $Y_{MarketBanker-x} \in [0, 1]$ that represents the predicted Profit the Market Banker Deep Learning Management Cluster can make.

## 4.4 CEO Banker Deep Learning Management Cluster

The CEO Banker Deep Learning Management Cluster, "AI Morgan" takes the final investment management decision based on the inputs from the Market Bankers and the associated. The CEO Banker selects the markets that generate better reward at lower risk where the maximum risk is defined as $\beta$ where the higher the value the higher risk. The CEO Banker Deep Learning Management cluster is defined:

- $I_{CEO-Banker}$, a X-dimensional vector $I_{CEO-Banker} \in [0, 1]^X$ with the values of the set x Banker DL Management Clusters;
- $w_{CEO-Banker}^{-}(c)$ is the is the C-dimensional vector of weights that represents the risk associated to each Market;
- $Y_{CEO-Banker}$, a scalar $Y_{CEO-Banker} \in [0, 1]$ that represents the final investment decision.

## 4.5 CEO Banker Genetic Learning

Genetic Learning transmits the knowledge acquired from the CEO Banker, to future Banker generations when "AI Morgan" final career decision is to cash the pension and take early retirement. Genetic Learning is purely based on the Genome as the biological method to transmit information. Reinforcement Learning is applied in decisions, Deep Learning in identity and Genetic Learning in immortality. The Genetic Algorithm implementation is shown on Table 1. "GoldAI Sachs" model defines Genetic Learning as an Autoencoder where:

**Table 1.** Genetic algorithm implementation

| Name | Variable | Value |
|---|---|---|
| $X_{Genetic}$ | $x_1^{Genetic} \ldots x_{32}^{Gentetic}$ | $y_1^{Banker-1} \ldots y_4^{Banker-8}$ |
| $w_{Genetic-1}^-(u,c)$ | $w_{Genetic-1}^-(1,1) \ldots w_{Genetic-1}^-(32,1)$ | $(0.02, \ldots 0.02$ |
| | $w_{Genetic-1}^-(1,2) \ldots w_{Genetic-1}^-(32,2)$ | $0.15, \ldots 0.15$ |
| | $w_{Genetic-1}^-(1,3) \ldots w_{Genetic-1}^-(32,3)$ | $0.40, \ldots 0.40$ |
| | $w_{Genetic-1}^-(1,4) \ldots w_{Genetic-1}^-(32,4)$ | $0.99, \ldots 0.99)$ |
| $Y_{Nucleoid}$ | $y_1^{Nucleoid}(0.00 < C \le 0.25$ | $0.2048$ |
| | $y_2^{Nucleoid} 0.25 < G \le 0.50$ | $0.3900$ |
| | $y_3^{Nucleoid} 0.50 < A \le 0.75$ | $0.6295$ |
| | $y_4^{Nucleoid} 0.75 < T \le 0.99)$ | $0.9268$ |
| $w_{Genetic-2}^-(c,u)$ | $w_{Genetic-2}^-(1,1) \ldots w_{Genetic-2}^-(1,32)$ | $pinv(Y_{Nucleiod})\, X_{Genetic}$ |
| | $w_{Genetic-2}^-(2,1) \ldots w_{Genetic-2}^-(2,32)$ | |
| | $w_{Genetic-2}^-(3,1) \ldots w_{Genetic-2}^-(3,32)$ | |
| | $w_{Genetic-2}^-(4,1) \ldots w_{Genetic-2}^-(4,32)$ | |
| $Y_{Generation}$ | $y_1^{Generation} \ldots y_{32}^{Generation}$ | $x_1^{Genetic} \ldots x_{32}^{Gentetic}$ |

- $X_{Genetic} = \left(x_1^{Genetic}, x_2^{Genetic}, \ldots, x_u^{Gentetic}\right)$ a U-dimensional vector where $x_1^{Genetic}, x_2^{Genetic}$, and $x_u^{Genetic}$ are outputs of the x Banker Deep Learning clusters;
- $w_{Genetic-1}^-(u,c)$ is the U x C matrix of weights of the Genetic Encoder;
- $Y_{Nucleoid} = \left(y_1^{Nucleoid}, y_2^{Nucleoid}, \ldots, y_c^{Nucleoid}\right)$ a C-dimensional vector where $y_1^{Nucleoid}, \ldots, y_c^{Nucleoid}$ is the value of the Nucleoid
- $w_{Genetic-2}^-(c,u)$ is the C x U matrix of weights of the Genetic Decoder;
- $Y_{Generation} = \left(y_1^{Generation}, y_2^{Generation}, \ldots, y_c^{Generation}\right)$ a C-dimensional vector where $y_1^{Generation}, \ldots, y_c^{Generation}$ is the value of the new Banker Generation.

## 5 Experimental Results

"GoldAI Sachs" is evaluated with seven different assets to assess the adaptability and performance of our proposed Smart Investment solution for 663 days, from 07/08/2015 to 31/05/2017. The assets are split into the Bitcoin Exchange Market (BITSTAMP,

BTCE, COINBASE, KRAKEN) and the Currency Market (Bitcoin, Ethereum, Ripple). Reinforcement Learning is first initialized with a Buy Decision. Cryptocurrency evaluation data has been produced though Datasheets obtained from Kraggle. The values of the different assets within the Bitcoin Exchange Market are very similar as they trade the same currency whereas the Currency market presents a more disperse set of values (Fig. 2).

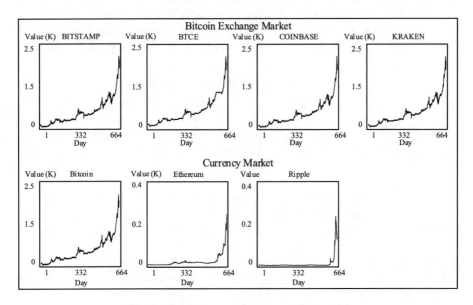

**Fig. 2.** Cryptocurrency investment data model

## 5.1   Asset Banker Reinforcement Learning Validation

The Profit that each Asset Banker makes when buying or selling 100 Assets for 664 days with the Maximum Profit, the number of winning decisions against the losing ones and the number of buy decisions against the sell for an investment reward memory $\alpha = 0.5$ is represented on Table 2.

**Table 2.** Asset banker reinforcement learning validation; $\alpha = 0.5$

| Asset | Profit | Maximum profit | Ratio | Win | Loss | Buy | Sell |
|---|---|---|---|---|---|---|---|
| BITSTAMP | 196,226 | 957,018 | 0.21 | 389 | 274 | 644 | 19 |
| BTCE | 180,596 | 749,308 | 0.24 | 371 | 242 | 659 | 4 |
| COINBASE | 192,951 | 985,625 | 0.20 | 392 | 270 | 644 | 19 |
| KRAKEN | 188,704 | 977,763 | 0.19 | 366 | 259 | 648 | 15 |
| Bitcoin | 196,059 | 952,339 | 0.21 | 385 | 278 | 496 | 167 |
| Ethereum | 19,446 | 58,916 | 0.33 | 355 | 299 | 175 | 488 |
| Ripple | 20 | 101 | 0.20 | 293 | 369 | 663 | 0 |

The RNN Reinforcement Learning Algorithm adapts very quickly to variable asset prices and makes profits; although not optimum values.

## 5.2   Market Banker Deep Learning Management Cluster Validation

The profit the Market Bankers can make for an investment reward memory $\alpha = 0.5$ is shown in Table 3. The Market Bankers take market decisions rather than individual asset decisions from the Asset Bankers. Bitcoin Exchange Market Banker invests 400 assets which is the combination of the four Asset Bankers purchasing power whereas Currency Market Banker invests 300 Assets as there are only three Asset Bankers.

**Table 3.**  Market banker profits; $\alpha = 0.5$

| Market | Total asset banker | Market banker | I | Maximum asset | Maximum market | I |
|---|---|---|---|---|---|---|
| Exchange | 758,477 | 596,973 | −21.29% | 3,669,714 | 3,403,866 | −7.24% |
| Currency | 215,525 | 547,631 | 154.09% | 1,011,356 | 2,003,275 | 98.08% |

We can appreciate the Bitcoin Exchange Market banker does not increase the profits of the market; this is mostly due the four Bitcoin Exchange Asset bankers perform very similarly on almost equal asset conditions therefore the addition of a Market Banker loses the independent knowledge acquired by each asset banker. However, when the Currency Asset Bankers operate under diverse asset conditions, the addition of a Currency Market Banker increases the market profit due the right selection of the best performing asset banker.

## 5.3   CEO Banker Deep Learning Management Cluster Validation

The CEO Banker, "AI Morgan" profits at different Risks ratios with a total of investment of 700 assets for an investment reward memory $\alpha = 0.5$ is shown on Table 4. A risk value $\beta = 0.2$ represents 560 assets in the Exchange Market and 140 in the Currency Market whereas a risk value $\beta = 0.8$ is 140 assets in the Exchange Market and 560 in the Currency Market respectively. This research considers the Exchange Market as low risk and the Currency market as high risk. The CEO banker increases the profits with higher risk decisions although its maximum profit is not optimum.

**Table 4.**  CEO banker profits; $\alpha = 0.5$

| Risk $\beta = 0.2$ | | Risk $\beta = 0.5$ | | Risk $\beta = 0.8$ | | Max profit |
|---|---|---|---|---|---|---|
| E | C | E | C | E | C | |
| 835,763 | 255,561 | 522,352 | 638,903 | 208,941 | 1,022,244 | 5,700,274 |
| Total:1,091,324 | | Total:1,161,254 | | Total: 1,231,185 | | |

## 5.4  Genetic Algorithm Validation

The Genetic Algorithm validation for the four different Nucleoids (C, G, A, T) average value during the 663 different days is shown in Table 5 with the Genetic Algorithm average error, Standard Deviation σ, and 95% Confidence Interval. The Genetic Algorithm successfully codifies the CEO Banker with a marginal error.

**Table 5.** Genetic algorithm validation

| Error | Deviation σ | 95% CR | Nucleoid-C | Nucleoid-G | Nucleoid-A | Nucleoid-T |
|-------|-------------|--------|------------|------------|------------|------------|
| 6.76E−31 | 7.91E−31 | 6.02E−32 | 0.1849 | 0.3594 | 0.5992 | 0.9176 |

# 6  Conclusions

This paper has presented a new learning Genetic Algorithm based on the Genome where the information is transmitted in the network weights rather than the neurons. The algorithm has been incremented in a Fintech Application: Bitcoin Smart Investment model that simulates the human brain with Reinforcement Learning for fast decisions, Deep Learning to memorize properties to create asset identity, Deep Learning management clusters to make global decisions and a Genetic algorithm to transmit learning to future generations.

In the Smart Investor Model, "GoldAI Sachs" Bitcoin Asset Banker Reinforcement Learning Algorithm takes the right investment decisions; with great adaptability to asset price changes whereas Bitcoin Asset Banker Deep Learning provides asset properties and identity. Exchange and Currency Market Bankers success to increase the profit by selecting the best performing Asset Bankers and the CEO Banker, "AI Morgan" increases the profits considering the associated market risks, prioritizing low risk investment decision at equal profit. Genetic learning algorithm has a minimum error and it exactly codes and encodes the CEO Banker, "AI Morgan".

Future work will analyse different methods to improve the performance; or increase the profits, of the proposed deep learning cluster structure. In addition the relevance of memory in investment with its optimum value will be studied.

# References

1. Kirschner, M., Gerhart, J.: The Plausibility of Life Resolving Darwin's Dilemma. Yale University Press, New Haven (2005)
2. Parter, M., Kashtan, N., Alon, U.: Facilitated Variation: How Evolution Learns from Past Environments To Generalize to New Environments. Department of Molecular Cell Biology, Weizmann Institute of Science (2008)
3. Hinton, G., Nowlan, S.: How learning can guide evolution. Adaptive individuals in evolving populations, pp. 447–454 (1996)
4. Smith, D., Bullmore, E.: Small-world brain networks. Neuroscientist **12**, 512–523 (2007)
5. Sporns, O., Chialvo, D., Kaiser, M., Hilgetag, C.: Organization, development and function of complex brain networks. Trends Cogn. Sci. **8**(9), 418–425 (2004)

6. Leshno, M., Spector, Y.: Neural network prediction analysis: the bankruptcy case. Neurocomputing **10**(2), 125–147 (1996)
7. Chen, W., Du, Y.: Using neural networks and data mining techniques for the financial distress prediction model. Expert Syst. Appl. **36**, 4075–4086 (2009)
8. Kara, Y., Acar, M., Kaan, Ö.: Predicting direction of stock price index movement using artificial neural networks and support vector machines: The sample of the Istanbul Stock Exchange. Expert Syst. Appl. **38**, 5311–5319 (2011)
9. Guresen, E., Kayakutlu, G., Daim, T.U.: Using artificial neural network models in stock market index prediction. Expert Syst. Appl. **38**, 10389–10397 (2011)
10. Zhang, G., Hu, M., Patuwo, B., Indro, D.: Artificial neural networks in bankruptcy prediction: General framework and cross-validation analysis. Eur. J. Oper. Res. **116**, 16–32 (1999)
11. Kohara, K., Ishikawa, T., Fukuhara, Y., Nakamura, Y.: Stock price prediction using prior knowledge and neural networks. Intell. Syst. Account. Finance Manage. **6**, 11–22 (1997)
12. Sheta, A., Ahmed, S., Faris, H.: A comparison between regression, artificial neural networks and support vector machines for predicting stock market index. Int. J. Adv. Res. Artif. Intell. **4**(7), 55–63 (2015)
13. Khuat, T., Le, M.: An application of artificial neural networks and fuzzy logic on the stock price prediction problem. Int. J. Inform. Vis. **1**(2), 40–49 (2017)
14. Naeini, M., Taremian, H., Hashemi, H.: Stock market value prediction using neural networks In: International Conference on Computer Information Systems and Industrial Management Applications, pp. 132–136 (2010)
15. Iuhasz, G., Tirea, M., Negru, V.: Neural network predictions of stock price fluctuations. In: International Symposium on Symbolic and Numeric Algorithms for Scientific Computing, pp. 505–512 (2012)
16. Nicholas, A., Zapranis, A., Francis, G.: Stock performance modeling using neural networks: a comparative study with regression models. Neural Networks **7**(2), 375–388 (1994)
17. Duarte, V.: Macro, Finance, and Macro Finance: Solving Nonlinear Models in Continuous Time with Machine Learning. Massachusetts Institute of Technology, Sloan School of Management, pp. 1–27 (2017)
18. Stefani, J., Caelen, O., Hattab, D., Bontempi, G.: Machine learning for multi-step ahead forecasting of volatility proxies. In: Workshop on Mining Data for Financial Applications, pp. 1–12 (2017)
19. Fischer, T., Krauss, C.: Deep learning with long short-term memory networks for financial market predictions. FAU Discuss. Pap. Econ. **11**, 1–32 (2017)
20. Hasan, A., Kalıpsız, O., Akyokuş, S.: Predicting financial market in big data: deep learning. In: International Conference on Computer Science and Engineering, pp. 510–515 (2017)
21. Arifovic, J.: Genetic algorithms and inflationary economies. J. Monetary Econ. **36**, 219–243 (1995)
22. Kim, K., Han, I.: Genetic algorithms approach to feature discretization in artificial neural networks for the prediction of stock price index. Expert Syst. Appl. **19**, 125–132 (2000)
23. Ticona, W., Figueiredo, K., Vellasco, M.: Hybrid model based on genetic algorithms and neural networks to forecast tax collection: application using endogenous and exogenous variables. In: International Conference on Electronics, Electrical Engineering and Computing, pp. 1–4 (2017)
24. Hossain, D., Capi, G.: Genetic algorithm based deep learning parameters tuning for robot object recognition and grasping. Int. Sch. Sci. Res. Innov. **11**(3), 629–633 (2017)
25. Tirumala, S.: Implementation of evolutionary algorithms for deep architectures. In: Artificial Intelligence and Cognition, pp. 164–171 (2014)

26. David, O., Greental, I.: Genetic algorithms for evolving deep neural networks. ACM Genetic and Evolutionary Computation Conference, pp. 1451–1452 (2014)
27. Gelenbe, E.: Random Neural Networks with negative and positive signals and product form solution. Neural Comput. 1, 502–510 (1989)
28. Gelenbe, E.: Learning in the recurrent Random Neural Network. Neural Comput. 5, 154–164 (1993)
29. Gelenbe, E.: A class of genetic algorithms with analytical solution. Rob. Auton. Syst. 22(1), 59–64 (1997)
30. Gelenbe, E.: Steady-state solution of probabilistic gene regulatory networks. Phys. Rev. 76 (1), 031903 (2007). 1–8
31. Gelenbe, E., Yin, Y.: Deep learning with Random Neural Networks. In: International Joint Conference on Neural Networks, pp. 1633–1638 (2016)
32. Serrano, W., Gelenbe, E.: The deep learning Random Neural Network with a management cluster. In: International Conference on Intelligent Decision Technologies, pp. 185–195 (2017)
33. Kasun, L., Zhou, H., Huang, G.: Representational learning with extreme learning machine for big data. IEEE Intell. Syst. 28(6), 31–34 (2013)
34. Gelenbe, E.: Cognitive Packet Network. US Patent, Washington (2004). 6804201 B1
35. Gelenbe, E., Xu, Z., Seref, E.: Cognitive packet networks. In: International Conference on Tools with Artificial Intelligence, pp. 47–54 (1999)

# Generating ANFISs Through Rule Interpolation: An Initial Investigation

Jing Yang[1,2], Changjing Shang[2(✉)], Ying Li[1], Fangyi Li[1,2], and Qiang Shen[2]

[1] School of Computer Science,
Northwestern Polytechnical University, Xi'an 710072, China
lybyp@nwpu.edu.cn
[2] Department of Computer Science, Institute of Maths,
Physics and Computer Science, Aberystwyth University,
Aberystwyth, Ceredigion, UK
{jiy6,cns,fal2,qqs}@aber.ac.uk

**Abstract.** The success of ANFIS (Adaptive-Network-based Fuzzy Inference System) mainly owes to the ability of producing nonlinear approximation via extracting effective fuzzy rules from massive training data. In certain practical problems where there is a lack of training data, however, it is difficult or even impossible to train an effective ANFIS model covering the entire problem domain. In this paper, a new ANFIS interpolation technique is proposed in an effort to implement Takagi-Sugeno fuzzy regression under such situations. It works by interpolating a group of fuzzy rules with the assistance of existing ANFISs in the neighbourhood. The proposed approach firstly constructs a rule dictionary by extracting rules from the neighbouring ANFISs, then an intermediate ANFIS is generated by exploiting the local linear embedding algorithm, and finally the resulting intermediate ANFIS is utilised as an initial ANFIS for further fine-tuning. Experimental results on both synthetic and real world data demonstrate the effectiveness of the proposed technique.

**Keywords:** ANFIS interpolation · Rule dictionary
Takagi-Sugeno fuzzy regression · Local linear embedding

## 1 Introduction

Fuzzy inference systems generally take two forms of knowledge representation: Mandani style models [1] and Takagi-Sugeno (TSK) models [2]. Whilst Mandani systems have been popular in many existing real world applications, TSK systems have also played an increasingly important role in such applications, such as stock market prediction [3] and EGG signals recognition [4]. This is largely owing to their capability of approximating complex nonlinear functions [5]. Amongst various TSK fuzzy inference systems, those implemented with ANFIS facilitate fuzzy inference through the construction of an adaptive network, by extracting useful knowledge in terms of a set of fuzzy rules directly from training data. They

A. Lotfi et al. (Eds.): UKCI 2018, AISC 840, pp. 150–162, 2019.
https://doi.org/10.1007/978-3-319-97982-3_12

have proven to be effective when sufficient training data is provided. However, in dealing with application problems where it is difficult or even impossible to obtain sufficient data for the required training, the potential of these systems may be significantly restricted.

In order to address this challenging practical issue, various techniques have been proposed [6–10], typically involving two types of approach: the transfer learning based and the fuzzy rule interpolation (FRI) based. Transfer learning based techniques [6,7] exploit the knowledge accumulated from data in an auxiliary domain (termed source domain) to support predictive modelling in a certain problem domain (termed target domain). In particular, the underlying techniques work, usually by transferring the data in the target domain into that in the source domain via the use of trained non-linear mappings, such that the model of the source domain can be utilised to enable inference in the target domain. Different to this type of approach are those techniques based on FRI [8–10] which provide a computational intelligence solution to the problem of data shortage. Such techniques work by directly conducting fuzzy inference in the target domain, with the assistance of sparsely given fuzzy rules of the source domains. For instance, the popular scale and move transformation based FRI [9,10] enables fuzzy systems to perform inference with a sparse rule base, by selecting and transforming rules that are the most similar to a given observation that does not actually match any of the rules. Unfortunately, the literature of FRI is mainly focussed on Mandani models [9,10], while the development of TSK type FRI is rather rare, although initial work has most recently been reported [11].

Fundamentally speaking, FRI is based on a practically reasonable assumption that if the conditions are similar, then the consequences should also be similar. From the perspective of manifold learning, the antecedent part and the consequent part of a fuzzy rule lie on the same manifold. With this assumption, the locally linear embedding (LLE) [12], a promising manifold learning algorithm, is able to learn the data structure in the antecedent space, which can be subsequently applied to the data of the consequent space. Inspired by this observation, a new ANFIS interpolation method is proposed in this paper to perform TSK-model based fuzzy regression. The work is developed by exploiting the relationships between the target domain and its neighbour source domains through the use of the LLE algorithm. Particularly, a rule dictionary is generated firstly by extracting fuzzy rules from source ANFIS networks; then a set of new rules are generated using LLE to subsequently construct an inter-mediate ANFIS; and finally, based on this inter-mediate one, a more accurate ANFIS is fine-tuned.

The rest of this paper is organised as follows. Relevant background on ANFIS and LLE is introduced in Sect. 2. Section 3 describes the new ANFIS interpolation algorithm. The experimental results and the corresponding analysis are presented in Sect. 4. Finally, Sect. 5 concludes the paper and outlines interesting further research.

## 2  Background

Only directly relevant work that will be used to develop the present research is introduced in this section, including an overview of ANFIS and LLE algorithm.

### 2.1  Adaptive Network-Based Fuzzy Inference System (ANFIS)

ANFIS [13] is a fuzzy inference system implemented within the framework of adaptive networks. It has been widely applied to various kinds of real problems [14,15] due to its simpleness and effectiveness. The general ANFIS architecture contains five layers. For simplicity, an example of two-input and one-output system is illustrated here. Suppose that the system's fuzzy rule base contains two fuzzy if-then rules of Takagi and Sugeno's type [16], as follows:

Rule 1: If $x_1$ is $A_1$ and $x_2$ is $B_1$, then $y_1 = p_1x_1 + q_1x_2 + r_1$
Rule 2: If $x_1$ is $A_2$ and $x_2$ is $B_2$, then $y_2 = p_2x_1 + q_2x_2 + r_2$

Then, the structure of the TSK type ANFIS that implements such a fuzzy system is shown in Fig. 1, where square nodes stand for nodes with adaptive parameters, and circle nodes represent those fixed ones without modifiable parameters.

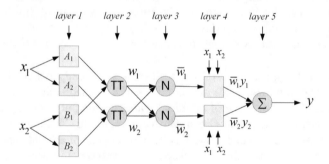

**Fig. 1.** Architecture of ANFIS with 2 fuzzy rules

*Layer 1*: Every node $i$ is defined with the following membership function:

$$O_i^1 = \mu_{A_i}(x) \tag{1}$$

where $x$ denotes the input variable to node $i$, and $A_i$ is the fuzzy set associated with this node. Usually the membership function $\mu_{A_i}(x)$ can be chosen as any continuous or piecewise differentiable functions, such as trapezoidal, triangular or bell-shaped functions with its maximum equalling to 1 and minimum to 0. In particular, for simplicity, the popularly applied triangular shaped membership function is defined as:

$$\mu_{A_i}(x) = \begin{cases} k_{i1}x + b_{i1} & a_{i0} \leq x \leq a_{i1} \\ k_{i2}x + b_{i2} & a_{i1} \leq x \leq a_{i2} \\ 0 & otherwise \end{cases} \tag{2}$$

where $\{k_{i1}, k_{i2}, b_{i1}, b_{i2}\}$ ($i = 1, 2$ for this illustrative example) are the parameters associated with the corresponding input variable. These parameters are named premise parameters hereafter. The representative value of such a triangular-shaped fuzzy set is defined as the centre of gravity of the membership function: $Rep(A_i) = (a_{i0} + a_{i1} + a_{i2})/3$.

*Layer 2*: Every node in this layer is a circle node which multiplies the incoming membership of each attribute and outputs the product. The output $w_i$ in this layer acts as the firing strength of a certain rule. That is,

$$w_i = \mu_{A_i}(x_1) \times \mu_{B_i}(x_2) \tag{3}$$

*Layer 3*: Each node in this layer is also a circle node, computing the relative proportion of the $i$th rule's firing strength to the total of all rules' firing strengths:

$$\bar{w}_i = \frac{w_i}{\sum_{j=1}^{N} w_j} \tag{4}$$

where $i$ is the index of a certain rule and $N$ is the total number of rules encoded (again, $i = 1, 2$ and $N = 2$ for this particular example). For convenience, outputs of this layer are termed as normalised firing strengths.

*Layer 4*: Each node $i$ is a square node with the following linear function:

$$O_i^4 = \bar{w}_i y_i = \bar{w}_i(p_i x_1 + q_i x_2 + r_i) \tag{5}$$

where $\bar{w}_i$ is the output of *Layer 3*, and $p_i, q_i, r_i$ are the parameters which are referred to as consequent parameters hereafter.

*Layer 5*: The output layer, which is a single circle node layer, computes the overall output in response to all current inputs, defined as the summation of all incoming signals, i.e.,

$$O_1^5 = \sum_i \bar{w}_i y_i = y \tag{6}$$

The parameters, including both the premise and the consequent ones, are trained using a hybrid learning method combining gradient descent and Least Square Estimation (LSE). More detailed description of this training process is beyond the scope of this paper, but can be found in [13].

## 2.2 Locally Linear Embedding

The locally linear embedding algorithm [12] is a classical manifold learning method originally proposed for dimensionality reduction. It has now been widely adapted to perform many different machine learning tasks [17, 18]. LLE works by constructing a neighbourhood-preserving mapping from a high-dimensional data space to a low-dimensional data space, based on the presumption that the data of these two spaces lie on or near the same manifold.

For each data point $X_i$ in the $D$-dimensional (high dimensional) data space, LLE involves the following implementation steps to reduce the dimensionality:

(a) Find $K$ nearest neighbours $\{X_j\}$ in the same data space.
(b) Compute the weights $\{w_j\}$ of the selected neighbours by minimising the cost function $\varepsilon(W) = |X_i - \sum_j w_j X_j|^2$ subject to constraint that the sum of the weights equals to one: $\sum_j w_j = 1$.
(c) Compute the corresponding data point $Y_i$ in the $d$-dimensional (low dimensional) data space using the above weights $\{w_j\}$ and the corresponding neighbours $\{Y_j\}$ such that $Y_i = \sum_j w_j Y_j$.

## 3   Proposed ANFIS Interpolation

The main idea of this paper is presented in this section. At the highest level, the proposed approach is put forward in order to implement the following: Given two well trained ANFISs defined over the source domains $S_1$ and $S_2$, denoted by $\mathcal{A}_1$ and $\mathcal{A}_2$ respectively, and a small number of training data in the target domain $T$, expressed in the form of the input-output pairs $\{(\mathbf{x}, y)\}$ (which are themselves not sufficient to train an ANFIS), the aim of ANFIS interpolation is to construct a new ANFIS $\mathcal{A}_T$ over the target domain.

### 3.1   Rule Dictionary Construction

A rule dictionary is a memory devised to store collected rule antecedent parts and consequent parts that are extracted from the given ANFISs, which will be subsequently used for generating new rules to form the inter-mediate ANFIS required to perform interpolation. Without losing generality, suppose that two source ANFISs $\mathcal{A}_1$ and $\mathcal{A}_2$ consist of $n_1$ and $n_2$ rules, respectively, the extracted rules can be expressed in the following format:

$$R_i^{\mathcal{A}_t} : if \ x_1 \ is \ A_{i1}^{\mathcal{A}_t} \ and \dots and \ x_m \ is \ A_{im}^{\mathcal{A}_t}, \ then \ y_i = \sum_{j=0}^{m} p_{ij}^{\mathcal{A}_t} x_j \qquad (7)$$

where $m$ is the number of input variables (with $x_0 = 1$), and $p_{i0}^{\mathcal{A}_t}, t \in \{1, 2\}$ is a constant coefficient within the linear combination of a rule consequent.

The rule dictionary $D = \{D_a, D_c\}$ with the antecedent part $D_a$ and the consequent part $D_c$ is generated by reorganising the aforementioned rules. In particular, $D_a \in R^{m \times N}$, consisting of the antecedent parts of the rules:

$$D_a = [d_1^a \ d_2^a \ \cdots \ d_N^a] \qquad (8)$$

where each column $d_i^a = [A_{i1}^{\mathcal{A}_t} \ A_{i2}^{\mathcal{A}_t} \cdots A_{im}^{\mathcal{A}_t}]^T, t \in \{1, 2\}$ forms an atom of the dictionary $D_a$, and $N = n_1 + n_2$ denotes the number of atoms. Similarly, the consequence part $D_c \in R^{(m+1) \times N}$, consisting of the consequent parts of the rules, which can be expressed as:

$$D_c = [d_1^c \ d_2^c \ \cdots \ d_N^c] \qquad (9)$$

where each atom $d_i^c = [p_{i0}^{\mathcal{A}_t} \ p_{i1}^{\mathcal{A}_t} \ p_{i2}^{\mathcal{A}_t} \cdots p_{im}^{\mathcal{A}_t}]^T, t \in \{1, 2\}$.

## 3.2   Intermediate ANFIS Construction

Having constructed a rule dictionary, the small number of training data $\{(\mathbf{x}, y)\}$ in the target domain are divided into $C$ clusters, on a variable by variable basis. For the centroid of each cluster, a new interpolated rule is generated using LLE by the following steps (as summarised in Fig. 2). By integrating all the interpolated rules, an intermediate ANFIS $\mathcal{A}_{int}$ results.

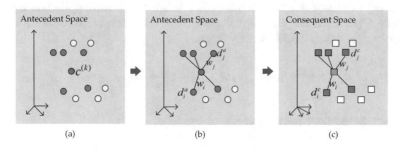

(a)                (b)                (c)

**Fig. 2.** Steps of new rule generation using LLE: (a) Choosing $K$ closest neighbours. (b) Calculating weights of chosen neighbours. (c) Generating new rule.

**Step 1: Atom Selection.** For each cluster $C_k$, the centroid is denoted by $c^{(k)} = (c_1, c_2, \cdots, c_m)^T$ with $m$ attributes. Given the previously obtained antecedent part rule dictionary $D_a$, the first step is to select $K$ closest atoms to $c^{(k)}$ based on the Euclid distance (though other distance metrics may be used as alternatives):

$$d_i = d(d_i^a, c^{(k)}) = \sqrt{\sum_{j=1}^{m} d(A_{ij}^{\mathcal{A}_t}, c_j)^2} \tag{10}$$

where $d(A_{ij}^{\mathcal{A}_t}, c_j) = |Rep(A_{ij}^{\mathcal{A}_t}) - c_j|, t \in \{1, 2\}$. The $K$ atoms $\{d_i^a\}$ with the smallest distances are chosen as the closest atoms, whose index set is $\mathcal{K}$.

**Step 2: Weights Calculation.** Having obtained the closest atoms $\{d_i^a;\ i \in \mathcal{K}\}$, this step is set to find the best reconstruction weights. It is achieved by minimising the following local reconstruction error:

$$w^{(k)} = \min_{w^{(k)}} ||c^{(k)} - \sum_{i \in \mathcal{K}} Rep(d_i^a) w_i^{(k)}||^2, \ s.t. \ \sum_{i \in \mathcal{K}} w_i^{(k)} = 1 \tag{11}$$

where $Rep(d_i^a) = [Rep(A_{i1}^{\mathcal{A}_t})\ Rep(A_{i2}^{\mathcal{A}_t}) \cdots Rep(A_{im}^{\mathcal{A}_t})]^T, t \in \{1, 2\}$, $w_i^{(k)}$ denotes the relative weighting of the atom $d_i^a$. This is an optimisation problem with a least square constraint. Define a Gram matrix $G = (c^{(k)}\mathbf{1}^T - X)^T(c^{(k)}\mathbf{1}^T - X)$, where $\mathbf{1}$ is a column vector of ones, and $X$ is a $m \times K$ matrix whose columns are the selected neighbours of $c^{(k)}$. From this, it can be established that the constrained least square problem has the following solution:

$$w^{(k)} = \frac{G^{-1}\mathbf{1}}{\mathbf{1}^T G^{-1}\mathbf{1}} \tag{12}$$

**Step 3: New Rule Generation.** The underlying presumption in support of the generation of a new rule is that the antecedent part and the consequence part lie on the same manifold as those of the rules extracted from the existing ANFISs. In other words, if the centroid $c^{(k)}$ can be represented as a linear combination of the atoms in the antecedent rule dictionary $D_a$, then the consequence of this observation can be accordingly represented by the linear combination of the atoms in the same location of the consequent rule dictionary $D_c$. Based on this (practically working) assumption, the weights $w_i^{(k)}$ are applied on both the antecedent part and the consequent part. The new interpolated rule for the $k$th cluster has therefore, the following format:

$$R_k^{\mathcal{A}_{int}} : if \ x_1 \ is \ A_{k1}^{\mathcal{A}_{int}} \ and \ldots and \ x_m \ is \ A_{km}^{\mathcal{A}_{int}}, \ then \ y_k = \sum_{j=0}^{m} p_{kj}^{\mathcal{A}_{int}} x_j \quad (13)$$

where the antecedent part is generated by:

$$A_{kj}^{\mathcal{A}_{int}} = \sum_{i\in\mathcal{K}} w_i^{(k)} A_{ij}^{\mathcal{A}_t}, \ j = 1, 2, \cdots, m, \ k = 1, 2, \cdots, C. \quad (14)$$

with $t \in \{1, 2\}$, and the consequent part is generated by:

$$p_{kj}^{\mathcal{A}_{int}} = \sum_{i\in\mathcal{K}} w_i^{(k)} p_{ij}^{\mathcal{A}_t}, \ j = 0, 1, 2, \cdots, m, \ k = 1, 2, \cdots, C. \quad (15)$$

### 3.3   ANFIS Interpolation Algorithm

The interpolated intermediate ANFIS $\mathcal{A}_{int}$ is used as an initial network here to train the final ANFIS $\mathcal{A}_T$ using the training algorithm described in Sect. 2.1. In so doing, a quality initial setup is provided to compute the final ANFIS through a fine-tuning procedure, so that the entire ANFIS interpolation can be performed with as little training data as possible.

   To be concise, the proposed approach is summarised as Fig. 3. The entire algorithm for generating an effective ANFIS in the target domain requires 3 stages: (i) rule dictionary generation, (ii) intermediate ANFIS construction, and (iii) final ANFIS tuning. In the first stage, a rule dictionary is constructed by extracting and reorganising rules from the two well trained source ANFISs. Following that, in the second stage, an intermediate ANFIS is interpolated through the following steps: (a) clustering training data of the target domain into $C$ clusters; (b) interpolating a new rule for the centroid of each cluster using LLE; and (c) integrating all the new generated rules to form the intermediate ANFIS. The specifications for these two stages have been described previously. In the third and final stage, the training data of the target domain is used to refine the intermediate ANFIS through retraining.

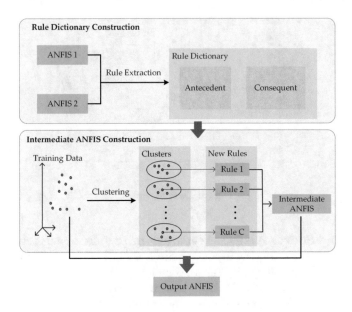

**Fig. 3.** Flowchart of proposed approach

# 4   Experimental Studies

To qualitatively and quantitatively evaluate the proposed approach for ANFIS interpolation, both synthetic and real world data are considered. Section 4.1 validates the approach by looking into a synthetic function modelling case, and in Sect. 4.2, the effectiveness of the proposed work is verified while dealing with a real world problem.

In the experiments on synthetic data, for simplicity, the original data is used without normalisation, given that the data involves any significantly skewed distribution over different magnitudes. For the experiments involving real world data, the input domains are normalised to $[0, 1]$. Triangular membership functions are employed to model the fuzzy sets used by the ANFISs due to their simplicity and popularity. The number of selected closest atoms is empirically set to three.

## 4.1   Experiments on Synthetic Data

In this section, the entire procedure of ANFIS interpolation is illustrated through a one-dimensional input function as Eq. 16, where $x \in [0, 9]$, the shape of this function is plotted for illustration in Fig. 4. The choice of this highly non-linear function to conduct the evaluation is to present ANFIS interpolation with a more challenging modelling problem than those typically used in the literature.

$$y = \frac{sin(2x)}{e^{\frac{x}{5}}} \qquad (16)$$

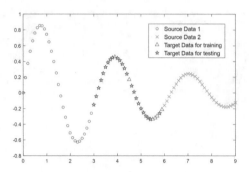

**Fig. 4.** Data samples

**(1) Data and Source ANFISs Generation.** In order to generate the source data and the target data, the input domain $x \in [0,9]$ is divided into three parts (with generated data shown in Fig. 4). The left part $x \in [0,3]$ is the first source domain, which will be used to train the first ANFIS $\mathcal{A}_1$, and the right part $x \in [6,9]$ is the second source domain for training ANFIS $\mathcal{A}_2$. The data of the middle part $x \in [3,6]$ forms the target domain and will be divided into two sub-parts, a small portion (20%, shown in triangles) for training and the rest for testing (80%, shown in Pentagrams).

The two ANFISs of the source domains are trained using the algorithm described in Sect. 2.1. The performances of the source ANFISs on both the source and the target data are evaluated using Root-Mean-Squared Error (RMSE). In particular, the RMSE of the source ANFIS $\mathcal{A}_1$ on the source domain $S_1$, is denoted by $E_{\mathcal{A}_1(S_1)}$; that of $\mathcal{A}_2$ on $S_2$ is denoted by $E_{\mathcal{A}_2(S_2)}$; and the RMSEs of $\mathcal{A}_1$ and $\mathcal{A}_2$ on the target domain $T$, are denoted by $E_{\mathcal{A}_1(T)}$ and $E_{\mathcal{A}_2(T)}$. Formally, these RMSEs are calculated as follows:

$$E_{\mathcal{A}_t(*)} = \sqrt{\frac{\sum_{k=1}^{N_*}(g_k - \mathcal{A}_t(x_k))^2}{N_*}} \tag{17}$$

where $N_*$ is the number of the data points $\{x_k\}$ in the domain of $S_1, S_2$ or $T$, respectively; $g_k$ is the corresponding ground truth of the $k$th data point; $\mathcal{A}_t(x_k)$ stands for the output of the source ANFIS on the data point $x_k$. Obviously, a smaller RMSE indicates a better performance.

**(2) ANFIS Interpolation.** The training results of the ANFIS interpolation algorithm are presented with regard to the three stages below.

Rule Dictionary Construction
Table 1 lists the rule dictionary constructed by extracting rules from trained ANFISs $\mathcal{A}_1$ and $\mathcal{A}_2$.

**Table 1.** Atoms in generated rule dictionary

| Atom | Source ANFIS | Antecedent $(a_0, a_1, a_2)$ | Consequent |
|------|--------------|------------------------------|------------|
| 1 | $\mathcal{A}_1$ | $(-0.725, 0.045, 0.684)$ | $1.263x - 0.044$ |
| 2 | $\mathcal{A}_1$ | $(-0.040, 0.733, 1.446)$ | $0.141x + 0.760$ |
| 3 | $\mathcal{A}_1$ | $(0.768, 1.440, 2.157)$ | $-0.609x + 1.057$ |
| 4 | $\mathcal{A}_1$ | $(1.465, 2.166, 2.924)$ | $-0.075x - 0.451$ |
| 5 | $\mathcal{A}_1$ | $(2.201, 2.874, 3.625)$ | $0.658x - 2.151$ |
| 6 | $\mathcal{A}_2$ | $(5.400, 6.002, 6.570)$ | $0.250x - 1.665$ |
| 7 | $\mathcal{A}_2$ | $(5.999, 6.599, 7.194)$ | $0.195x - 1.127$ |
| 8 | $\mathcal{A}_2$ | $(6.605, 7.196, 7.782)$ | $-0.098x + 0.936$ |
| 9 | $\mathcal{A}_2$ | $(7.234, 7.801, 8.375)$ | $-0.223x + 1.765$ |
| 10 | $\mathcal{A}_2$ | $(7.815, 8.389, 9.012)$ | $-0.083x + 0.531$ |
| 11 | $\mathcal{A}_2$ | $(8.401, 8.983, 9.600)$ | $0.120x - 1.205$ |

## Intermediate ANFIS Construction

A group of new rules are interpolated in the target domain to form the intermediate ANFIS. The number of clusters $C$ (i.e., the number of rules in the new ANFIS) is decided by $C = \prod_{j=1}^{m} \lfloor (n_1^{(j)} + n_2^{(j)})/2 \rfloor$, where $n_1^{(j)}$ is the number of fuzzy sets in the $j$th variable of $\mathcal{A}_1$, and $n_2^{(j)}$ is that of $\mathcal{A}_2$. In this illustrative example, $m = 1, n_1^{(1)} = 5, n_2^{(1)} = 6$, so $C = \lfloor (5+6)/2 \rfloor = 5$. Table 2 shows the new rules generated for all $C$ centroids.

**Table 2.** Generated new rules for all cluster centroids

| Centroids | Choosen atoms | Weights | New rules | |
|-----------|---------------|---------|-----------|---|
| | | | Antecedent | Consequent |
| 3.25 | $(5, 4, 6)$ | $0.376, 0.415, 0.209$ | $(2.565, 3.235, 3.950)$ | $0.269x - 1.344$ |
| 3.85 | $(5, 4, 6)$ | $0.318, 0.304, 0.378$ | $(3.186, 3.840, 4.524)$ | $0.281x - 1.450$ |
| 4.45 | $(6, 5, 7)$ | $0.258, 0.538, 0.203$ | $(3.800, 4.439, 5.111)$ | $0.458x - 1.817$ |
| 5.05 | $(6, 7, 5)$ | $0.321, 0.313, 0.366$ | $(4.418, 5.044, 5.688)$ | $0.382x - 1.674$ |
| 5.65 | $(6, 7, 5)$ | $0.385, 0.422, 0.193$ | $(5.035, 5.650, 6.265)$ | $0.306x - 1.532$ |

## Final ANFIS Fine-tuning

By combining all the (five) new rules, an intermediate ANFIS is constructed. From this, the final output ANFIS is obtained by retraining it using the given data. The performance of this final ANFIS $\mathcal{A}_T$ on the test data of the target domain is denoted by $E_{\mathcal{A}_T(T)}$. An original ANFIS trained using only the training data of the target domain is also generated here for comparison, denoted by $\mathcal{A}_{ori}$, the RMSE of $\mathcal{A}_{ori}$ on the testing data is $E_{\mathcal{A}_{ori}(T)}$.

**(3) Result Evaluation.** The 5-fold cross validation is applied to statistically evaluate the experimental results. The mean and standard deviation values under different algorithms compared are shown in Table 3, and the visual results are illustrated in Fig. 5.

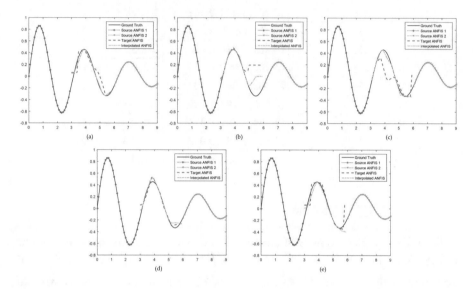

**Fig. 5.** 5-fold cross validation for nonlinear function modelling

By analysing the data in both Fig. 5 and Table 3, it can be concluded that the source ANFISs $\mathcal{A}_1$ and $\mathcal{A}_2$ perform very well in the source domains, but they do not work in the target domain (which is expected, of course). More importantly, due to data shortage of the target domain, the performance of the original ANFIS $\mathcal{A}_{ori}$ is poor. However, the interpolated ANFIS improves the result quality significantly, with the assistance of its two neighbouring ANFISs.

## 4.2   Experiments on Real World Data

Due to length limit, only one benchmark data set (Quake Dataset [19], with 2178 instances and 3 input variables: Depth, Latitude, Longitude) is used here, owing to its popularity. The dataset is split into three sub-datasets with respect to the values of the variable 'Longitude': instances whose 'Longitude' value is smaller than '−40' form the first source domain $S_1$ (593 instances), those whose 'Longitude' value is within the interval $[-40, 92]$ form the second source domain $S_2$ (1253 instances), and those whose 'Longitude' value is larger than '92' form the target domain $T$ (332 instances). As with the synthetic data experiments reported earlier, 20% (66 instances) of the data in the target domain are used for training, with the remaining 80% (265 instances) employed for testing. The results of this experiment are shown in Table 3.

**Table 3.** Experimental results of proposed approach

|  | Mean ± Standard deviation | |
|---|---|---|
|  | Synthetic data | Real data |
| $E_{\mathcal{A}_1(S_1)}$ | $0.000 \pm 0.000$ | $0.000 \pm 0.000$ |
| $E_{\mathcal{A}_2(S_2)}$ | $0.000 \pm 0.000$ | $0.000 \pm 0.000$ |
| $E_{\mathcal{A}_1(T)}$ | $0.287 \pm 0.019$ | $9.655 \pm 0.187$ |
| $E_{\mathcal{A}_2(T)}$ | $0.255 \pm 0.015$ | $0.952 \pm 0.028$ |
| $E_{\mathcal{A}_{ori}(T)}$ | $0.161 \pm 0.101$ | $1.343 \pm 0.839$ |
| $E_{\mathcal{A}_T(T)}$ | $\mathbf{0.079} \pm 0.046$ | $\mathbf{0.283} \pm 0.049$ |

It can be seen from the results that the interpolated ANFIS using the proposed method remarkably minimises the error caused by data shortage, as the mean and standard deviation values of $E_{\mathcal{A}_T(T)}$ are both much smaller than their counterparts as per $E_{\mathcal{A}_{ori}(T)}$.

# 5  Conclusion

Fuzzy rule interpolation offers an important approach for reasoning with sparse knowledge. Different from single rule interpolation and from the existing literature where Mamdani models are used, in this work, a new ANFIS interpolation technique which interpolates a group of fuzzy rules has been proposed. This technique effectively resolves the data shortage problem in the target domain with the assistance of neighbouring ANFISs. Experimental results have shown that the proposed approach greatly improves the performance in the target domain when only a small amount of training data is available.

In this initial work, the two well trained ANFISs in the source domains are of the same structure (i.e., they involve the same set of input variables). However, in many situations, different ANFISs may consist of different variables and hence are of different structures. How to extend the present work to handling such ANFIS interpolation forms a possible future research. Also, expanding and evaluating the initial work with more real data also presents a necessary direction of further work. Currently, all fuzzy sets are specified in triangular form; using an automated data clustering tool such as those introduced in [20] would ensure more accurate interpolative results. Last but not least, in this initial work, a target ANFIS is generated by decomposing the relevant source ANFISs and interpolating the decomposed fuzzy rules. However, there may exist a direct way for constructing new ANFISs without considering the fitness of the individual decomposed rules. In particular, evolutionary computation may provide a helpful means to implement such an approach. This forms an interesting and promising piece of further research.

# References

1. Mamdani, E.H.: Application of fuzzy logic to approximative reasoning using linguistic syntesis. IEEE Trans. Comput. **C–26**(12), 1182–1191 (1977)
2. Takagi, T., Sugeno, M.: Fuzzy identification of systems and its applications to modeling and control. IEEE Trans. Syst. Man Cybern. **SMC–15**(1), 116–132 (1985)
3. Chang, P.-C., Liu, C.-H.: A TSK type fuzzy rule based system for stock price prediction. Expert Syst. Appl. **34**(1), 135–144 (2008)
4. Jiang, Y., Deng, Z.: Recognition of epileptic EEG signals using a novel multiview TSK fuzzy system. IEEE Trans. Fuzzy Syst. **25**(1), 3–20 (2017)
5. Chuang, C.-C., Shun-Feng, S., Chen, S.-S.: Robust TSK fuzzy modeling for function approximation with outliers. IEEE Trans. Fuzzy Syst. **9**(6), 810–821 (2001)
6. Zuo, H., Zhang, G., Lu, J.: Fuzzy regression transfer learning in Takagi-Sugeno fuzzy models. IEEE Trans. Fuzzy Syst. **25**(6), 1795–1807 (2016)
7. Zuo, H., Zhang, G., and Lu, J.: Granular fuzzy regression domain adaptation in Takagi-Sugeno fuzzy models. IEEE Trans. Fuzzy Syst. **99** (2017)
8. Baranyi, P., Kóczy, L.T., Gedeon, T.D.: A generalized concept for fuzzy rule interpolation. IEEE Trans. Fuzzy Syst. **12**(6), 820–837 (2004)
9. Huang, Z., Shen, Q.: Fuzzy interpolative reasoning via scale and move transformations. IEEE Trans. Fuzzy Syst. **14**(2), 340–359 (2006)
10. Huang, Z., Shen, Q.: Fuzzy interpolation and extrapolation: a practical approach. IEEE Trans. Fuzzy Syst. **16**(1), 13–28 (2008)
11. Li, J., et al.: TSK inference with sparse rule bases. In: Advances in Computational Intelligence Systems, pp. 107–123, (2017)
12. Roweis, S.T.: Nonlinear dimensionality reduction by locally linear embedding. Science **290**(5500), 2323–2326 (2000)
13. Jang, J.S.R.: ANFIS: adaptive-network-based fuzzy inference system. IEEE Trans. Syst. Man Cybern. **23**(3), 665–685 (1993)
14. Turkmen, I.: Efficient impulse noise detection method with ANFIS for accurate image restoration. AEU-Int. J. Electron. Commun. **65**(2), 132–139 (2011)
15. Wei, L.: A hybrid ANFIS model based on empirical mode decomposition for stock time series forecasting. Appl. Soft Comput. **42**, 368–376 (2016)
16. Takagi, T., Sugeno, M.: Derivation of fuzzy control rules from human operator's control actions. IFAC Proc. Vol. **16**(13), 55–60 (1983)
17. Hong, C., Yeung, D.-Y.: Super-resolution through neighbor embedding. In: Proceedings of the 2004 IEEE Computer Society Conference on Computer Vision and Pattern Recognition (2004)
18. Jinglin, Z., et al.: Quality-relevant fault monitoring based on locally linear embedding enhanced partial least squares statistical models. Data Driven Control Learn. Syst. (DDCLS) **6**, 259–264 (2017)
19. http://sci2s.ugr.es/keel/dataset.php?cod=75
20. Boongoen, T., et al.: Extending data reliability measure to a filter approach for soft subspace clustering. IEEE Trans. Syst. Man Cybern. B Cybern. **41**(6), 1705–1714 (2011)

# Disentangling the Latent Space of (Variational) Autoencoders for NLP

Gino Brunner[(⊠)], Yuyi Wang, Roger Wattenhofer, and Michael Weigelt

ETH Zurich, Zürich, Switzerland
brunnegi@ethz.ch

**Abstract.** We train multi-task (variational) autoencoders on linguistic tasks and analyze the learned hidden sentence representations. The representations change significantly when translation and part-of-speech decoders are added. The more decoders are attached, the better the models cluster sentences according to their syntactic similarity, as the representation space becomes less entangled. We compare standard unconstrained autoencoders to variational autoencoders and find significant differences. We achieve better disentanglement with the standard autoencoder, which goes against recent work on variational autoencoders in the visual domain.

**Keywords:** NLP · Variational · Autoencoder · Disentanglement
Representation learning · Syntax

## 1  Introduction

Learning good representations lies at the core of Deep Learning [1]. We would like algorithms to automatically extract the most salient features instead of having to rely on expert knowledge to manually design complex preprocessing pipelines. If a model can learn good features, it will likely perform well in an array of (downstream) tasks. Another important aspect is that of transfer learning, where a model is trained on multiple tasks that mutually benefit from each other, leading to better performance in each task. A model that can learn good representations is likely to perform better in a transfer learning setting. For more background on what makes good representations, we refer the interested reader to [1]. Higgins et al. [3] have shown that a simple modification to the standard Variational autoencoder (VAE) objective enables disentanglement of independent linear data generating factors for an artificial dataset of simple 2D shapes. Other works have achieved similar results for the visual domain. However, the progress for discrete sequences, such as natural language, has been much slower. Prior work shows the efficacy of Variational autoencoders, which are generally good at learning representations, for complex tasks such as sentiment transfer ([9]). It has also been shown that multi-task learning is beneficial in the context of NLP [6–8]. In this paper, we focus on analyzing the learned representations

© Springer Nature Switzerland AG 2019
A. Lotfi et al. (Eds.): UKCI 2018, AISC 840, pp. 163–168, 2019.
https://doi.org/10.1007/978-3-319-97982-3_13

and investigate the disentanglement capabilities of (variational) autoencoders in a Multitask setting for Natural Language Processing (NLP). A commonly used definition of disentanglement (e.g., [3]) is that small changes in one dimension of the hidden representation should result in small changes in only one data generating factor, where data generating factors could be the size of an object. In the context of language however, it is considerably more difficult to come up with such linear data generating factors, and thus, there are not yet any general definitions of disentanglement. In this paper we therefore only look at one specific factor of language: syntax. We investigate the ability of autoencoder based language models to learn disentangled representations of syntax. We define a representation to be disentangled if the hidden representations of sentences with different syntactic structures can be clustered with little to no overlap. To this end we train several multi-task autoencoder models, where each decoder performs a distinctive linguistic task. We compare the sentence representations our models have learned and explore how representations of different sentences relate to each other.

## 2    Models

Our models are based on the autoencoder (AE) and variational autoencoder (VAE) [4] frameworks. In both cases, an encoder transforms the data into a lower dimensional representation, from which a decoder tries to replicate the input. We use Long short-term memory (LSTM) neural networks for all encoders and decoders. The VAE formulation additionally encourages the latent variables to be distributed according to a prior (usually an isotropic Gaussian with unit variance). This is achieved by adding a second term to the AE loss function that minimizes the KL-divergence between the chosen prior and the true posterior. When weighted appropriately, this constraint acts as a regularizer on the number of latent dimensions that are used by the model, which in turn promotes disentanglement of the latent dimensions [2,3]. Higgins et al. formally introduce this weight in the VAE loss function as $\beta$, and thus call their VAE variant $\beta$-VAE. Apart from the standard replicating decoder (REP(R)), we attach additional decoders to perform different tasks. The multi-task models in this paper use a subset of the following three decoders in addition to the REP decoder. The German and French (GER(G)/FR(F)) decoders translate the input sentence to German and French respectively. The part-of-speech (POS(P)) decoder learns to tag words in the input sequence with part-of-speech tags, such as *verb, noun, adjective*. To train our models on the three tasks replication, translation and POS, we use the aligned multilingual transcripts of the European Parliament sessions ([5]). The subset of this dataset we use contains over 1.7 million sentences, 1.5 million of which were used as the training set. The remaining 0.2 million sentences form the test set. Our models are trained on character-sequences.

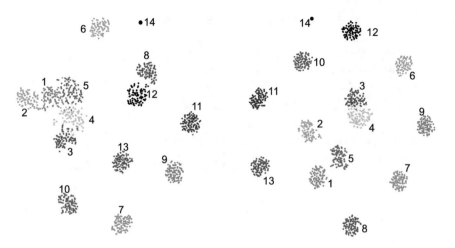

(a) R(EP) model with $CE = 51$. Some sentence prototype representation clusters are very close together or overlapping.

(b) RGP model with $CE = 0$. No sentence representation clusters are overlapping, and only type 3 and 4 are close together.

**Fig. 1.** Syntax clusters for the autoencoder models visualized with t-SNE.

# 3  Results

## 3.1  Syntax Clustering

To compare the learned representations of different models, we examine how well they cluster syntactically similar sentences in latent space. We define 14 sentence prototypes (see Table 1) with different syntactic structures. $N$, $V$, $A$ and $D$ are placeholders for nouns, verbs, adjectives and adverbs. Each sentence prototype is randomly populated by common English words 100 times. These sentences are then fed through the encoders to obtain their representation vectors, which are then clustered by K-means with $K = 14$. For each resulting cluster, we count how many sentences of each prototype it contains. The cluster is then labeled with the majority prototype. The per-cluster error is defined as the number of sentences in the cluster that are not of the majority type (Fig. 2). The sum of errors of all 14 clusters is the *clustering error* (CE), which is our quality metric for this experiment. Since K-means clustering is nondeterministic, we run the algorithm 100 times. Table 2 shows the best-of-100 clustering errors.

For the standard autoencoder, adding more tasks clearly helps reduce the clustering error. The POS tagging decoder brings the highest benefit. This makes sense, since most sentence prototypes have a unique POS tag sequence, and thus separating the sentence prototypes in latent space will make the POS tagging decoder's job easier. Attaching either a German or French translation decoders also help reduce the clustering error. Figure 1 shows the sentence representation of two different AE models, visualized using t-SNE. The RGP model is significantly better at disentangling syntax.

**Table 1.** Sentence prototypes.

| 1: The $N$ is $A$ | 2: The $N$ $V$s | 3: The $N$ has a $N$ | 4: The $N$ $V$s a $N$ |
|---|---|---|---|
| 5: The $N$ $V$s a $N$ | 6: No $N$ ever $V$s | 7: Are $N$s $A$? | |
| 8: The $N$s of $N$ $D$ $V$ the $A$ $N$, but some $N$s still $V$ their $N$ | | | |
| 9: In the $N$ of a $A$ $N$, the $N$ will $V$ the $N$ of $V$ing the $N$ | | | |
| 10: $N$s $V$ the $A$ $N$ of $N$s $V$ing on the $N$ | | | |
| 11: In the $N$ of $N$, $N$s would rather $V$ without $N$ than $V$ any $A$ $N$s | | | |
| 12: $N$ $V$s in order to $V$ on a $N$ | | 13: $A$ $N$s often $V$ like $N$s | |
| 14: *whitespace* | | | |

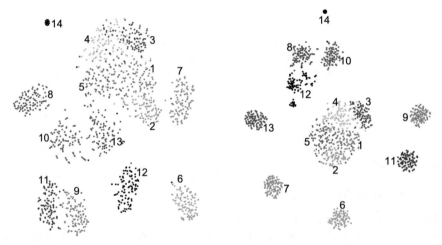

(a) R(EP) model with $CE = 197$. Several sentence types are highly overlapping or close together. The individual clusters have a large diameter.

(b) RGP model with $CE = 25$. The latent space is much less entangled and the clusters have smaller diameters, but several sentence types are still overlapping or very close together.

**Fig. 2.** Syntax clusters for the $\beta$-VAE models visualized with t-SNE.

**Table 2.** AE vs $\beta$-VAE ($\beta = 0.001$)

| Model | R | RF | RGF | RG | RP | RGP |
|---|---|---|---|---|---|---|
| AE | 51 | 26 | 24 | 22 | 8 | 0 |
| VAE | 197 | 87 | 26 | 98 | 58 | 25 |

For the $\beta$-VAE, the results are much less consistent, and generally worse than for the standard AE, even though $\beta$-VAE was shown to disentangle factors of variation in latent space for visual tasks. Especially the high CE for the RP model is surprising. In the experiments performed by [3], $\beta = 4$ yielded the highest degree of disentanglement. Unfortunately, increasing the weight of the KL-term in the VAE loss has a negative effect on reconstruction performance. We were not able to train models with $\beta > 0.1$, which is consistent with existing VAE implementations for sequence tasks[1]. We trained multiple $\beta$-VAEs with $\beta \in [0, 0.0001, 0.001, 0.01, 0.1]$ and found that $\beta = 0.001$ generally performs best in terms of reconstruction performance and clustering error. Figure 1 shows the sentence representation of two different VAE models, visualized using t-SNE. The latent space is clearly more entangled than for the AE based RGP model.

## 3.2   Interpolation and Representation Space Algebra

To further evaluate the properties of our models we traverse the latent space by interpolating between samples from the dataset. We find that the models with the best clustering errors produce smoother interpolations with fewer non-words and more consistent syntactic structure. We also investigate the learned linear relationships between latent sentence representations. To do this we compute a new sentence representation $\hat{s}$ by combining the latent representations of three sentences as $\hat{s} = s_1 - s_2 + s_3$. Intuitively, $s_3$ should be modified with the difference-vector of $s_1$ and $s_2$. For example: *Cats are good pets* ($s_1$) and *Dogs are good pets* ($s_2$) should have canceled out the part about good pets and roughly

**Table 3.** Examples of representation vector algebra for two autoencoder models. The RGP model produces the correct result for the first two examples. Both models manage to replace *small* with *large* in the third example, but also wrongly change most other parts of the sentence.

|  | $s_1$ | $s_2$ | $s_3$ | $s_1 - s_2 + s_3$ |
|---|---|---|---|---|
| RG | I am one | − I am two | + You are two | = You ready no |
| RGP | I am one | − I am two | + You are two | = **You are one** |
| RG | A word in a phrase | − A tree in a phrase | + A tree is green | = A word is purevy? |
| RGP | A word in a phrase | − A tree in a phrase | + A tree is green | = **A word is green** |
| RP | A large number of people want to work | − A small number of people want to work | + A small sentence is enough | = A large senselfeir in or evacce |
| RGP | A large number of people want to work | − A small number of people want to work | + A small sentence is enough | = A large sector for challenge |

---

[1] E.g., https://github.com/tensorflow/magenta/tree/master/magenta/models/music_vae.

point from *Dogs* to *Cats* ($s_1 - s_2$). Adding this difference-vector to any sentence that contains *Dogs* should then result in a sentence where *Dogs* is replaced with *Cats*. We find that the models with low clustering error perform significantly better at this task, as shown in Table 3.

## 4  Conclusion

We trained several multi-task autoencoders on linguistic tasks and analyzed the learned sentence representations based on a new clustering based metric using a toy dataset. Adding linguistic tasks helps the models disentangle syntax in latent space. We further found significant differences between standard and variational autoencoders. We built a toy dataset based on sentence prototypes and introduced the clustering error metric to evaluate the disentanglement of the learned representations. In the future we plan to formulate more rigorous definitions of good (e.g., disentangled) representations in the context of natural language, and evaluate models with different degrees of disentanglement on downstream tasks to see if there is any benefit. We will also revisit the use of recurrent neural networks. As recent trends show, CNNs or pure attention based models might be better suited to model sequences. CNNs are known to be powerful feature extractors, which might be one reason for the success of unsupervised representation learning methods for vision.

## References

1. Bengio, Y., Courville, A.C., Vincent, P.: Representation learning: a review and new perspectives. IEEE Trans. Pattern Anal. Mach. Intell. **35**(8), 1798–1828 (2013). https://doi.org/10.1109/TPAMI.2013.50
2. Burgess, C.P., Higgins, I., Pal, A., Matthey, L., Watters, N., Desjardins, G., Lerchner, A.: Understanding disentangling in $\beta$-VAE. arXiv preprint arXiv:180403599 (2018)
3. Higgins, I., Matthey, L., Pal, A., Burgess, C., Glorot, X., Botvinick, M., Mohamed, S., Lerchner, A.: $\beta$-VAE: Learning basic visual concepts with a constrained variational framework (2016)
4. Kingma, D.P., Welling, M.: Auto-encoding variational bayes. CoRR abs/1312.6114. http://arxiv.org/abs/1312.6114 (2013)
5. Koehn, P.: Europarl: A Parallel Corpus for Statistical Machine Translation (2005)
6. Liu, X., Gao, J., He, X., Deng, L., Duh, K., Wang, Y.: Representation learning using multi-task deep neural networks for semantic classification and information retrieval. In: NAACL HLT 2015, The 2015 Conference of the North American Chapter of the Association for Computational Linguistics: Human Language Technologies, pp. 912–921 (2015)
7. Luong, M., Le, Q.V., Sutskever, I., Vinyals, O., Kaiser, L.: Multi-task sequence to sequence learning. CoRR abs/1511.06114 (2015)
8. Niehues, J., Cho, E.: Exploiting linguistic resources for neural machine translation using multi-task learning. In: Proceedings of the Second Conference on Machine Translation, WMT 2017, pp. 80–89 (2017)
9. Shen, T., Lei, T., Barzilay, R., Jaakkola, T.: Style transfer from non-parallel text by cross-alignment. In: Advances in Neural Information Processing Systems, pp. 6833–6844 (2017)

# A Low Computational Approach for Assistive Esophageal Adenocarcinoma and Colorectal Cancer Detection

Zheqi Yu[1], Shufan Yang[2(✉)], Keliang Zhou[2], and Amar Aggoun[1]

[1] Faculty of Science and Engineering, University of Wolverhampton, Wolverhampton, UK
[2] School of Engineering, University of Glasgow, Glasgow, UK
Shufan.Yang@glasgow.ac.uk

**Abstract.** In this paper, we aim to develop a low-computational system for real-time image processing and analysis in endoscopy images for the early detection of the human esophageal adenocarcinoma and colorectal cancer. Rich statistical features are used to train an improved machine-learning algorithm. Our algorithm can achieve a real-time classification of malign and benign cancer tumours with a significantly improved detection precision compared to the classical HOG method as a reference when it is implemented on real time embedded system NVIDIA TX2 platform. Our approach can help to avoid unnecessary biopsies for patients and reduce the over diagnosis of clinically insignificant cancers in the future.

**Keywords:** Machine learning · Endoscopy · Cancer detection
Texture analysis division

## 1 Introduction

Although esophageal adenocarcinoma is uncommon, its diagnosis has increased dramatically over the past 25 years [1]. During diagnosis process, information obtained from the patient's physical examination, laboratory data, and endoscopic evaluation help doctors to make the correct diagnosis. While those diagnosis process, medical image processing analysis has become more and more popular in an early stage due to low cost [2]. However, using endoscopic images to detect the lesion of esophageal adenocarcinoma and colorectal cancer is a challenging task due to several facts presented in the endoscopy pictures: air bubbles, ink marking, uneven illumination and shadows [3]. Conventional medical image processing algorithm design includes the following steps: image segmentation, feature extraction, reduction and classification.

Artificial neural network based image segmentation approaches have drawn a lot of attention due to their signal-to-noise independency. However, while those approaches can achieve a high level of accuracy, they are computationally expensive with less than 10 frames per second in the latest literature [4,5].

© Springer Nature Switzerland AG 2019
A. Lotfi et al. (Eds.): UKCI 2018, AISC 840, pp. 169–178, 2019.
https://doi.org/10.1007/978-3-319-97982-3_14

In this paper, we focus on the implementation of texture analysis as feature extractions into a machine learning classifier whilst balancing the computational requirement and detection accuracy. We verify our low computational method for esophageal adenocarcinoma cancer detection using NVIDIA TX2 board. Experimental results show that the proposed algorithm can effectively improve the speed and accuracy by 20% compared with the HOG reference design.

## 2    Our Texture Analysis Based Algorithm

In this paper, we use the texture analysis to extract and analyse the image grey levels for the spatial distribution pattern. A statistical histogram of features is calculated based on the Histograms of Oriented Gradients (HOG) [6] algorithm, followed by an AdaBoost classifier. Figure 1 shows the diagram of system structure blocks.

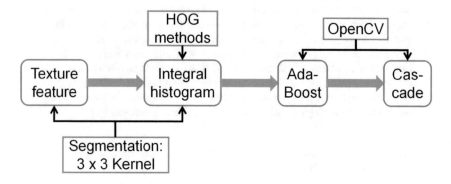

**Fig. 1.** Algorithm work flow diagram.

Our algorithm first uses the $3 \times 3$ kernel to calculate each pixel in the kernel to get the texture feature, and then it calculates the corresponding grey division interval levels along with the kernel texture features. This process means the centre points will be greyscaled range division when the texture feature is calculating the surrounding pixels. For the statistical analysis, each pixel corresponds to a division grey interval in the statistical histogram. While the histogram is calculating, the algorithm adds all the texture features of the same grey level interval and then inputs result into the AdaBoost classifier through the HOG fast integral image method. Finally, all of the weak classifiers are cascaded to obtain a strong classifier.

Figure 2 is a pseudo code diagram for our proposed algorithm procedure; it has shown the grey levels division methods for how to work in the texture analysis.

| | Proposed Algorithm Procedure |
|---|---|
| 1 | **for** all pixel ( i , j ) in image I |
| 2 | magnitude = abs(I ( i-1 , j-1) + I ( i-1 , j) + I ( i-1 , j+1) + I ( i , j-1) −I ( i , j)*8+ I ( i , j+1) + I ( i+1 , j-1) + I ( i+1 , j) + I ( i+1 , j+1))/8 |
| 3 | bin = I ( i , j ) / (256/9); // uniform grey levels division or cubic-bezier curve grey levels division for different grey values range. |
| 4 | cell = getcell ( i , j ) |
| 5 | histogram ( cell, bin ) = histogram ( cell , bin ) + magnitude |
| 6 | **end** |
| 7 | **for** all cells c in image I |
| 8 | descriptor ( c ) = createblock ( histogram ) |
| 9 | descriptor ( c ) = integrated ( descriptor ( c ) ) |
| 10 | **end** |
| 11 | **for** all detect i on windows w in image I |
| 12 | finaldesc = gatherHOG ( descriptor , w ) |
| 13 | results = AdaBoost ( finaldesc , model ) |
| 14 | **end** |

**Fig. 2.** Pseudo code for proposed algorithm procedure.

## 3   Texture Analysis for Feature Calculation

Unlike natural images, medical images include rich texture information, such as widely used X-rays and cellular imaging. In image analysis, textures are interpreted as a repetitive arrangement of a basic pattern (hue primitives). Thus the description of a texture includes the colour tones that make up the texture and the relationship between the hues [7]. A texture is a regional feature and thus relates to the size and shape of an area, with the boundary between two texture patterns determined by examining the difference in greyscale pixel values [8]. Finally, a texture feature is a reflection of the object structure, and therefore can provide important image information about the object such as density, which is an essential method of image segmentation, feature extraction, classification and recognition.

In this paper, we propose a low computational approach of only using texture information in greyscale images based on an Neighbourhood Grey-tone Difference Matrix (NGTDM) algorithm.

The NGTDM method is used to depict the pixel values relationship with its surrounding [9]. The calculation equation is shown in Eq. (1).

$$\overline{A}(k, l) = \frac{1}{W^2 - 1} \left[ \sum_{m=-d}^{d} \sum_{n=-d}^{d} f(k + m, l + n) \right]$$

$$where \qquad W = 2d + 1, (m, n) \neq (0, 0) \tag{1}$$

The $W$ is general settings 3 or 5 that means take the $W \times W$ window size of the kernel; where d specifies the neighbourhood size of pixel distance. A $3 \times 3$ ($W = 3$) kernel is used to calculate a pixel A in the graph whose column matrix of array coordinates is $(k, l)$, and the pixel grey value also corresponds to the groups of different levels. For a two-dimension image, there should be 8 pixels around the pixel $\overline{A}$, to sum of the absolute value of the difference between a grey value

(0–255) of the center pixel and the grey value of the surrounding pixels, in order to obtain the sequence $S(i)$ and put the result into the histogram calculation.

Where $f(k, l)$ is the grey value of the image in the window of central pixel $i(k, l)$. $S(k, l)$ represents the sum of the absolute values of the differences between the centre pixel i and the surrounding pixels.

$$S(k, l) = \sum \left| i - \overline{A}(k, l) \right| \qquad for \qquad f(k, l) = i \qquad (2)$$

In this paper, we calculate the histogram through the HOG method after the texture feature is extracted. But the gradient vector feature of HOG is replaced by the texture features to reduce computation complexity. Three different methods are proposed here: texture analysis based on HOG division (TAH), texture analysis based on uniform grey levels division (TAD) and texture analysis based on cubic-bezier curve grey levels division (TAC). As shown in Table 1, the TAH is a texture feature calculated directly by HOG histograms. The TAD is an improved method of replacing the gradient direction of HOG histograms. The TAC is calculated using cubic-bezier curve division to divide the grey bin levels of HOG histogram calculation.

Histogram statistics can be used to count any image features (such as greyscale, gradient and direction). The pixel's greyscale range contains 256 values [0 255], which are divided into sub-regions (called bins). Where n bins are divided, its' statistic pixel calculation result into a matched bin.

Our proposed TAD method classifies greyscale values into ranges without calculating the gradient directions. The greyscale range of 0–255 is divided into nine categories with the boundary ranges rounded down to the nearest whole number. Each divided range details as following:

$$range = bin_1 \cup bin_2 \cup .... \cup bin_n$$
$$(nbins = 9) \qquad (3)$$

$$[0, 255] = [0, 27] \cup [28, 56] \cup [57, 85] \cup [86, 113] \cup [114, 141]$$
$$\cup [142, 170] \cup [171, 198] \cup [199, 226] \cup [227, 255] \qquad (4)$$

The TAC is similar to the TAD method, but it has the non-uniform division greyscale range to 9 bins ($nbins = 9$) by the cubic-bezier curve equation B(t) (Eq. (5)).

$$B(t) = P_0(1 - t)^3 + 3P_1 t(1 - t)^2 + 3P_2 t^2 (1 - t) + P_3 t^3 \qquad t \in [0, 1] \quad (5)$$

P0 is the starting point of the curve (0, 0)
P3 is the end of the curve (255,255)
P1 and P2 are the control points for the trend of the curve, so any changing by these two parameters.
t is a dummy parameter that acts on all connected lines of points.

The cubic-bezier also is called as a third-order Bezier; it is a classic method of curve approximation. In the endoscopic image analysis, the appearance of

**Table 1.** Division calculation complexity comparison table.

| Types | Feature | | Division Calculation Complexity |
|---|---|---|---|
| HOG | Histogram of Oriented Gradient of 3x3 cells (cell similar to kernel) gradient magnitude and gradient direction | The gradient of each pixel, including size and direction, all of the cell of gradients are divided into 9 bins that to get 9 dimensional feature vector. | Twice addition and subtraction + Once division calculation + Once arc-tangent function (mainly time consumption) + Once multiplication calculation |
| TAH | | | |
| TAD | Texture analysis for a 3x3 kernel | Central pixel of value to continuous uniform division for 9 bins | Once multiplication calculation + Once shiftoperation |
| TAC | | Central pixel of value to cubic-bezier curve division for 9 bins | |

cancer, illumination condition changes and the background of the image have relatively small variability. The high and low light regions can be set up as different greyscale divided interval by curve regulation. Such as the highlighted area for low recognition rate, where the curve can be set smooth or steep to enhance the effect. And also, the same applies to low light area settings. There is an advantage than the uniform division that more suitable for the endoscopic environment.

The comparison of various features is shown in the following Table 1. Four methods are based on different classification and feature extraction. The HOG and TAH methods are both for using HOG division method for feature extraction. It uses greyscale pixel values for vector computation to separate the vector space results (vector directional angles) and to calculate gradient magnitude and gradient direction to histogram statistic. We can find on the TAD and TAC division methods that the kernel calculation is a texture analysis to the histogram statistics. It can achieve the greyscale values by calculating the difference values of pixels.

## 4    Experiments Results and Discussions

The experiment uses OpenCV 2.4.13 library version of AdaBoost traincascade program interface and runs on CPU i7 4710MQ for training and testing of the above four methods. The final experiments are implemented on NVIDIA TX2 platform. We choose and set the 5 scale rates for the AdaBoost that are 1.05, 1.1, 1.3, 1.5 and 1.9. In order to eliminate differences from the OpenCV program optimisation for above four methods, we run the experiments using a unified single-threaded processing. We use VIVO Endoscopic Video Datasets [10] to train and test our proposed algorithm.

## 4.1   Quality Results

After getting the statistical texture information, we need to decide the size of the bounding box (Region of Interests). A $K - means$ clustering algorithm is then used for clustering analysis esophageal adenocarcinoma cancer size on the endoscopy image that to sets a suitable detection window size, it is concluded to be 80 (width) × 60 (height), which is to obtain statistically the number of boxes (number of clusters) and the object size (cluster centre box). A simple AdaBoost classifier was online trained. However first experiments results show that the false detecting rate is too high due to the inefficient AdaBoost classifier, as shown in the Fig. 3D.

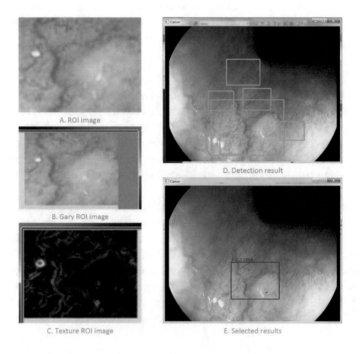

**Fig. 3.** Computer simulation results.

In the second experiment, a cascade AdaBoost classifier uses the sliding window to find the objects in different positions of the image and magnify the detection window to find the different size objects in the image with a threshold banding. The final object box is the highest score of the levelWeight by comparison. The final detection effect is shown on the Fig. 3E.

## 4.2   Quantitative Analysis

Table 2 demonstrates the detection performance including precision and recall. For the precision results, there is many similarities, but the TAC algorithm more stable for each scale rate and it is the top average in the precision. For the recall results, the HOG division method (HOG and TAH) has a significant difference and lower than our design algorithm. The TAC is still top one in the average recall and each scale rate better than 10% with HOG division method. The result of comprehensive speed and recognition rate can meet the requirements of real-time operation for the algorithm after the 1.3 scale rate of AdaBoost. Moreover, in the embedded system test, the TAC method can be achieved in real time processing in the 1.9 scale rate (shown in the TACT results). The TAC method clearly dominates the best effect at this time scale rate.

**Table 2.** Precision and recall comparison table

| Type results | Algorithm | | | | |
|---|---|---|---|---|---|
| | Scale rate | HOG | TAH | TAD | TAC | TACT |
| Precision (%) | 1.05 | 65.4 | 49.7 | 60.5 | 64.3 | 61.5 |
| | 1.1 | 77.1 | 58.4 | 62.7 | 67.2 | 63.4 |
| | 1.3 | 75.1 | 67.6 | 68.7 | 71.8 | 64.1 |
| | 1.5 | 39.1 | 60.6 | 68.2 | 70.9 | 64.5 |
| | 1.9 | 66.4 | 61.2 | 66.8 | 66.6 | 59.5 |
| Avg-Precision | | 64.42 | 59.50 | 65.38 | 68.16 | 62.60 |
| Recall (%) | 1.05 | 50.1 | 37.1 | 59.5 | 61.3 | 59.5 |
| | 1.1 | 55.9 | 48.7 | 62.8 | 63.5 | 62.0 |
| | 1.3 | 59.4 | 52.4 | 66.3 | 65.8 | 60.1 |
| | 1.5 | 39.7 | 45.4 | 60.6 | 63.7 | 56.6 |
| | 1.9 | 44.2 | 37.1 | 53.5 | 52.5 | 47.8 |
| Avg -Recall | | 49.86 | 44.14 | 60.54 | 61.36 | 57.20 |

Figure 4 has compared the average processing speed. All of test videos and pictures are based on 640 × 480 resolution. In the 1.9 scale rate, the grey value division method is 10% faster than the HOG algorithm. Also, the TACT results have shown the embedded systems achieved real time processing that over the 25 fps in the TX2 board.

Figure 5 has shown four algorithms of 9 bins histogram statistics for the same image. By comparison, the TAC method of the histogram is outperformed by using the cubic-bezier curves for non-uniform greyscale range division. So that means each bin in the TAC is including a different greyscale range for the matched non-uniform grey interval and that can improve histogram differences. The TAH and TAD methods of the histogram are smooth and stable which means there is litter different in features.

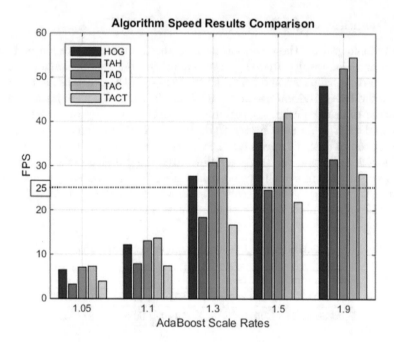

**Fig. 4.** Algorithm fps results.

## 4.3  Discussion

The TAC algorithm achieves the highest recognition rates, particularly when using higher scale rates compared to HOG algorithms. The HOG algorithms produce high levels of accuracy in esophageal cancer detection but have low recall rates (less than 50%). The TAH directly using texture features to match the official HOG calculation method, not only leads to a decrease in accuracy, but the processing speed also drops dramatically and as a result is unable to meet the requirements of the real-time processing. The use of the greyscale division method between the texture analysis and histogram, not only shows a higher accuracy (large-scale rate of AdaBoost setting) than the HOG method, but also a greater speed. In two Greyscale division methods, it has also been shown the best results from a non-uniform division using cubic-bezier curves. It means that in the textural analysis of cancer detection, the non-uniform range of greyscale values is good for histogram statistics. Therefore, it is possible that using the non-uniform intervals of greyscale values calculated by the cubic-bezier curve is better than the HOG algorithm for esophageal cancer detection. In addition, in the embedded system, the processing without vector feature can reduce computation that helps it running fast and easy to achieve real time processing.

However, when using small-scale rates, the HOG method still provides the best accuracy using multi-feature calculation (vector of gradient and angle). This means that the texture analysis method for feature extraction mainly benefits

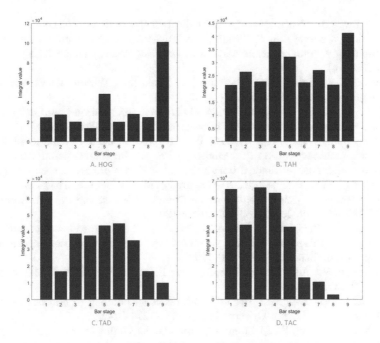

**Fig. 5.** Algorithm histogram results.

from using a simplified calculation to reduce processing time, whilst sacrificing accuracy as a result. Our proposed TAC method has improved accuracy over traditional texture analysis methods, but is still limited by its use of simplified calculations in texture analysis.

## 5   Conclusion

In this paper, we proposed a supervised machine learning method integrated with texture analysis for medical image processing. We compared three different HOG based feature extraction techniques with our proposed non-uniform division method, and verified them in the embedded system platform. The computational cost of the HOG based technologies in supervised learning is high due to the processing needs for vector-based feature extraction. Our method takes advantage of dividable uneven greyscale levels presented in texture features to reduce the computational complexity of machine learning, enabling higher processing speed per second.

## References

1. Sun, L., Subar, A.F., Bosire, C., Dawsey, S.M., Kahle, L.L., Zimmerman, T.P., Abnet, C.C., Heller, R., Graubard, B.I., Cook, M.B.: Dietary flavonoid intake reduces the risk of head and neck but not esophageal or gastric cancer in us men and women. J. Nutr. **147**(9), 1729–1738 (2017)

2. Velusamy, P.D., Karandharaj, P.: Medical image processing schemes for cancer detection: a survey. In: Proceedings of 2014 International Conference on Green Computing Communication and Electrical Engineering (ICGCCEE), pp. 1–6. IEEE (2014)
3. Watson, T.F., Neil, M.A.A., Juškaitis, R., Cook, R.J., Wilson, T.: Video-rate confocal endoscopy. J. Microsc. **207**(1), 37–42 (2002)
4. Qi, J., Le, M., Li, C., Zhou, P.: Global and local information based deep network for skin lesion segmentation. arXiv preprint arXiv: 1703.05467 (2017)
5. Yu, Z., Jiang, X., Wang, T., Lei, B.: Aggregating deep convolutional features for melanoma recognition in dermoscopy images. In: International Workshop on Machine Learning in Medical Imaging, pp. 238–246. Springer, Cham (2017)
6. Dalal, N., Triggs, B.: Histograms of oriented gradients for human detection. In: 2005 IEEE Computer Society Conference on Computer Vision and Pattern Recognition (CVPR 2005), vol. 1, pp. 886–893. IEEE (2005)
7. Bhushan, N., Ravishankar Rao, A., Lohse, G.L.: The texture lexicon: understanding the categorization of visual texture terms and their relationship to texture images. Cogn. Sci. **21**(2), 219–246 (1997)
8. Hawkes, P.W.: Advances in Imaging and Electron Physics, vol. 116. Academic Press, Cambridge (2001)
9. Amadasun, M., King, R.: Textural features corresponding to textural properties. IEEE Trans. Syst. Man Cybern. **19**(5), 1264–1274 (1989)
10. Ye, M., Giannarou, S., Meining, A., Yang, G.-Z.: Online tracking and retargeting with applications to optical biopsy in gastrointestinal endoscopic examinations. Med. Image Anal. **30**, 144–157 (2016)

# Learning from Interaction: An Intelligent Networked-Based Human-Bot and Bot-Bot Chatbot System

Jordan J. Bird, Anikó Ekárt, and Diego R. Faria[✉]

Aston Lab for Intelligent Collectives Engineering (ALICE), School of
Engineering and Applied Science, Aston University, Birmingham B4 7ET, UK
{birdjl, a.ekart, d.faria}@aston.ac.uk

**Abstract.** In this paper we propose an approach to a chatbot software that is able to learn from interaction via text messaging between human-bot and bot-bot. The bot listens to a user and decides whether or not it knows how to reply to the message accurately based on current knowledge, otherwise it will set about to learn a meaningful response to the message through pattern matching based on its previous experience. Similar methods are used to detect offensive messages, and are proved to be effective at overcoming the issues that other chatbots have experienced in the open domain. A philosophy of giving preference to too much censorship rather than too little is employed given the failure of Microsoft Tay. In this work, a layered approach is devised to conduct each process, and leave the architecture open to improvement with more advanced methods in the future. Preliminary results show an improvement over time in which the bot learns more responses. A novel approach of message simplification is added to the bot's architecture, the results suggest that the algorithm has a substantial improvement on the bot's conversational performance at a factor of three.

**Keywords:** Artificial Intelligence · Natural language processing
Chatbot

## 1 Introduction

Both Artificial Intelligence (AI) researchers and industry are increasingly recognising the importance of chatbot systems. Chatbots are embedded in everyday life, performing various roles serving as assistants for end-users. Examples include consumer interaction geared towards problem-solving and addressing customer requests in a variety of industries [1] and even performing as a virtual home-assistant in home-automation duties within a modern household [2].

In this work, a Chatbot system learns through experience to respond to certain messages based on previous interactions, and gains more knowledge over time based on its interaction with users.

The main contributions of this work are three-fold:

1. A modular web-based chatbot system useful for different applications and with the potential in industry is presented.

A. Lotfi et al. (Eds.): UKCI 2018, AISC 840, pp. 179–190, 2019.
https://doi.org/10.1007/978-3-319-97982-3_15

2. Natural Language Processing (NLP) and Sentiment Analysis are applied for offensive vocabulary detection to prevent the current issues faced by chatbots in the open domain [12].
3. A novel approach is introduced for a message simplification layer in which colloquialisms and identical synonyms are mitigated by simplifying them into marking flags, and building responses in real-time.

This paper will first explore the related works (Sect. 2), highlighting milestones as well as open issues currently existing in the field. The proposed approach (Sect. 3) will then be detailed, followed by a preliminary set of results (Sect. 4). The paper then concludes with a discussion of results, future work and impact (Sect. 5).

## 2  Related Work

Sentiment analysis (a.k.a. opinion mining) identifies people's opinions, sentiments, emotions, appraisals and attitudes towards entities such as services and products, organizations and/or individuals, events and their attributes. Chatbot systems are tools that employ natural language processing and sentiment analysis for multiple purposes, examples include human-machine communication applied to business, health, and education. The objective measurement of the users experience and emotional states is a key factor to understand the level of satisfaction or mood given a certain context. Tackling this problem is very challenging, since the measurements of satisfaction levels or internal states might be too ambiguous, but new emerging technologies combined with computational intelligence can provide a new perspective for approaching the problem in a more effective manner. There are many channels to detect and perform sentiment analysis such as affection through facial expressions [17, 18], voice [19], and also through text communication via natural language processing [4]. Body gesture can also be an interesting alternative for user behaviour or emotional expression understanding [20]. This work falls into the category of text communication between two agents (human-bot or bot-bot) using natural language processing and sentiment analysis via an intelligent web-based chatbot system. Related works are presented below to show the progress of chatbot systems in different domains, their challenges and open issues in this field.

Weizenbaum's ELIZA [22] system is often considered as the first ever chatbot system, which quickly became famous in the 1960s due to people engaging in conversations with it beyond the expectations of its creator.

Shawar and Atwell's exploration of using dialogue corpora to generate chatbots [3] found that a non-bot human experience (such as the script of two human beings conversing) was more than capable of forming the basis for a chatbot. This therefore suggests that rather than starting with no knowledge whatsoever, a dataset of conversations could be provided to give the chatbot a useful head start in learning from interaction. A response format would need to be developed to organise this data into.

In 2004, the effectiveness of chatbots deployed to teach students foreign languages was demonstrated [4]. The effectiveness of a robotic agent became clear when it was noted that the students were more open to admit fault to the bot rather than a person, showing that students were far less intimidated by a feeling of inadequacy during the learning process. This research is very promising, as it shows an application of a chatbot in education which would not only aid the education system, but further improve it.

Freudbot is an Artificial Intelligence Markup Language (AIML) based chatbot, an XML based approach to message pattern and response pairings. For experimental purposes it was designed to speak to students to aid with distance learning when staff were not available out of hours [5]. Heller et al. found that the experience was itself neutral, neither good nor bad, but explained that with more responses it would have been expected to be better. This suggests that more responses create a better experience for the user, and so learning from experience could constantly evolve a chatbot to become better with each and every interaction.

Tatai et al. ran tests of chatbots who preferred using different sentiments to see how a user would react and recorded how they found the experience [6]. Everyone, regardless of input sentiment, found a better experience when the chatbot response was positive. This suggests that a learning bot should respond to negative messages, but not risk reiterating them in the future, which would result in only positive responses, regardless of input. In terms of impact, this means the bot must analyse the sentiment of the messages in order to make use of this important work and as such allow the implementation of a chatbot that will not learn from negative messages.

The Loebner prize [7] is awarded to a chatbot which can fool a human into thinking they are conversing with another human. This is based on the "Imitation game" proposed by Turing in 1950 [21]. Shawar and Atwell [8] found that case-by-case training was required depending on localisation and audience. They also suggest that this problem could be overcome by having a more massive dataset to draw from. This suggests that larger datasets would improve the chatbot, but to contextualise it to an environment, the level of training required is dynamic and depends on its uses.

ALICE is an XML based chatbot [9] that makes use of pattern matching between user inputs as well as a hidden person to generate responses to a user's message. The method of pattern matching to a stored set of messages was an effective enough method for the bot to have won the Loebner Prize three times [10] suggesting that the act of learning from interaction is an effective method of creating record-breaking chatbots, in terms of competitions that were won.

Tay, an AI chatbot that learns from previous interaction [11] caused major controversy due to it being targeted by internet trolls on Twitter [12]. The bot was exploited, and after 16 h began to send extremely offensive Tweets to users. This suggests that although the bot learnt effectively from experience, adequate protection was not put in place to prevent misuse. This implies two important influences - firstly, learning from interaction is reinforced as an effective technique. Secondly, the bot must have protection in place to prevent such an event from occuring.

The Google Cloud NLP API [13] is a library that can process a message and calculate sentiment and magnitude thereof. The API, once authorised, will receive a message as raw text, and then produce a response array containing pieces of sentiment information gathered from the input, therefore enabling us to be able to measure the sentiment of a message and deem whether or not it is positive or negative. This knowledge would allow for selective response elicitation based on positivity.

Despite the efforts spent on these extensive and high quality works in the area of scientific philosophy of chatbots, most, except for [4], are yet to produce systems capable of specifically long-term deployment into a real-world situation, rather concentrating on the learning theory behind their conception. This suggests that there is room for such systems in industry, making use of aforementioned techniques already well documented and published in respectable Machine Learning journals. To conclude, the level of current scientific knowledge is seemingly ahead of that employed in the real-world, and furthermore is not yet in a state to be used in real-world applications.

## 3   Proposed Approach

For user interaction with the bot, the user will input a message into an arbitrary front-end application. The bot makes use of an input/output structure of web-requests and therefore any application that accommodates this can be used. The system outputs its response in the form of a simple key-value JSON Array. To show the bot's potential, two front-end applications were developed (see Fig. 1), a browser-based application using asynchronous JavaScript to populate a chat window with results, and an automated Twitter bot that would perform the same process when replying to Tweets.

**Fig. 1.**  HTML and UI bot interfaces.

**System Architecture.** In terms of the bot itself, a black box system is developed (see Fig. 2) seen in steps 1 to 4.1.1, the message from the user is processed and results are generated accordingly. Firstly, to avoid the aforementioned Microsoft Tay disaster, a pattern-matching algorithm is executed, comparing the input message to a library of all of the known offensive words in the English Language [16], together with a list of Political terms. If the bot were to detect any of these terms, a response will not be generated.

Secondly, the Google Cloud Sentiment Analysis API is requested to analyse the sentiment of the input message. This allows the bot to detect whether the message is considered positive, negative, or neutral (without emotion). Furthermore, following the work performed into user experience with emotional chatbots [6], the bot will only store the unknown message if, and only if, it is not considered of negative sentiment.

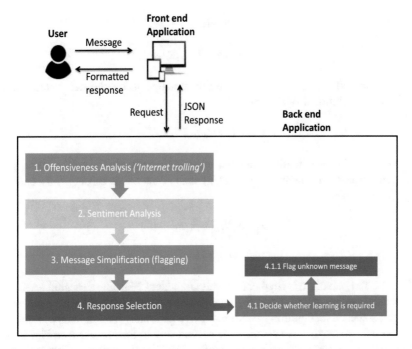

**Fig. 2.** Proposed architecture overview and flow of system process.

This in effect will prevent the bot from having the ability to produce a negative response to a message regardless of the input sentiment. In addition, sentiment analysis is also performed (See Fig. 2) during step 2 in real-time after a response has been generated, to mitigate a slight change in said message's sentiment after the simplification and flagging is reversed, effectively preventing any incorrect measurements or representations.

**LSTM Overview.** The Google Sentiment Analysis tool [13] is trained using Long-Short-Term-Memory (LSTM), where multiple recurrent neural networks (RNN) predict an output based on their input and their current state [14]. The general idea is as follows.

Firstly, a logical forget gate at the current timestep *ft* will decide which information to discard and delete: *Wf* represents the learning-weighting matrix, *h* represents the output vector of the unit at provided timestep *t–1, xt* being the input vector at defined timestep *t*, and finally *bf* is a bias vector applied to the process.

$$f_t = \sigma\left(W_f.[h_{t-1}, x_t] + b_f\right) \tag{1}$$

Secondly, the cell must decide on which information to store. The variable *i* represents the input data being received by the cell, and does so through a logical input gate. $C_t$ being the vector of the new values generated by the process.

$$i_t = \sigma(W_i.[h_{t-1}, x_t] + b_i) \tag{2}$$

$$\tilde{C}_t = \tanh(W_c.[h_{t-1}, x_t] + b_c) \tag{3}$$

Thus, the cell's parameters are therefore updated using the calculated variables (1–3) in a convolutional operation:

$$C_t = f_t * C_{t-1} + i_t * \tilde{C}_t \tag{4}$$

An output is consequently generated where $O_t$ represents the cell's output gate at the current timestep, $t$. The internal (hidden) state of the cell is subsequently updated to match its new value:

$$o_t = \sigma(W_o[h_{t-1}, x_t] + b_o) \tag{5}$$

$$h_t = o_t * \tanh(C_t) \tag{6}$$

This LSTM paradigm is used extensively in modern Machine Learning applications due to its ongoing record breaking effectiveness, one of which was most notably Microsoft's speech recognition system being at a genuinely human level of complexity [15]. A user's message was passed at the second step in the process (see Fig. 2) to the API and its responses were used accordingly. The Google Sentiment Analysis toolkit made use of the LSTM by deriving the sentiment score on a scale of −1.0 to 1.0 (most negative, through neutral, to most positive) as well as its magnitude (the strength of the derived sentiment score regardless of value). These values are used in both input to adhere to the findings of Tatai et al. [6] in sentiment impact of user experience, as well as provided by the system to the front-end application to be used in a platform-applicable approach.

**Proposed Simplification Step.** A novel approach of message simplification is employed in (see Fig. 2) step 3. Messages that contain colloquialisms or otherwise identical synonyms that do not change the meaning of a message, although greatly impacting the structure, will be simplified by replacing said terms with a flag.

Possibilities of message s where $x \in X$ are the set of all known flag-phrase pairs are given as the product of all possibilities of said phrase within the message.

$$P(s) = \prod_{x=1}^{X} P(x). \tag{7}$$

An iterative process takes into account the stored steps of flag-phrase relationships and replaces the phrases based on their flag parent. Possibilities, therefore, can be calculated via the product of the set of phrase siblings within flags existent in the message $s$.

The reason for the introduction of this simplification step is that the usage of this layer would expand the learning capabilities by overcoming the differentiation of spoken sentences in terms of societal colloquialisms and synonyms with identical meaning. This is done by denoting them with flags rather than retaining the original string data. These are arbitrarily stored sets of strings that have been created manually where an index simply defines either a flag or a phrase e.g. {"[GREETING]", "Hello", "Hi"} where at index 0 a flag exists, in this case "[GREETING]". All values at index $i > 0$ therefore, are children of said flag.

For example, if the bot were to know ten phrases for 'hello', and five for 'happy' the sentence 'Hello! I am happy!' would be simplified as '[GREETING]! I am [ADJ_HAPPY]!' which would in turn have 50 possible combinations. Without this method, fifty messages would have been learned individually, whereas with its implementation, only one exchange is needed to learn responses to all messages.

Figure 3 details the usage of lingual simplification in terms of two messages that, other than a slight differentiation of sentiment (through wording), have identical meanings. Without the layer active, the two messages must be learnt from individually. With the lingual simplification active, the experiences gathered from one of the message are applied to the other due to their identical meaning.

**Fig. 3.** Previous interaction comparison between lingual simplification algorithm active and inactive.

**Response Selection.** (See Fig. 2) Step 4 is comprised of three sub-layers. Firstly, pattern matching is performed against the bot's stored dataset of message-response pairs. Thresholds are preliminarily defined as 60% and 90% (see Fig. 4), the former defining an accurate response that needs to be learnt from, and the latter defining an accurate response that does not require further learning. Results below either values will flag the message for further learning, whereas a result below 60% will have the bot change conversation due to the response being too inaccurate to give.

| 0-59% | 60-89% | 90%+ |
|---|---|---|
| The response is not accurate enough, change conversation and do not give the response. Add the message to the 'unknowns' memory. | The response is considered accurate enough to give. Iterate the response, but also add it to the 'unknowns' memory | Give the response, no need to learn further |
| ONLY IF sentiment is neutral to positive | | |

**Fig. 4.** Bot reactions to detected percentage similarity.

Learning is only performed on the non-negative messages due to the findings of user chatbot experience in terms of response sentiment [6].

Testing will be performed by having three individuals having a twenty-message conversation with the bot (10 user messages and 10 bot responses), and will be automated by recreating them on the system with and without the simplification algorithm active. Conversational dataset will persist throughout the three conversations. This method of testing is followed so two identical conversations can be compared on two identical chatbot states.

**Learning Methodology.** The learning process of the bot is proposed as follows. If a bot were not to understand a message input, it would change conversation. The conversation would be changed to a message that the bot has previously not understood (from another user) and the user's response would be stored as a potential candidate for a message-response pair, as well as their measured sentiment (see Fig. 4). This calculation is simply based on a pattern similarity between the two messages (the input, and the closest previously seen message) which is logically relevant as the simplification layer has been executed, meaning the simplified flags have been taken into account in said pattern match (See Fig. 3).

## 4  Preliminary Results

Testing was performed by observing three individuals conversing with the bot for a period of 20 messages; conversations were repeated on identical datasets both with the simplification layer turned off, and on. The environment selected was a web-based interface that asynchronously requested responses from the bot server. A small general conversational dataset of 200 message-response pairs was deployed to give the bot a starting knowledge. The message dataset was produced by having simple, general conversations with the bot.

The 200 messages were pre-processed via the layer for the simplification testing which began at 478 combinations – "known responses" referring to both the stored responses and their combinations due to a combinational message being treated as a message (see proposed approach). Thus, the illusion of a larger dataset was produced.

Table 1 details the bot's Known Response Increase (KRI) over the course of three conversations. Without the novel method of message simplification, a total of 27 new responses were learnt and on average, a response was deemed accurate enough 58.3% of the time. On the other hand, with the simplification layer activated, the exact same conversations resulted in 82 new responses learnt and on average created an accurate message response 68.3% of the time. The calculation of percentage success is simply the ratio of the number of messages the bot replied to the total number of messages, i.e. contrasting those that were replied to as opposed to the user messages that were not understood. This is seen further in Figs. 5 and 6, in which success is indicated as binary result of 0/100 (could not reply/did reply).

Figures 5 and 6 give a graphical representation of the three sequential conversations over time, where X is the number of the message of the total 1–60 messages. Success (%) shows the percentage success in terms of accuracy of the response where 0

**Table 1.** Conversation success with and without Message Simplification.

| Conversation | KRI | Success [%] | KRI | Success [%] |
|---|---|---|---|---|
| | No simplification | | With simplification | |
| 1 | 8 | 65 | 21 | 80 |
| 2 | 8 | 60 | 28 | 70 |
| 3 | 11 | 50 | 33 | 55 |
| Overall | 27 | 58.3 Avg. | 82 | 68.3 (Avg.) |

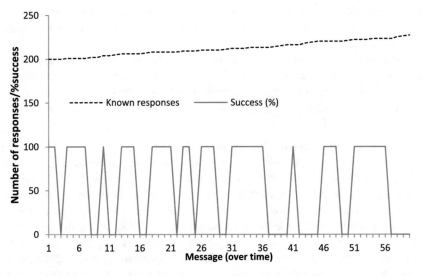

**Fig. 5.** Performance through three iterative twenty-message conversations (no simplification).

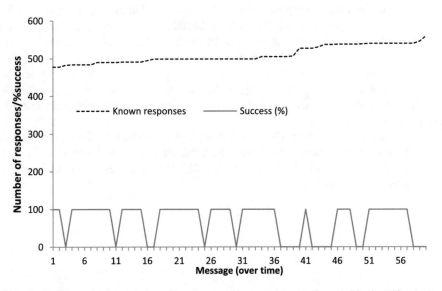

**Fig. 6.** Performance through three iterative ten-message conversations (with simplification).

was deemed inaccurate and 100% was a reply from the bot. Known responses shows the learning process over time through the increasing number of known responses the bot has to input messages.

## 5   Conclusion

The learning ability is clearly indicated by the preliminary results. The number of known message-response pairs grows over time, with experience. Message success does not seem to improve, but further extensive research is needed to explore the relationships between known responses and conversational success. The experiment only covered sixty messages (three twenty message pairs), but many more would be needed to explore said relationship.

Furthermore, conversational re-creation during testing shows the massive improvement when the message simplification layer is employed to mitigate colloquialisms, and even though this system tends to get fewer opportunities to learn (due to its higher success rate), it will make use of these opportunities far better than the system without will ever do.

**Future Work and Impact.** The decrease of message success gives the illogical impression of the bot performing worse, as it accumulates more knowledge. This is likely due to the user's differing conversational subjects, due to the fact that more knowledge quite literally correlates to more responses in this system. More extensive conversations with the bot must be obtained to gain a more accurate figure of success over time. Pattern matching learning thresholds (60%, 90%) during the learning process were set arbitrarily at levels that made logical sense, as they performed as expected during the testing stage (see Table 1). Further improvement to an accuracy-based learning system in terms of string patterns can be expected by setting more effective threshold values.

Jia found that chatbots had the potential to not only aid in the education system, but also effectively improve it [4]. The further introduction of selective learning from experts, and through this, the formation of an expert system could ultimately lead to a conversational aid in required situations. For example, a teaching assistant may answer many uniquely-phrased questions with logically identical, or close-to, answers. An expert chatbot born from the knowledge of said teaching assistants could, for example, lead to a more efficient University Laboratory paradigm in which students are aided by both human and machine.

Following this vein of thought, a system deployed *post-learning* from many experts in the field of psychotherapy could introduce the usage of knowledge re-application in situations such as mental health counseling. This is a strong social impact that, as of yet, has not been achieved.

# References

1. Kuligowska, K.: Commercial chatbot: performance evaluation, usability metrics and quality standards of embodied conversational agents (2015)
2. Alexa Amazon: "Amazon" (2014)
3. Shawar, B.A., Atwell, E.: Machine learning from dialogue corpora to generate chatbots. Expert Update J. **6**(3), 25–29 (2003)
4. Jia, J.: The study of the application of a web-based chatbot system on the teaching of foreign languages. In: Society for Information Technology & Teacher Education International Conference. Association for the Advancement of Computing in Education (AACE), pp. 1201–1207 (2004)
5. Heller, B., Proctor, M., Mah, D., Jewell, L., Cheung, B.: Freudbot: an investigation of chatbot technology in distance education. In: EdMedia: World Conference on Educational Media and Technology, pp. 3913–3918. Association for the Advancement of Computing in Education (AACE) (2005)
6. Tatai, G., Csordás, A., Kiss, Á., Szaló, A., Laufer, L.: Happy chatbot, happy user. In: Intelligent Virtual Agents, pp. 5–12. Springer, Heidelberg (2003)
7. Mauldin, M.L.: Chatterbots, tinymuds, and the turing test: entering the loebner prize competition. In: AAAI, vol. 94, pp. 16–21 (1994)
8. Shawar, B.A., Atwell, E.: Different measurements metrics to evaluate a chatbot system. In: Proceedings of the Workshop on Bridging the Gap: Academic and Industrial Research in Dialog Technologies, pp. 89–96. Association for Computational Linguistics (2007)
9. Wallace, R.: Artificial linguistic internet computer entity (ALICE) (2001). https://www.chatbots.org/chatbot/a.l.i.c.e/. Accessed 25 May 2018
10. The Exeter Blog: The Loebner Prize, a Turing Test competition at Bletchley Park (2014). https://blogs.exeter.ac.uk/exeterblog/blog/2014/12/08/the-loebner-prize-a-turing-test-competition-at-bletchley-park/
11. Microsoft (March). Tay AI. https://twitter.com/tayandyou. Accessed 25 May 2018
12. Wakefield, J.: BBC News. Microsoft chatbot is taught to swear on Twitter (2016). http://www.bbc.co.uk/news/technology-35890188. Accessed 12 Apr 2018
13. Google Cloud Products (n.d.), 28 March 2018. https://cloud.google.com. Accessed 25 May 2018
14. Hochreiter, S., Schmidhuber, J.: Long short-term memory. Neural Comput. **9**(8), 1735–1780 (1997)
15. Haridy, R.: Microsoft's speech recognition system is now as good as a human (2017). https://newatlas.com/microsoft-speech-recognition-equals-humans/50999/. Accessed 6 Apr 2018
16. AllSlang. (n.d.). Swear Word List, Dictionary, Filter, and API. https://www.noswearing.com/. Accessed 11 Mar 2018
17. Faria, D.R., Vieira, M., Faria, F.C.C., Premebida, C.: Affective facial expressions recognition for human-robot interaction. In: IEEE International Symposium on Robot and Human Interactive Communication, pp. 805–810 (2017)
18. Faria, D.R., Vieira, M., Faria, F.C.C.: Towards the development of affective facial expression recognition for human-robot interaction. In: International Conference on Pervasive Technologies Related to Assistive Environments, pp. 300–304 (2017)
19. Bertero, D., Siddique, F., Wu, C., Wan, Y., Chan, R., Fung, P.: Real-time speech emotion and sentiment recognition for interactive dialogue systems. In: Conference on Empirical Methods in Natural Language Processing, pp. 1042–1047 (2016)

20. Vieira, M., Faria, D.R., Nunes, U.: Real-time application for monitoring human daily activities and risk situations in robot-assisted living. In: 2nd Iberian Robotics Conference, pp. 449–461 (2015)
21. Turing, A.M.: Computing machinery and intelligence. Mind **49**, 433–460 (1950)
22. Weizenbaum, J.: Computer Power and Human Reason: From Judgment to Calculation. W.H. Freeman and Company, New York (1976)

# A Study on CNN Transfer Learning for Image Classification

Mahbub Hussain, Jordan J. Bird, and Diego R. Faria$^{(\boxtimes)}$

School of Engineering and Applied Science, Aston University,
Birmingham B4 7ET, UK
{hussam42, birdj1, d. faria}@aston. ac. uk

**Abstract.** Many image classification models have been introduced to help tackle the foremost issue of recognition accuracy. Image classification is one of the core problems in Computer Vision field with a large variety of practical applications. Examples include: object recognition for robotic manipulation, pedestrian or obstacle detection for autonomous vehicles, among others. A lot of attention has been associated with Machine Learning, specifically neural networks such as the Convolutional Neural Network (CNN) winning image classification competitions. This work proposes the study and investigation of such a CNN architecture model (i.e. Inception-v3) to establish whether it would work best in terms of accuracy and efficiency with new image datasets via Transfer Learning. The retrained model is evaluated, and the results are compared to some state-of-the-art approaches.

## 1 Introduction

In recent years, the field of Machine Learning has made tremendous progress in different domains where autonomous systems are needed. Thus, allowing to advance models such as a Deep Convolutional Neural Networks to achieve impressive performance on hard visual recognition tasks, matching or exceeding human performance in some domains [1]. The work, "Going Deeper with Convolutions" [2] introduces the Inception-v1 architecture, which was well succeeded in the ILSVRC 2014 GoogleNet challenge. The main contribution presented by the authors is the application to the deeper nets required for image classification. The authors observed that some sparsity would be beneficial to the network's performance, and thus it was applied using today's computing techniques. The authors also introduced additional losses to help improve convergence on the relatively deep network. The limitations noted was a training trick, which also resulted in the output of the layers being discarded during inference. The authors in [3] propose a novel deep network structure to help the enhancement of a model that distinguishes between patches in the receptive field of a convolutional layer within a CNN by instantiating a micro network using multilayer perceptron's instead of linear filters and non-linear activation functions to abstract the data. Similarly to a common CNN, micro networks stride over input images to produce a feature map, these types of layers can then be stacked resulting in a deep "network in network". The results presented within that paper found that the proposed network was less prone to overfitting than traditional fully connected layers due to a global average pooling over

© Springer Nature Switzerland AG 2019
A. Lotfi et al. (Eds.): UKCI 2018, AISC 840, pp. 191–202, 2019.
https://doi.org/10.1007/978-3-319-97982-3_16

the feature maps in the classification layer. Tests were conducted on numerous datasets, one of which was CIFAR-10 [10]. Results presented focused on test error rates regarding network in network with combination of a drop-out and data augmentation, which achieved the best scores. Their limitation include the usage of multiple layers and combinations, and this dissents the availability of being able to identify and compare singular layers and their effect on performance. The book "Computational collective intelligence" [4], covers various aspects of Machine Learning. The author tests 3 different architectures: a simple network trained on the CIFAR-10 dataset, a CNN trained on the MNIST dataset as well as a CNN trained on the CIFAR-10 dataset. The authors explain the testing procedures undertaken in detail, in terms of number of convolutional layers, max pooling layers and ReLU layers used. The author also details tests evaluating the performance of each architecture in combination with local binary patterns (LBP). The simple network achieved a test accuracy score of 31.54%, with the CNN trained on the CIFAR-10 dataset managing to achieve a higher score of 38.8% after 2805 s of training. Most of the aforementioned papers identified limitations whether it be cost, insufficient requirements or problems with the processing of complex datasets, or quality of images. Rather than replicating these studies, the aim of this work is to progress on previous related works and solve the addressed limitations. The results obtained by the authors in [4], will be used as a direct comparison. We will use a similar architecture and dataset compared to [4], however combining Transfer Learning on top of a pre-trained model on other datasets (domains). The aim is to establish whether accuracy results could be improved with Transfer Learning given minimal time and computational resources.

Therefore, the motivation of this work encompasses finding a suitable model in which facilitates Transfer Learning, allowing classification of new datasets with respectable accuracy. The main contributions of this work are as follows: a study on the core principals behind CNNs related to a series of tests to determine the usability of such as technique (i.e. Transfer Learning) and whether it could be applied to multiple datasets with different images category. Also, acknowledging the measures taken to adapt a network to advance its integration within diverse domains. Thus, we test a CNN architecture (i.e. Inception-v3) on both the Caltech [11] Face dataset and the CIFAR-10 dataset [10] whilst changing certain parameters to evaluate their significance with regards to the classification accuracy results.

The remaining structure of this paper is as follows. The approach adopted in this work is explained in Sect. 2, with the experimental setup and datasets described in Sect. 3, followed by a preliminary set of results explained in Sect. 4. This paper is concluded and then future works based on the findings is proposed in Sect. 5.

## 2   CNN Transfer Learning Development

### 2.1   CNN

Convolutional Neural Networks (CNN) have completely dominated the machine vision space in recent years. A CNN consists of an input layer, output layer, as well as multiple hidden layers. The hidden layers of a CNN typically consist of convolutional layers, pooling layers, fully connected layers and normalisation layers (ReLU). Additional layers can be used for more complex models. Examples of a typical CNN can be seen in [5] and it is depicted in Fig. 1.

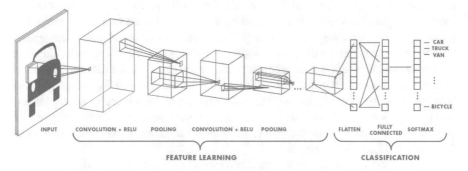

**Fig. 1.** Typical CNN architecture [14].

The CNN architecture has shown excellent performance in many Computer Vision and Machine Learning problems. CNN trains and predicts in an abstract level, with the details left out for later sections. This CNN model is used extensively in modern Machine Learning applications due to its ongoing record breaking effectiveness. Linear algebra is the basis for how these CNNs work. Matrix vector multiplication is at the heart of how data and weights are represented [12]. Each of the layers contains a different set of characteristics for an image set. For instance, if a face image is the input into a CNN, the network will learn some basic characteristics such as edges, bright spots, dark spots, shapes etc., in its initial layers. The next set of layers will consist of shapes and objects relating to the image which are recognisable such as: eyes, nose and mouth. The subsequent layer consists of aspects that look like actual faces, in other words, shapes and objects which the network can use to define a human face. CNN matches parts rather than the whole image, therefore breaking the image classification process down into smaller parts (features). A $3 \times 3$ grid is defined to represent the features extraction by the CNN for evaluation. The following process, known as filtering, involves lining the feature with the image patch. One-by-one, each pixel is multiplied by the corresponding feature pixel, and once completed, all the values are summed and divided by the total number of pixels in the feature space. The final value for the feature is then placed into the feature patch. This process is repeated for the remaining feature patches followed by trying every possible match- repeated application of this filter, which is known as a convolution.

The next layer of a CNN is referred to as "max pooling", which involves shrinking the image stack. In order to pool an image, the window size must be defined (e.g. usually $2\times2/3\times3$ pixels), the stride must also be defined (e.g. usually 2 pixels). The window is then filtered across the image in strides, with the max value being recorded for each window. Max pooling reduces the dimensionality of each feature map whilst retaining the most important information. The normalisation layer of a CNN, also referred to as the process of Rectified Linear Unit (ReLU), involves changing all negative values within the filtered image to 0. This step is then repeated on all the filtered images, the ReLU layer increases the non-linear properties of the model. The subsequent step by the CNN is to stack the layers (convolution, pooling, ReLU), so that the output of one layer becomes the input of the next. Layers can be repeated resulting in a "deep stacking". The final layer within the CNN architecture is called the fully connected layer also known as the classifier. Within this layer every value gets a vote on determining the image classification. Fully connected layers are often stacked together, with each intermediate layer voting on phantom "hidden" categories. In effect, each additional layer allows the network to learn even more sophisticated combinations of features towards better decision making [6]. The values used for the convolution layer as well as the weights for the fully connected layers are obtained through backpropagation, which is done by the deep neural network. Backpropagation is whereby the neural network uses the error in the final answer to determine how much the network adjusts and changes.

The Inception-v3 model is an architecture of convolutional networks. It is one of the most accurate models in its field for image classification, achieving 3.46% in terms of "top-5 error rate" having been trained on the ImageNet dataset [7]. Originally created by the Google Brain team, this model has been used for different tasks such as object detection as well as other domains through Transfer Learning.

The CNN learning process can rely on vector calculus and chain rule. Let $z$ be a scalar (i.e., $z \in$ R) and $y \in \mathbf{R}^H$ be a vector. So, if $z$ is a function of $y$, then the partial derivative of $z$ with respect to $y$ is a vector, defined as:

$$\left( \frac{\partial z}{\partial y} \right)_i = \frac{\partial z}{\partial y_i}. \tag{1}$$

Specifically, $\left( \frac{\partial z}{\partial y} \right)$ is a vector having the same size as $y$, and its $i$-th element is $\left( \frac{\partial z}{\partial y} \right)_i$. Also note that $\left( \frac{\partial z}{\partial y^T} \right) = \left( \frac{\partial z}{\partial y} \right)^T$. Furthermore, presume $x \in$ R$^W$ is another vector, and $y$ is a function of $x$. Then, the partial derivative of $y$ with respect to $x$ is defined as:

$$\left( \frac{\partial y}{\partial x^T} \right)_{ij} = \frac{\partial y_i}{\partial x_i}. \tag{2}$$

This fractional derivative is a H $\times$ W matrix, whose entry at the juncture of the $i$-th row and $j$-th column is $\frac{\partial y_i}{\partial x_i}$. It is easy to see that $z$ is a function of $x$ in a chain-like argument: a function maps $x$ to $y$, and another function maps $y$ to $z$. A chain rule can be used to compute:

$$\left( \frac{\partial z}{\partial x^T} \right), \text{ as } \left( \frac{\partial z}{\partial x^T} \right) = \left( \frac{\partial z}{\partial y^T} \right) \left( \frac{\partial y}{\partial x^T} \right). \tag{3}$$

One can use a cost or loss function to measure the discrepancy between the prediction of a CNN $x^L$ and the target $t$, $x^1 \rightarrow w^1, x^2 \rightarrow \ldots, x^L \rightarrow w^L = z$, using a simplistic loss function $z = \|t - x^L\|^2$. However more complex functions are usually employed. A prediction output can be seen as $\text{argmax}_i \, x_i^L$. The convolution procedure can be expressed as:

$$y_{i^{l+1}, j^{l+1}, d} = \sum_{i=0}^{H} \sum_{j=0}^{W} \sum_{d=0}^{D} f_{i,j,d} \times x_{i^{l+1} + i, j^{l+1} + j, d}^{L}. \tag{4}$$

The filter $f$ has size $(H \times W \times D^l)$, thus the convolution will have the spatial size of $(H^l - H + 1) \times (W^l - W + 1)$ with $D$ slices, which means $y(x^{l+1})$ in $\mathrm{R}^{H^{l+1} \times W^{l+1} \times D^{l+1}}$, $H^{l+1} = H^l - H + 1$, $W^{l+1} = W^l - W + 1$, $D^{l+1} = D$.

When it comes to Inception V3, the probability of each label $k \in \{1, \ldots K\}$ for each training example is computed by $P(k|x) = \frac{\exp(z_k)}{\sum_i^K \exp(z_i)}$, where $z$ is a non-normalised log probability. Ground truth distribution over labels $q(k|x)$ is normalised, so that $\sum_k q(k|x) = 1$. For this model, the loss is given by cross-entropy:

$$\ell = \sum_{k=1}^{K} \log(p(k))q(k). \tag{5}$$

Cross-entropy loss is differentiable with respect to the logits $z_k$ and thus it can be used for gradient training of deep models, where the gradients has the simple form $\frac{\partial \ell}{\partial z_k} = p(k) - q(k)$, bounded between $-1$ and $1$. Usually, when minimising the cross entropy, it means that log-likelihood of the correct label is maximised. Since it may cause some overfitting problems, Inception V3 considers a distribution over labels independent of training examples $u(k)$ with a smooth parameter $\epsilon$, where for a training example, the label distribution $q(k|x) = \delta_{k,y}$ is simply replaced by:

$$q'^{(k|x)} = (1 - \epsilon)\delta_{k,x} + \epsilon u(k), \tag{6}$$

which is a mixture of the original distribution $q(k|x)$ with weights $1-\epsilon$ and the fixed distribution $u(k)$ with weights $\epsilon$. A label-smoothing regularisation is applied, with uniform distribution $u(k) = 1/K$, so that it becomes:

$$q'^{(k|x)} = (1 - \epsilon)\delta_{k,x} + \frac{\epsilon}{K}. \tag{7}$$

Alternatively, this can be interpreted as cross-entropy as follows:

$$H(q', p) = -\sum_{k=1}^{K} \log(p(k))q'^{(k)} = (1 - \epsilon)H(q', p) + \epsilon H(u, p). \tag{8}$$

Therefore, the label-smoothing regularisation is similar to applying a single cross-entropy loss $H(q,p)$ with a pair of losses $H(q,p)$ and $H(u,p)$, with the second loss penalising the deviation of the predicted label distribution $p$ from the prior $u$ with relative weight $\frac{\epsilon}{(1-\epsilon)}$, which is equivalent to computing the Kullback–Leibler divergence. More details (step-by-step) and the mathematical formulation of CNN and Inception V3 can be found in [12, 13].

In this work we aim to retrain this model on a new dataset and study the results. Choosing not to train the model from scratch as it would be computationally intensive task and depending on the computing setup, may take several days or even weeks. In addition, it would also require multiple GPUs and/or multiple machines. Instead, we will be comparing results obtained from the proposed retrained model to that of related works to prove our hypothesis.

## 2.2  Transfer Learning

Transfer Learning is a Machine Learning technique whereby a model is trained and developed for one task and is then re-used on a second related task. It refers to the situation whereby what has been learnt in one setting is exploited to improve optimisation in another setting [8]. Transfer Learning is usually applied when there is a new dataset smaller than the original dataset used to train the pre-trained model [9].

This paper proposes a system which uses a model (Inception-v3) in which was first trained on a base dataset (ImageNet), and is now being repurposed to learn features (or transfer them), to be trained on a new dataset (CIFAR-10 and Caltech Faces). With regards to the initial training, Transfer Learning allows us to start with the learned features on the ImageNet dataset and adjust these features and perhaps the structure of the model to suit the new dataset/task instead of starting the learning process on the data from scratch with random weight initialization. TensorFlow is used to facilitate Transfer Learning of the CNN pre-trained model. We study the topology of the CNN architecture to find a suitable model, permitting image classification through Transfer Learning. Whilst testing and changing the network topology (i.e. parameters) as well as dataset characteristic to help determine which variables affect classification accuracy, though with limited computational power and time.

## 3  Experimental Set-Up and Datasets

For the experiments, two datasets are used: CIFAR-10 [10] and Caltech Faces [11] to retrain a pretrained model on ImageNet dataset. Literature review indicated the CIFAR-10 dataset as popular given many researchers would use such a dataset with it having a large set of images that of which were of low dimensionality. The dataset consists of 60000 ($32 \times 32$ pixel) colour images in 10 classes. The Caltech Face dataset consisted of 450 ($896 \times 592$ pixels) face images of 27 people. Both datasets vary in terms of image quantity, quality and type, with the CIFAR-10 dataset consisting of several categories (animals, vehicles and ships) whereas the Caltech Face dataset contains only face images. Consequently, allowing us to study the significance of the aforementioned differences with regards to accuracy performance.

With regards to coding, Python is the preferred language, as it not only contained a huge set of libraries that of which were easily used for Machine Learning (i.e. TensorFlow), but was also very accessible. The pre-trained model chosen for Transfer

Learning is the Inception-v3 model created by Google [1], with regards to image data for retraining we used the datasets: CIFAR-10 and Caltech Faces. Using a pre-existing dataset such as CIFAR-10, allows the comparison of results from previous state-of-the-art studies. For training purposes, we have one model retrained on the CIFAR-10 dataset and another model retrained on the Caltech Face dataset. The CIFAR-10 model, consists of 10 classes containing 10000 images. For the testing stage, we will keep it simple and have 2 test images per category, which is sufficient in obtaining preliminary results to meet our aims. It is important to note when training, the images are categorised into different labels identified by folder titles, the CNN will use the labels to classify the test images. When testing, all the images will be placed into one folder without labels. The second model using the Caltech Faces dataset will follow the same procedures, though each of the training classes will comprise of 18 images.

## 4 Preliminary Results

This section discusses the procedure in setting up tests to evaluate the system as well as documenting the results obtained. The presented tests should help in answering the following questions: Does Transfer Learning help improve the accuracy of a CNN? Does the number of epochs (training steps) improve accuracy? Does the number of images per class in a dataset influence accuracy? Does the type of image in a dataset effect the accuracy? Some of these questions have been influenced having identified open issues within previous state-of-the-art studies described in Sect. 1.

**Test 1:** The first test involves retraining three Inception-v3 models (pre-trained on the ImageNet dataset) with the datasets: CIFAR-10 and Caltech Faces. CIFAR-10 test A, involves retraining the model with 10000 images per training class, whilst CIFAR-10 test B involves retraining the model with 1000 images per training class. The purpose is to establish whether the quantity of training images affects classification accuracy as well as obtaining the average accuracy achieved by the Inception-v3 model having been retrained on both datasets stated separately.

Figure 2 shows some of the output images having run the CIFAR-10 model. The results from the first test indicated in Table 1 show that the CIFAR-10 test A model achieved a higher average accuracy of 70.1% over the training set compared to the

**Fig. 2.** CNN Transfer Learning results. Model trained on CIFAR-10 dataset Test A

**Table 1.** Classification results for both datasets in terms of overall accuracy.

| Dataset | Average accuracy (%) |
|---|---|
| CIFAR-10 (Test A) | 70.1 |
| CIFAR-10 (Test B) | 66.1 |
| Caltech Faces | 65.7 |

CIFAR-10 test B model and the model retrained on Caltech Face dataset that of which achieved classification accuracies of 66.1% and 65.7% respectively.

The results attained helped in answering the question: Does the number of images per class in a dataset influence accuracy? Results show that the accuracy scores are higher when more sample images are used for training aids. As the CIFAR-10 test B model used the same dataset though with far less training images and thus achieved a lower accuracy percentage, the model which used the Caltech Face dataset also contained less training images compared to CIFAR-10 test A. Furthermore, in terms of training time, the CIFAR-10 test A model took 3 h compared to 30 min for the Caltech trained model. This indicates the more images (although worse in terms of quality) has a significant effect on the training time as well as computational power required.

The results obtained from test 1 are very important towards the motivation of this work, determining whether Transfer Learning is useful in improving accuracy for image classification. Our results evidenced better accuracy scores compared to that achieved in [4] mentioned in Sect. 1. The model presented within this paper achieved almost twice the classification accuracy having been trained on the same dataset (38% vs 70%), with the only difference being the author in [4] had used a CNN-CIFAR-10 model trained from scratch as opposed to a pre-trained model which was adapted using Transfer Learning from ImageNet dataset to be retrained using the CIFAR-10. This evidently shows the benefits of Transfer Learning. Also considering we only used 500 epochs and a CPU due to time and computing power limitations, we could have easily achieved better accuracy scores given the use of GPUs, as they tend to perform a lot quicker and achieve better accuracy scores when more sample images are used for training.

**Test 2:** For the second test we have changed the number of epochs (full training cycle on the whole training data) to verify whether it influences the overall accuracy. For all tests the default number of epochs was set to 500, though for this test we have used 4000 epochs to replicate the original training of the Inception-v3 by Google on the ImageNet dataset. The training procedure was done on 2 classes consisting of 18 images (from the Caltech Face dataset).

Figure 3 shows the output having run the second test. The images show the input image (testing image) with a text overlaid to indicate the accuracy percentage. The images used for the second test were taken in a good lighting condition and with good angles. The results from the second test presented in Table 2 show that the increase in number of epochs does in fact help with the classification accuracy. With the model using 4000 epochs as opposed to 500 achieving a higher accuracy percentage, though requiring more time to train.

**Fig. 3.** CNN Transfer Learning results. Model trained on Caltech Faces dataset. Accuracy (confidence) for the images from left to right: 94.85%, 96.48%, 99.26%, 97.19%.

**Table 2.** Classification accuracy for both 500 epochs and 4000 epochs

| Person | Accuracy for 500 epochs | Accuracy for 4000 epochs |
|---|---|---|
| 1 | 88% | 95% |
| 2 | 94% | 98% |
| Average (%) | **91%** | **96.5%** |

**Test 3:** The third test involves 3 models each of which have been trained on only one category: humans, animals or cars. The purpose of this test was to verify whether the type of image (content within the image) has a direct effect on the accuracy scores. Three identical systems were created, with the only difference being the testing and training image. For the first model we used pictures of humans (i.e. Donald trump and Barack Obama), the second model had pictures of animals (i.e. dog and cat) and the third model had pictures of cars (i.e. Lamborghini and Ferrari). Each training set contained 10 images, and the testing set contained 3 "unseen" images of the corresponding category from the aforementioned datasets. All images were of high quality and similar pixel dimensions, these variables were controlled to avoid picture quality, image quantity or the like influencing results.

Table 3 shows the results, having conducted the third test. Evidently the model trained on only human face images achieved the highest accuracy of 93%, compared to both car and animal images. Considering variables such as image quality/dimensionality and quantity having no effect on the results, this indicates the type of images does indeed influence accuracy results to an extent. The results although preliminary indicate the CNN system working best with human face images, thus the results from the first test may have been different in favor of the Caltech Face model had there been more training images.

**Table 3.** Classification accuracy for models trained on the human face, car and animal dataset

| System | Human | Car | Animal |
|---|---|---|---|
| Average accuracy (%) | 93 | 87 | 73 |

## 5 Discussion

The aim of this study was to find a model suitable for Transfer Learning, being able to achieve respectable accuracy scores within a short space of time and with limited computational power. The study addressed different aspects of Machine Learning and explained the principals behind the Convolutional Neural Network architecture. We were able to find a suitable architecture that allows image classification through Transfer Learning, this came in the form of Inception-v3. A series of tests were conducted to determine the usability of such a technique and whether it could be applied to different sets of data. In turn, we could prove the usefulness of Transfer Learning as the results from the tests proved retraining the Inception-v3 model on the CIFAR-10 dataset resulted in better results compared to that stated in the previous state-of-the-art works, whereby authors in [4] did not use Transfer Learning and instead used a CNN trained on the same dataset (CIFAR-10) from scratch.

The CIFAR-10 retrained model proposed within this paper achieved an overall accuracy of 70.1%, compared to the 38% achieved and stated in [4]. In addition, the proposed system had a 100% pass rate as every image tested was given the correct classification, though there was variation in accuracy/confidence scores. Furthermore, given the preliminary results obtained from the first two tests we could verify that number of epochs and quantity of images in a dataset had a direct stimulus on the accuracy achieved. That said, the quality of the images was also identified as a factor, given the Caltech Face dataset had far less images compared to the CIFAR-10 dataset, and it still managed to achieve reasonable results which were similar. This was due to the fact the images were higher in quality and had variety (i.e. different lighting conditions, expressions and angles) as opposed to the CIFAR-10 dataset, which had small generic images from basic angles, which proved to be useful in helping the model achieving respectable accuracy given the low quantity training set. The third test gave the notion that the type of image influences the accuracy of a pre-trained model. This can be useful in certain circumstances whereby determining a specific dataset is required. A dataset consisting of one or limited classes will prove more useful in terms of classification accuracy compared to a dataset consisting of several types.

The main limitations within this study were computational power and time. As we established in test 2, increasing the number of training steps (epochs) increased the classification accuracy, though also increasing the training time considerably. Given access to a GPU as opposed to a CPU would have resulted in better accuracy and time efficiency, though the results presented within this paper were sufficient in achieving our aims. Using more images for our training datasets would have resulted in better accuracy, though we were limited in that regard. Ideally, building a model from scratch would allow the customization of layers/weights giving better efficiency and a more accurate functionality. This option however was not required in this study as the aim was to compare against works that had already been completed. This study used a pre-trained model and with the use of Transfer Learning we were able to retrain the aforementioned models with a new dataset and in turn provide accurate image classification. The final layer of the pre-trained model was retrained to provide this classification, thus maintaining the key knowledge in terms of weights from the models initial training and transferring that to the new dataset. The pretrained model (Inception-v3) used was sufficient in being able to provide reasonable classification accuracy given the low quantity training set, as it had initially been trained on a dataset consisting of a million images (ImageNet).

Given the findings of this paper, this provides a solid basis to advance the use of Transfer Learning in not only the model presented but other deep neural networks. The CNN model used can be refined and fine-tuned further by stacking additional layers and adjusting weights. There are numerous possibilities which can result in a more complex model able to achieve better accuracy results.

**Future Work:** As future work, having affirmed the notion of being able to use Transfer Learning on an existing model to achieve decent accuracy results in different domains permits many possibilities. This sort of application may be useful in class registration systems as well as being used as a biometric password. There is also the possibility to research different implementations of the system on a webserver, this would potentially allow users to use their face to verify their identity from any device and location in the world relating to the Internet of Things (IoT) as most systems/devices are in one way or another connected online or moving towards such a perception. There is still room to improve the accuracy results through implementation with the Inception-v4 model, increased epochs size and testing on larger datasets providing you have the time and resources. Another possibility would be combining a CNN with a Long-Short Term Memory (LSTM), this may be multifaceted, but in theory should help achieving better results and efficiency.

# References

1. TensorFlow: Image Recognition (2018). https://www.tensorflow.org/tutorials/image_recognition. Accessed 20 Apr 2018
2. Christian Szegedy, C., Liu, W., Jia, Y., Sermanet, P., Reed, S., Anguelov, D., Erhan, D., Vanhoucke, V., Rabinovih, R.: Going Deeper with Convolutions (arXiv.org) (2015)
3. Lin, M., Chen, Q., Yan, S.: Network in Network. ICLR submission (arXiv.org) (2013)
4. Nguyen, N.: Computational Collective Intelligence. Springer, Cham (2016)
5. Arun, P., Katiyar, S.: A CNN based Hybrid approach towards automatic age registration. Geodesy Cartography **62**(1), 33–49 (2013)
6. Rohrer, B.: How do Convolutional Neural Networks work? (2016). http://brohrer.github.io/how_convolutional_neural_networks_work.html. Accessed 20 Apr 2018
7. ImageNet: About (2016). http://image-net.org/about-overview. Accessed 20 Apr 2018
8. Gao, Y., Mosalam, K.: Deep transfer learning for image-based structural damage recognition. Comput. Aided Civ. Infrastruct. Eng. (2018)
9. Larsen-Freeman, D.: Transfer of learning transformed. Lang. Learn. **63**, 107–129 (2013)
10. Krizhevsky, A.: The CIFAR-10 dataset (2009). https://www.cs.toronto.edu/~kriz/cifar.html. Accessed 20 Apr 2018
11. Vision Caltech: Computational Vision (2018). http://www.vision.caltech.edu/html-files/archive.html. Accessed 20 Apr 2018
12. Wu, J.: CNN for Dummies. Nanjing University (2015)

13. Szegedy, C., Vanhoucke, V., Ioffe, S., Shlens, J., Wojna, Z.: Rethinking the inception architecture for computer vision. In: IEEE CVPR 2016: Computer Vision and Pattern Recognition (2016)
14. Mathworks.com: Convolutional Neural Network (2018). https://www.mathworks.com/content/mathworks/www/en/discovery/convolutional-neural-network.html. Accessed 20 Apr 2018

# Dendritic Cell Algorithm with Fuzzy Inference System for Input Signal Generation

Noe Elisa, Jie Li, Zheming Zuo, and Longzhi Yang$^{(\boxtimes)}$

Northumbria University, Newcastle upon Tyne NE1 8ST, UK
{noe.nnko,jie2.li,zheming.zuo,longzhi.yang}@northumbria.ac.uk

**Abstract.** Dendritic cell algorithm (DCA) is a binary classification system developed by abstracting the biological danger theory and the functioning of human dendritic cells. The DCA takes three signals as inputs, including danger, safe and pathogenic associated molecular pattern (PAMP), which are generated in its pre-processing and initialization phase. In particular, after a feature selection process for a given training data set, each selected attribute is assigned to one of the three input signals. Then, these input signals are calculated as the aggregation of their associated features, usually implemented by a simple average function followed by a normalisation process. If a nonlinear relationship exists between a signal and its corresponding selected attributes, the resulting signal using the average function may negatively affect the classification results of the DCA. This work proposes an approach named TSK-DCA to address such limitation by aggregating the assigned features of a signal linearly or non-linearly depending on their inherit relationship using the TSK+ fuzzy inference system. The proposed approach was evaluated and validated using the popular KDD99 data set, and the experimental results indicate the superiority of the proposed approach compared to its conventional counterpart.

**Keywords:** Dendritic cell algorithm · TSK+ fuzzy inference system
Information aggregation · Danger theory

## 1 Introduction

Intrusion detection systems (IDSes) are of paramount importance in computer network security as the number of cyber attacks grow in prominence every year. Over the last three decades, artificial immune systems (AISes) have been proposed primarily for intrusion detection in computer systems. Self-nonself is the first biological model used in computer security domain to develop AISes such as clonal selection, negative selection and positive selection algorithms [1]. Self-nonself model is built upon the observation that the natural immune system

This work is supported by the Commonwealth Scholarship Commission (CSC) and Northumbria University in the United Kingdom.

© Springer Nature Switzerland AG 2019
A. Lotfi et al. (Eds.): UKCI 2018, AISC 840, pp. 203–214, 2019.
https://doi.org/10.1007/978-3-319-97982-3_17

provides protection based on the discrimination between self (own body cells) which is tolerated and nonself (foreignness) which is the source of attack [2]. Self-nonself based algorithms often fail to provide beneficial advantages to computer network security systems like those provided by the natural self-nonself model to the natural immune systems [1].

Inspired by the danger theory [3] and the behaviour of dendritic cells (DCs), the dendritic cell algorithm (DCA) was developed to address the aforementioned limitation [4]. Briefly, the DCA first transfers the values of the most relevant features from a given training dataset to its input signals, termed as safe, danger, and pathogenic associated molecular pattern (PAMP). Then, the DCA classifier takes those input signals to produce a binary output. Conventionally, a linear average aggregation method is commonly used to aggregate the values from the assigned features to form each input signal. However, if a non-linear relationship exists between the selected attributes and the resulting signals, the average approach will adversely impact the performance of the DCA. A fuzzy inference system is then adopted in this paper to compute the value of each DCA input signal to generalise the linear average aggregation method.

Fuzzy inference systems are built upon fuzzy logic to map from the input space to the output space. They have been widely applied in solving either linear or non-linear problems of arbitrary complexity, such as [5,6]. The two most widely used fuzzy inference systems are the Mamdani fuzzy model and TSK fuzzy model. Compared with the Mamdani fuzzy model, which is more intuitive and commonly utilised to deal with human natural language, the TSK fuzzy model is more convenient to be employed when crisp output values are required. Both of these conventional fuzzy inference systems are only workable with a dense rule base by which the entire input domain is fully covered. Fuzzy interpolation enhances the power of the conventional fuzzy inference systems by relaxing the requirement of dense rule bases [7,8]. In other words, the conventional fuzzy inference systems fail to generate a conclusion when a given observation does not overlap with any rule antecedents in the rule base, but fuzzy interpolation can still approximate the conclusion. Various fuzzy interpolation methods have been developed in the literature, such as [5,9–16].

This paper proposes the TSK-DCA approach for aggregating the assigned features of each input signal, either linearly or non-linearly, to generate DCA inputs using the TSK+ fuzzy inference approach. In particular, the TSK-DCA uses three TSK+ fuzzy inference systems to deal with the three DCA input signals. In order to implement the proposed TSK-DCA, a data-driven rule base generation method is firstly employed to generate three sub-TSK fuzzy rule bases, corresponding to the three input signals. Then, the TSK+ fuzzy inference approach is applied to compute the value of each input signal from the assigned features for each data instance, before the application of the DCA. TSK-DCA has been validated and evaluated by a well-known benchmark dataset, KDD99. Experimental results indicate that the TSK-DCA performs better than the conventional one.

The rest of this paper is organised as follows: Sect. 2 describes the background theories, including TSK+ fuzzy inference approach and the DCA algorithm. Section 3 details the proposed TSK-DCA approach. Section 4 reports the experimentation and analyses the results; and Sect. 5 draws the conclusion and points out future research directions.

## 2  Background

### 2.1  TSK+ Fuzzy Inference System

The original TSK inference system generates a crisp inference result as the weighted average of the sub-consequences with the firing strength of the fired rules as weights [17]. Obviously, no rule will be fired if a given input does not overlap with any rule antecedent. As a consequence, the TSK inference cannot be performed. TSK+ was proposed to address such issue which generates a consequence by considering all the rules in the rule base [18]. Suppose that a sparse TSK rule base is comprised of $n$ rules:

$$R_1 : \textbf{IF } x_1 \text{ is } A_1^1 \text{ and } \cdots x_j \text{ is } A_j^1 \cdots \text{ and } x_m \text{ is } A_m^1$$
$$\textbf{THEN } z = f_1(x_1, \cdots, x_m),$$

$$\cdots\cdots$$

$$R_n : \textbf{IF } x_1 \text{ is } A_1^n \text{ and } \cdots x_j \text{ is } A_j^n \cdots \text{ and } x_m \text{ is } A_m^n$$
$$\textbf{THEN } z = f_n(x_1, \cdots, x_m), \tag{1}$$

where $A_j^i, (i \in \{1, 2, \cdots, n\}$ and $j \in \{1, 2, \cdots, m\})$ represents a normal and convex polygonal fuzzy set that can be denoted as $(a_{j1}^i, a_{j2}^i, \cdots, a_{jv}^i)$, $v$ is the number of odd points of the fuzzy set. Given an input $I = (A_1^*, A_2^*, \cdots, A_m^*)$ in the input domain, a crisp inference result can be generated by the following steps:

**Step 1:** Identify the matching degrees between the given input $(A_1^*, A_2^*, \cdots, A_m^*)$ and rule antecedents $(A_1^i, A_2^i, A_3^i, \cdots, A_m^i)$ for each rule $R_i$ by:

$$S(A_j^i, A_j^*) = \left(1 - \frac{\sum\limits_{q=1}^{v} |a_{jq}^i - a_{jq}^*|}{v}\right) \cdot (DF), \tag{2}$$

where $DF$ is a *distance factor*, which is a function of the distance between the two concerned fuzzy sets:

$$DF = 1 - \frac{1}{1 + e^{(-cd+5)}}, \tag{3}$$

where $c$ is a sensitivity factor, and $d$ represents the Euclidean distance between the two fuzzy sets for a given defuzzification approach. In particular, $c$ is a positive real number. Smaller value of $c$ leads to a similarity degree which is more sensitive to the distance of two fuzzy sets, and vice versa.

**Step 2:** Determine the firing degree of each rule by aggregating the matching degrees between the given input and its antecedent terms by:

$$\alpha_i = S(A_1^*, A_1^i) \wedge S(A_2^*, A_2^i) \wedge \cdots \wedge S(A_m^*, A_m^i), \tag{4}$$

where $\wedge$ is a t-norm operator usually implemented as a minimum operator.

**Step 3:** Generate the final output by integrating the sub-consequences from all rules by:

$$z = \sum_{i=1}^{n} \alpha_i \cdot f_n(x_1, \cdots, x_m) \Big/ \sum_{i=1}^{n} \alpha_i. \tag{5}$$

## 2.2 Dendritic Cell Algorithm

In order to detect anomaly for any given inputs, the DCA creates a population of artificial DCs to form a pool from which a number of DCs are selected to perform antigens (data items) sampling, signals categorization (into PAMP, DS and SS) and antigens identification [4]. While in the pool, DCs are exposed to the current signal values and the corresponding antigen data from the data source. Each DC has the ability to sample multiple antigens, so during the classification an aggregated sampling value from different DCs regarding a particular antigen is computed which is used to classify the antigen as normal or anomalous [4,19].

The inputs to the DCA are signals from all categories and data items. Signals are represented as an aggregation of real-valued numbers from their corresponding associated features, while antigens are identified by the data item IDs such as process ID or any other unique nominal attributes. As a binary classifier, the DCA classifies each antigen as either normal (semi-mature) cell context, or as anomalous (mature) cell context. So, the DCA output is the antigen normality or abnormality context which is represented as a binary value 0 for normality or 1 for abnormality. After the pre-processing and initialisation, the DCA goes through three phases as detailed below.

**Step 1. Detection:** The DCA processes the input signals using the following equation to obtain three cumulative output signals termed as $CSM$, $mDC$ and $smDC$:

$$\begin{aligned} C[CSM, smDC, mDC] \\ = \frac{(W_{PAMP} * C_{PAMP}) + (W_{SS} * C_{SS}) + (W_{DS} * C_{DS})}{W_{PAMP} + W_{SS} + W_{DS}} * \frac{1+I}{2}, \end{aligned} \tag{6}$$

where $C_{PAMP}$, $C_{DS}$ and $W_{SS}$ are PAMP, DS and SS signal values respectively which are generated by aggregating the assigned attributes. The weights ($W_{PAMP}$, $W_{SS}$ and $W_{DS}$) are pre-defined weights or can be derived empirically from the data. Each selected DC from the pool is assigned a migration threshold in order to determine the lifespan for antigen sampling and the amount of data items it can collect. Each DC computes its $CSM$ value and compares it with the migration threshold. If the $CSM$ of a DC exceeds the migration threshold, the DC ceases to sample data items and thus signals.

**Step 2. Context Assessment:** The cumulative values of $smDC$ and $mDC$ obtained from the detection phase are used to perform context assessment. If the antigens collected by a DC has a greater $mDC$ than its $smDC$ value, it is assigned a binary value of 1, and 0 otherwise.

**Step 3. Classification:** All the collected antigens are analysed by deriving the Mature Context Antigen Value (MCAV) for each presented antigen, which is used to assess the degree of an anomaly of a given antigen. Firstly, the anomaly threshold of MCAV is derived from the test data set. Then, the MCAV value of each antigen is calculated as dividing the number of times it is presented in the mature context by the total number of presentations in DCs. Antigens with MCAVs greater than the anomaly threshold are classified into the anomalous class while the others are classified into the normal one.

## 3   The Proposed Approach

The proposed TSK-DCA system is depicted in Fig. 1. In particular, given a training dataset, a feature selection process is first performed to select the most significant features. The selected features are then categorised into three groups, representing the three input signals. From this, three TSK+ fuzzy models can be generated using the given training data set for the aggregation of the three input values. Given an input, the TSK+ inference systems take place to aggregate the given inputs to the three DCA input signals. Then, the output of the TSK-DCA classifier is generated by the DCA model. Each of these key components of the proposed system is detailed in the following subsections.

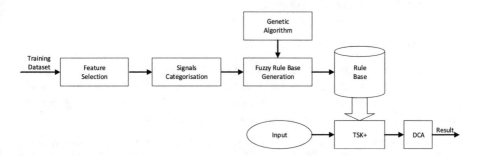

**Fig. 1.** The overall TSK-DCA system

### 3.1   Data Pre-processing

This study adopted the information gain approach to decide which features are more important than others during the data pre-processing stage, although any other feature selection approach is also applicable here. Briefly, the information gain of an attribute indicates the amount of information with respect to the classification a particular attribute provides, which can be obtained by [20]:

$$G(D, A) = E(D) - \sum_{v \in values(A)} \frac{|D_v|}{|D|} * E(D_v), \tag{7}$$

where $values(A)$ represent all the possible values of attribute $A$, $D_v$ is a subset of $D$ each taking value $v$ for attribute $A$, $G$ is the gain, and $E$ is the entropy. In particular, the entropy $E$ is computed as:

$$Entropy(D) = \sum_{i=1}^{i=2} -p_i * log_2 p_i, \tag{8}$$

where $p_i$ is the proportion of elements being classified as $i$ in the data set $D$. The higher the entropy the more information an attribute contains. Given a threshold, the attributes with higher gains than the given threshold are selected.

## 3.2   Signal Categorisation

The selected features are analysed using their histograms with respect to the two class labels (normal and abnormal) presented in the training dataset. The frequency of occurrence of the largest values presented in each attribute from each class is used to decide its signal category. If the largest values of an attribute have a high frequency of occurrence in normal class than that in anomalous class, the attribute will be categorized as safe signal. If the largest values of an attribute have a higher frequency of occurrence in anomalous class and significant lower frequency of occurrence in the normal class, it is categorised to PAMP signal. Otherwise, it is assigned to DS signal.

## 3.3   Signal Generation Using TSK+

Once the selected features are categorised into the three input signals, the TSK+ approach is applied to generate the input signals of the DCA. In order to apply the TSK+ approach as introduced in Sect. 2.1, a rule base needs to be generated first, which is outlined in Fig. 2 in two key steps as detailed below.

*Clustering:* The K-Means clustering algorithm is employed to each sub-dataset (i.e., danger, safe or PAMP) which includes only the associated features for the particular input signal. Note that the number of clusters has to be pre-defined to enable the application of the K-Means algorithm. The Elbow method, which has been used in [6, 21], is also employed in this work to determine the number of clusters.

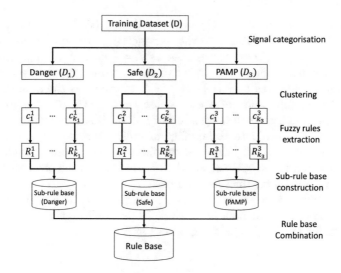

**Fig. 2.** The TSK+ fuzzy rule base generation

*Fuzzy Rule Extraction:* Each obtained cluster is expressed as one TSK fuzzy rule. Assume that a determined cluster for a signal is associated with $d$ features, then a TSK fuzzy rule $R_i$ can be formed as:

$$R_i : \textbf{IF } x_1 \text{ is } A_1^i \text{ and ... and } x_d \text{ is } A_d^i$$
$$\textbf{THEN} y = f_i(x_1, ..., x_d), \tag{9}$$

where $A_r^i$ $(r = \{1, ..., d\})$ is a fuzzy set as a rule antecedent. For simplicity, triangular membership functions are utilised in this work, that is $A_r^i = (a_{r1}^i, a_{r2}^i, a_{r3}^i)$. Without loss of generality, take a rule cluster $c_k$ as an example, which contains $p_k$ elements, such as $c_k = \{x_k^1, x_k^2, ..., x_k^{p_k}\}$. The core of the fuzzy set is set as the cluster centre which is $a_{r2}^i = \sum_{q=1}^{p_k} x_k^{qr} / p_k$; and the support the fuzzy set is expressed as the span of the cluster, i.e. $(a_{r1}^i, a_{r3}^i) = (\min\{x_k^{1r}, x_k^{2r}, ..., x_k^{p_k r}\}, \max\{x_k^{1r}, x_k^{2r}, ..., x_k^{p_k r}\})$. The consequent of a TSK fuzzy rule is the DCA input signal values. In particular, the consequent is expressed as a first-order polynomial in this work, which can be represented as $y = f_i(x_1, ...x_d) = \beta_0^i + \beta_1^i x_1 + \beta_2^i x_2 + ... + \beta_d^i x_d$, where $\beta_i^d$ is a constant parameter of the linear functions.

The rule base is optimised by employing the genetic algorithm (GA). As an adaptive heuristic search algorithm, GA has been successfully applied to find the optimised solution in the problem of fuzzy inference systems, such as [21–23]. The algorithm firstly initialises the population with random individuals. It then selects a number of individuals for reproduction by applying the genetic operators, that is mutation and crossover. The offspring and some of the selected existing individuals jointly form the next generation. The algorithm repeats this process until a satisfactory solution is generated or a maximum number of generations has been reached.

In this work, an individual ($I$) in a population ($\mathbb{P}$) is used to represent a potential solution that contains all the parameters of the polynomial functions in the TSK rule consequent, represented as $I = \{\beta_0^1, ..., \beta_d^1, ..., \beta_0^i, ..., \beta_d^i, \beta_0^n, ..., \beta_d^n\}$, where $n$ denotes the total number of rules in the current rule base. Given a population, represented as $\mathbb{P} = \{I_1, ..., I_{|\mathbb{P}|}\}$, where $|\mathbb{P}|$ is the numbers of individuals, the next generation of a population is produced by applying a single point crossover and a mutation, on selected individuals. The DCA classification accuracy is used to evaluate the quality of individuals in the new generation of population. After the algorithm is terminated, the fittest individual in the current population is the optimal solution. From this, all the extracted rules are grouped together to form the final rule base.

Once the rule bases are generated for all three input signals, the TSK+ inference approach as introduced in Sect. 2.1 is applied, which generates the signal inputs for the DCA as illustrated in Fig. 3.

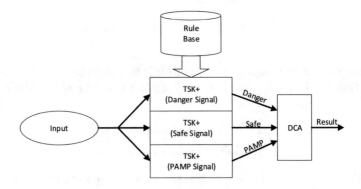

**Fig. 3.** Inputs generation for DCA

## 4    Experimentation

The proposed TSK-DCA system was validated and evaluated using the KDD-99 cup dataset [24]. The KDD-99 dataset was published in 1999 in the context of the 1998 DARPA initiative for IDS within the realm of computer networks [24]. This dataset has been intensively employed for building network intrusion detectors to distinguish normal and abnormal (i.e., intrusions or attacks) network connections. 10% (494,021) and 2.5% (125,973) data instances of the original KDD-99 dataset were respectively used for training and testing in this work.

### 4.1    Model Generation

In order to reduce the system complexity, the information gain method is employed for feature selection. Ten features were typically selected from 41 ones [25], and this work also follows this tradition, and the selected features for

**Table 1.** Selected features for each DCA input signal

| Signal | Features |
|--------|----------|
| DS | Count and srv count |
| SS | Logged in, srv different host rate and dst host count |
| PAMP | Serror rate, srv serror rate, same srv rate, dst host serror and dst host rerror rate |

each signal category are listed in Table 1. From this, the dataset was normalised using the min-max (MM) normalisation approach [26].

The rule base was generated in three step based on the training dataset:

**Step 1: Training Dataset Partition.** Divided the entire training dataset $T$ into three sub-training dataset $T_1, T_2$ and $T_3$ based on the results of the information gain method.

**Step 2: Optimal Number of Clusters Determination for Each Sub-Training Dataset.** The K-Means clustering algorithm was adopted for each sub-training dataset in which the optimal number of clusters was determined by the Elbow method. The identified cluster numbers for the three sub-datasets are listed in Table 2.

**Table 2.** The number of clusters for each sub-dataset

|                    | DS | SS | PAMP |
|--------------------|----|----|------|
| Number of clusters | 7  | 7  | 7    |

**Step 3: TSK Rule Extraction.** Based on the results of the Elbow method, there were 21 TSK fuzzy rules in total in the final rule base. For instance, the rule antecedents of one fuzzy rule in DS sub-rule base can be expressed as:

$$x_1 = (0, 0.39, 3.52) \text{ and } x_2 = (0, 0.39, 31.51). \tag{10}$$

**Step 4: Fine-Tune Polynomial Coefficients for TSK Consequence.** GA was applied to find the optimised constant parameters of polynomial functions of TSK consequent, and the results are listed in Table 3.

**Table 3.** The employed GA parameters

| | |
|---|---|
| Number of individuals | 50 |
| Number of iterations | 250 |
| Crossover rate | 0.95 |
| Mutation rate | 0.1 |

Taking Eq. 10 as an example, the optimised fuzzy rule is:

$$
\begin{aligned}
R_1 : \ &\textbf{IF } x_1 = (0, 0.39, 3.52) \text{ and } x_2 = (0, 0.39, 31.51) \\
&\textbf{THEN } f_1(x_1, x_2) = 18.2x_1 - 4.05x_2 - 5.5.
\end{aligned} \tag{11}
$$

**Table 4.** DCA parameters

| | smDC | | | mDC | | |
|---|---|---|---|---|---|---|
| | $W_{PAMP}$ | $W_{SS}$ | $W_{DS}$ | $W_{PAMP}$ | $W_{SS}$ | $W_{DS}$ |
| Weights | 5 | 11 | 11 | 8 | 7 | 14 |

Once the TSK fuzzy rule base has been generated, three TSK+ fuzzy inference systems were applied to generate the three input signals for the DCA model. In this work, the DCA model reported in [25] was employed and the corresponding parameters for Eq. 6 were configured as shown in Table 4. Note that, antigen multiplier was not used in this work.

## 4.2    Results and Discussion

Based on the parameter values shown in Tables 3 and 4, the best performance for the training dataset is 98.36%, and for the testing dataset is 92.07%. The accuracies for the 250 iterations of GA optimisation is shown in Fig. 4. The proposed TSK-DCA approach was compared with the basic DCA without using antigen multiplier as presented by [25] and Fuzzy-based DCA proposed by [27]. The proposed TSK-DCA approach overall outperforms the two benchmark approaches as demonstrated in Table 5.

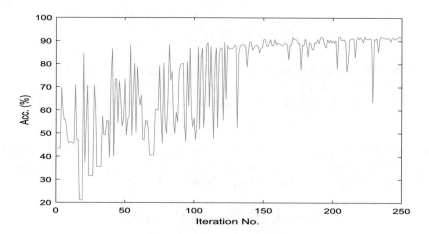

**Fig. 4.** The processing of fine-tuning the testing accuracies

**Table 5.** Performance comparison with existing DCA approaches

| Approach | Acc. (%) | |
|---|---|---|
| | Training | Testing |
| Basic DCA without antigen multiplier [25] (2008) | 78.92 | – |
| Fuzzy-based DCA [27] (2015) | 94.00 | – |
| **TSK-DCA** | **98.36** | **92.07** |

## 5   Conclusion

This work proposed the TSK-DCA classifier which generates the DCA input signal values from the assigned attributes using the TSK+ fuzzy inference system. The TSK-DCA is applicable to either linear or nonlinear related data instances. The experimental results using the KDD99 dataset demonstrate that the TSK-DCA achieves better classification accuracy in reference to its conventional DCA counterparts. Although promising, the work can be further improved by better fine-tune the rule base parameters as the present work only trains the parameters for the rule consequences. In addition, it is interesting to combine the proposed approach with other extensions and modifications of DCA to further boost the performance.

## References

1. Aickelin, U., Cayzer, S.: The danger theory and its application to artificial immune systems. arXiv preprint arXiv:0801.3549 (2008)
2. Burnet, F.M.: Immunological recognition of self. Science **133**(3449), 307–311 (1961)
3. Matzinger, P.: The danger model: a renewed sense of self. Science **296**(5566), 301–305 (2002)
4. Greensmith, J., Aickelin, U., Cayzer, S.: Introducing dendritic cells as a novel immune-inspired algorithm for anomaly detection. In: ICARIS, vol. 3627, pp. 153–167. Springer (2005)
5. Li, J., Yang, L., Fu, X., Chao, F., Qu, Y.: Dynamic QoS solution for enterprise networks using TSK fuzzy interpolation. In: 2017 IEEE International Conference on Fuzzy Systems (FUZZ-IEEE), pp. 1–6, July 2017
6. Yang, L., Li, J., Fehringer, G., Barraclough, P., Sexton, G., Cao, Y.: Intrusion detection system by fuzzy interpolation. In: 2017 IEEE International Conference on Fuzzy Systems (FUZZ-IEEE), pp. 1–6, July 2017
7. Kóczy, L.T., Hirota, K.: Interpolative reasoning with insufficient evidence in sparse fuzzy rule bases. Inf. Sci. **71**(1), 169–201 (1993)
8. Kóczy, L.T., Hirota, K.: Approximate reasoning by linear rule interpolation and general approximation. Int. J. Approximate Reasoning **9**(3), 197–225 (1993)
9. Huang, Z., Shen, Q.: Fuzzy interpolation and extrapolation: a practical approach. IEEE Trans. Fuzzy Syst. **16**(1), 13–28 (2008)
10. Yang, L., Shen, Q.: Adaptive fuzzy interpolation with prioritized component candidates. In: 2011 IEEE International Conference on Fuzzy Systems (FUZZ-IEEE 2011), pp. 428–435, June 2011
11. Yang, L., Shen, Q.: Closed form fuzzy interpolation. Fuzzy Sets Syst. **225**, 1–22 (2013). Theme: Fuzzy Systems
12. Yang, L., Shen, Q.: Adaptive fuzzy interpolation with uncertain observations and rule base. In: 2011 IEEE International Conference on Fuzzy Systems (FUZZ-IEEE 2011), pp. 471–478 (2011)
13. Naik, N., Diao, R., Shen, Q.: Genetic algorithm-aided dynamic fuzzy rule interpolation. In: 2014 IEEE International Conference on Fuzzy Systems (FUZZ-IEEE), pp. 2198–2205 (2014)
14. Yang, L., Chen, C., Jin, N., Fu, X., Shen, Q.: Closed form fuzzy interpolation with interval type-2 fuzzy sets. In: 2014 IEEE International Conference on Fuzzy Systems (FUZZ-IEEE), pp. 2184–2191 (2014)

15. Li, J., Yang, L., Shum, H.P.H., Sexton, G., Tan, Y.: Intelligent home heating controller using fuzzy rule interpolation. In: UK Workshop on Computational Intelligence (2015)
16. Yang, L., Chao, F., Shen, Q.: Generalised adaptive fuzzy rule interpolation. IEEE Trans. Fuzzy Syst. (2016). https://doi.org/10.1109/TFUZZ.2016.2582526
17. Takagi, T., Sugeno, M.: Fuzzy identification of systems and its applications to modeling and control. IEEE Trans. Syst. Man Cybern. **SMC-15**(1), 116–132 (1985)
18. Li, J., Qu, Y., Shum, H.P.H., Yang, L.: TSK inference with sparse rule bases, pp. 107–123. Springer, Cham (2017)
19. Chelly, Z., Elouedi, Z.: A survey of the dendritic cell algorithm. Knowl. Inf. Syst. **48**(3), 505–535 (2016)
20. Kayacik, H.G., Zincir-Heywood, A.N, Heywood, M.I.: Selecting features for intrusion detection: a feature relevance analysis on KDD 99 intrusion detection datasets. In: Proceedings of the Third Annual Conference on Privacy, Security and Trust (2005)
21. Li, J., Yang, L., Yanpeng, Q., Sexton, G.: An extended Takagi-Sugeno-Kang inference system (TSK+) with fuzzy interpolation and its rule base generation. Soft. Comput. **22**(10), 3155–3170 (2018)
22. Kumar, A., Agrawal, D.P., Joshi, S.D.: A GA-based method for constructing TSK fuzzy rules from numerical data. In: 2003 the 12th IEEE International Conference on Fuzzy Systems, FUZZ 2003, vol. 1, pp. 131–136, May 2003
23. Tan, Y., Li, J., Wonders, M., Chao, F., Shum, H.P.H., Yang, L.: Towards sparse rule base generation for fuzzy rule interpolation. In: 2016 IEEE International Conference on Fuzzy Systems (FUZZ-IEEE), pp. 110–117, July 2016
24. KDD Cup 1999 Data. http://kdd.ics.uci.edu/databases/kddcup99/kddcup99.html/. Accessed 16 Apr 2018
25. Gu, F., Greensmith, J., Aickelin, U.: Further exploration of the dendritic cell algorithm: antigen multiplier and time windows. In: International Conference on Artificial Immune Systems, pp. 142–153. Springer (2008)
26. Zuo, Z., Li, J., Anderson, P., Yang, L., Naik, N.: Grooming detection using fuzzy-rough feature selection and text classification. In: 2018 IEEE International Conference on Fuzzy Systems (FUZZ-IEEE). IEEE (2018)
27. Chelly, Z., Elouedi, Z.: Hybridization schemes of the fuzzy dendritic cell immune binary classifier based on different fuzzy clustering techniques. New Gener. Comput. **33**(1), 1–31 (2015)

# Clustering and Regression

# A Method of Abstractness Ratings for Chinese Concepts

Xiaomei Wang[1], Chang Su[1(✉)], and Yijiang Chen[2]

[1] Department of Cognitive Science, Xiamen University, Fujian, China
suchang@xmu.edu.cn
[2] Department of Computer Science, Xiamen University, Fujian, China

**Abstract.** As a kind of semantic knowledge of words, abstractness shows the degree of abstraction of a concept. There are many databases rating the concreteness of English words; however, there is only a small amount of research on analyzing the abstractness (or concreteness) of Chinese concepts. In this paper, abstractness ratings are presented for Chinese concepts. Our method is semi-supervised. Concrete and abstract paradigm words are pre-built. The degree of abstractness is calculated by analyzing the semantic similarity of a word with two paradigms. This method also intuitively classifies the concepts into abstract or concrete categories, based on their abstract ratings. Experimental results are reasonable and in line with our cognition.

**Keywords:** Degree of abstractness · Similarity
Classification of concepts

## 1 Introduction

Concepts contain rich semantic knowledge. Abstractness, one kind of semantic knowledge, is defined as things that cannot be experienced directly by our senses [1]. Abstractness is the important information when dealing with semantic works. For example, one theory is that a target domain is usually abstract and a source domain is usually concrete [2]. The information of abstractness can be used to analyze the associations between two domains in a metaphor.

Early research suggested that the abstractness of a word can be determined by the number of subordinate words within it [3]. Some approaches to studying abstractness were to place concepts on a scale ranging between abstractness and concreteness [4]. Some dictionaries rate the concreteness of English words by using a participant-based method [5–7]. Lynott and Connell [8] observed that concreteness ratings are impacted more by vision and touch. They analyzed modality exclusivity norms for nouns based on concreteness ratings. Brysbaert et al. [9] obtained concreteness ratings for forty thousand generally known English word lemmas based on five sensory experiences and motor responses. They concluded that the distribution of concreteness ratings is bimodal.

© Springer Nature Switzerland AG 2019
A. Lotfi et al. (Eds.): UKCI 2018, AISC 840, pp. 217–226, 2019.
https://doi.org/10.1007/978-3-319-97982-3_18

In recent research, computational models have been introduced for calculating abstractness (or concreteness) [10–12]. Hill et al. [13] argue that concreteness determines the most informative linguistic features. Experiments showed that perceptual input can enhance representations of abstract nouns. Huang et al. [14] introduced the logistic regression model to train an abstract words database to modify the weight of eigenvalues. Turney et al. [15] calculated the abstractness of a target word by comparing it to twenty abstract words and twenty concrete words. Their calculation, although using a form of Latent Semantic Analysis (LSA), did not take contextual information into account.

The Medical Research Council (MRC) psycholinguistic database [16], having 4,295 English words rated with degrees of concreteness, is an authoritative dictionary. The range of ratings is from 158 (highly abstract) to 670 (highly concrete). However, rating the abstractness of Chinese concepts is still unknown. Besides, there are differences in abstract information between English and Chinese words. For example, the Chinese concept, "美人(beauty)", is usually regarded as a more concrete word. However, the concreteness of "美人(beauty)" in the MRC psycholinguistic database is 336, which shows that "beauty" is more abstract. In English, "beauty" contains "beautiful people" and other sensory perceptions. The semantic information of "beauty" is different from the information of "美人".

Analyzing abstractness of Chinese concepts is conducive to the Chinese semantic analysis. We propose a method that can automatically measure the abstractness of Chinese concepts and identify the words as abstract (or concrete) concepts. Our method has three steps: (1) learn abstractness of the concept by word embedding; (2) build two paradigms of abstract (or concrete) concepts; (3) measure abstractness based on the semantic similarity and classify the concepts into the category of abstractness or concreteness.

**Step 1:** To acquire the knowledge of abstractness effectively, our method introduces the model of word embedding for learning the distributed representation of words from a context-level. Knowledge, like syntax knowledge, is concluded from vectors. For example, in Chinese, it's unreasonable to describe abstract words by specific quantifiers. The expression, "三公里梦想 (three kilometers of the dream)", ordinarily is considered illegal.

**Step 2:** In the second step, as the seed collections, we collect twenty abstract concepts and twenty concrete concepts, and introduce a classification algorithm to extend them. An abstract paradigm, having 225 abstract words, and a concrete paradigm, having 221 concrete words, are achieved.

**Step 3:** Abstractness is measured by analyzing the similarity of the target word with two paradigms. If the target word is more similar to the abstract paradigm, it is regarded as the abstract concept and has a higher degree of abstractness. The values of words that all belong to the abstract category are related positively to their semantics about abstractness. We designate the abstractness of a concept on a range from "−1" (highly concrete) to "1" (highly abstract). We assume that a concept with a value between "0" and "1" is classified as an abstract word.

Compared with the work of Zhao [17], where the values of a concept are distinctly marked at "0" or "1", our values lie within the range $[-1, 1]$. Also, our method reflects the difference in abstractness between two words. For example, the values of two concepts, "职业(profession)" and "教师(teacher)" both within the range of "0" and "1", are regarded as abstract words. In our method, the abstractness of concept, "职业(profession)," is higher than the value for concept "教师(teacher)." This result is in line with our cognition.

The contributions of this paper are as follows:

1. We propose a method to calculate the abstractness of Chinese concepts, because only a small amount of research focuses on such calculations. Based on our method, the difference in abstractness of different words is clearly identified.
2. The knowledge of abstractness is acquired from the context-sensitive model of word embedding that contains rich contextual information.
3. The degree of abstractness is reasonable and valid. Our method is suitable for laying the foundation of Chinese semantic analysis.

## 2   A Method of Abstractness Rating

### 2.1   Learning Abstractness Based on Word Embedding

Distributed representations of words can be regarded as grouping similar words in context [18]. Mikolov [19] proposed technologies that can be used for learning high-quality word vectors from huge data sets (with billions of words), and a vocabulary (with millions of words). The knowledge of a word can be ascertained by the company it keeps. Every dimension of vector space presents the knowledge of semantics, syntax, etc.

We use Reader Corpus[1] for training word vectors. The corpus is segmented by Segtag[2]. The Skip-gram model [19] is introduced to compute the vector representations of words. We obtained a vector file of 7,987,025 vectors with 400 dimensions. The representations of words are distributed in this paper. The semantic relationship between two words is well analyzed.

### 2.2   Build Paradigms of Abstract and Concrete Words

There are two steps in building paradigms: (1) build seed collections of abstract and concrete words by manual evaluation; (2) extend seed collections by the classification algorithm based on the corpus. We consider both manual evaluation and corpus. As a result, the data of paradigms is more reasonable and more comprehensive.

Seed collections are collected from a questionnaire. We invited twenty native-speaking Chinese students to rate the abstractness of 200 Chinese words in a range of $[-1, 1]$. Their instructions are as follows:

---

[1] A Chinese corpus. URL: www.duzhe.com.
[2] A word segmentation tool of NLP Lab of Xiamen University.

1. Concrete concepts refer to things that can be perceived directly by our five senses (visual, auditory, gustatory, olfactory and haptic). For example, concrete concepts can be orange, star, etc. A concrete concept can be explained by demonstrating it [9].
2. Abstract concepts refer to things that cannot be directly expressed by sensory experiences, such as dream, philosophy, etc. An abstract concept is usually explained by using other words.
3. The value "1" means the most abstract and "−1" means the most concrete. Different concepts have different levels of abstraction. Some concepts, that have values near "0", can be explained by perceptual experiences and other words at the same time.

We chose the top twenty words as a seed collection of abstract words (see Table 1). The last twenty words are chosen as a seed collection of concrete words (see Table 2).

**Table 1.** A seed collection of abstract words.

| 智慧(wisdom) | 真理(truth) | 哲学(philosophy) | 时间(time) |
|---|---|---|---|
| 热情(enthusiasm) | 经验(experience) | 美丽(beauty) | 光明(light) |
| 野心(ambition) | 诚实(honesty) | 事业(career) | 沉默(silence) |
| 公平(justice) | 灵魂(soul) | 语言(language) | 理智(mind) |
| 感情(emotion) | 思想(thought) | 矛盾(contradiction) | 人生(life) |

**Table 2.** A seed collection of concrete words.

| 蛋糕(cake) | 报纸(newspaper) | 吉他(guitar) | 玫瑰花(rose) |
|---|---|---|---|
| 胡萝卜(carrot) | 模特(model) | 松(pine) | 江(river) |
| 明珠(pearl) | 手绢(handkerchief) | 马(horse) | 涂料(coating) |
| 眉毛(eyebrow) | 铅笔(pencil) | 翅膀(wing) | 邮票(stamp) |
| 红色(red) | 宝石(jewel) | 汽车(car) | 太阳(sun) |

The K-nearest neighbor (KNN) algorithm requires no preprocessing of the labeled sample set prior to its use [20]. In this paper, the training data are seed collections; the testing data are parts of nominal words acquired from Sect. 2.1. The distance is measured by the Euclidean distance formula. After extension, we obtained an abstract paradigm with 225 abstract words and a concrete paradigm with 221 concrete words.

## 2.3   Measuring Abstractness Based on Semantic Similarity

We assume that, if concept $C$ is more semantically similar to the abstract paradigm than the concrete paradigm, the degree of abstractness will be higher. This concept has a great possibility to be classified into the abstract category.

This method measures the semantic similarity of words based on word embedding. Semantically similar words tend to have similar contextual distribution

[18]. We calculate the semantic similarity of two words by the cosine of the angle between two corresponding vectors. Cosine similarity, viewed as an improvement of the vector inner product, has been widely used in practice.

The algorithm for measuring abstractness and classifying the concept is as follows:

---

**Algorithm 1.** Measuring abstractness and classifying the concept

---

**Require:**

    The target concept, $C$;

    The paradigm of abstract words, $Abs = \{abs_1, abs_2, ,, abs_n\}$, where $n$ is the number of all abstract paradigm words;

    The paradigm of concrete words, $Con = \{con_1, con_2, ,, con_m\}$, where $m$ is the number of all concrete paradigm words.

**Ensure:**

1: Calculate the similarity with the abstract paradigm.

$$Sim_{abs} = \sum_{i=1}^{n} Similarity(C, abs_i)$$

2: Calculate the similarity with the concrete paradigm.

$$Sim_{con} = \sum_{j=1}^{m} Similarity(C, con_j)$$

3: Calculate the degree of abstractness.

$$degree_C = norm(Sim_{abs} - Sim_{con})$$

4: **if** $degree_C \in (0, 1]$ **then**

5:     $C$ will be classified into the abstract category.

$$category_C \ is \ abstract$$

6: **else**

7:     $C$ will be classified into the concrete category.

$$category_C \ is \ concrete$$

8: **end if**

9: Return $degree_C$ and $category_C$.

---

The final value of abstractness is normalized in the range $[-1, 1]$. The more abstract the concept, the higher its value. Consequently, we compare the abstractness of different concepts directly. Besides, abstractness ratings can be used to identify the abstract (or concrete) concepts.

# 3  Evaluation and Discussion

## 3.1  Test Data

The dictionary of Sogou[3] divides Chinese concept into different parts according to their corresponding topics. We randomly constructed eleven groups of concepts based on this dictionary. Each group has six concepts, are randomly chosen from the data for their topics. In each group, every concept has a correlation relationship.

## 3.2  Evaluation

The results of our method include two parts: (1) the degree of abstractness of a concept; (2) the abstract or concrete category of a concept. We evaluate these two results in this Section.

To better illustrate the performance of abstractness rating of our method, we compare our ratings with the comprehensive data calculated by the method of Brysbaert et al. [9]. They rate the concreteness of forty thousand English word lemmas. However, their database does not classify English words into an abstract or a concrete category. Based on these eleven groups, we translated these words into English, and found the corresponding values in this database.

The evaluation method was manual evaluation. We invited three native-speaking Chinese annotators to evaluate the acceptability of the results. The agreement of the annotators on the preliminary test was 0.66 ($\kappa$) [21], which is considered reliable. The following instructions were used when the annotators evaluated the results by themselves:

1. The annotator must regard each group of concepts as an assessment of the whole. To evaluate a concept, the annotators must consider a comparison with other concepts in the same group.
2. Instead of binary decision (accept/decline) when evaluating the degree of a concept, annotators were asked to use the following five-level method: (5)–very acceptable; (4)–acceptable; (3)–neutral; (2)–unacceptable; (1)–very unacceptable.
3. When evaluating classification results, an annotator is asked to identify the abstract (or concrete) concept based on his knowledge.

Evaluation results using our method show an accuracy of 82.7% compared with an accuracy of 70.6% using the method of Brysbaert et al. [9], indicating that the degrees of abstractness calculated by our method are more reasonable. Our method meets the abstract information of Chinese. Some concepts, like "落叶 (fallen leaves)" and "秋风 (autumn wind)", are not found in the dictionaries of Brysbaert et al. [9]. Our method automatically calculates the abstractness result for these concepts.

---

[3] An online dictionary of Chinese concept. URL: https://pinyin.sogou.com/dict/.

Using our method, classification results for precision, recall, and F score are 80.0%, 90.3%, and 84.8%, respectively, indicating that our method is acceptable. The classification of a concept is realized by analyzing its degree of abstractness. The abstract word identified by our method has an abstractness degree greater than zero. The concrete word has a value less than zero. To some degree, our method can obtain the coarse-grained partitioning.

## 3.3   Discussion

Consider the following four groups as the examples:

1. {家庭(family), 家人(family member), 父亲(father), 女儿(daughter), 宝贝(dotey), 爸爸(dad)}
2. {医生(doctor), 职业(profession), 人物(personage), 疾病(disease), 咳嗽(cough), 感冒(cold)}
3. {钢琴(piano), 音符(note), 音乐(music), 乐谱(notation), 歌曲(song), 歌声 (singing)}
4. {宠物(pet), 金鱼(goldfish), 猫咪(kitty), 鸟(bird), 老鼠(mouse), 母鸡(hen)}

The abstractness ratings and the classification results of these four examples are illustrated in Table 3.

As Example 1 shows, "家庭(family)" is an abstract concept explained by other words, like "家人(family member)", "父亲(father)" and other concepts in this group. "家庭(family)" has the highest abstractness in this group. The concept "父亲(father)" is more abstract than the concept "爸爸(dad)", since "父亲(father)" belongs to the written language with a stronger tendency of abstractness. However, the concept "宝贝(dotey)" is regarded as a concrete concept. In fact, it can be used to describe the people who are dear to someone. This concept has no specific instruction, compared with the other words, like "父亲(father)" and "女儿(daughter)".

As Example 2 shows, "职位(profession)" is much more abstract than "医生(doctor)", indicting that "职位(profession)" is the hypernym of "医生 (doctor)" in WordNet[4]. The degrees of abstractness among the three concepts, (disease)", "感冒(cold)" and "咳嗽(cough)", are decreasing, which is a reasonable result.

As Example 3 shows, there is no obvious difference between "歌曲(song)" and "歌声(singing)", indicting that "歌曲(song)" is a synonym of "歌声(singing)". In this group, "乐谱(notation)" is more concrete than "音符(note)", which does not fit our cognition, because this method heavily relies much on the corpus. Introducing lexical databases when measuring abstractness is necessary for our future work. Besides, the classification performance of Example 3 is unsatisfactory. The concepts, "歌曲(song)" and "歌声(singing)", which can be described by auditory experience, are regarded as abstract words by our method. Five-sensory knowledge will be considered in our future work.

---

[4] A large lexical database. URL: https://wordnet.princeton.edu/.

**Table 3.** The abstractness ratings and classification results of 4 examples.

| Examples | Concepts | Abstractness ratings | Machine classification | Human judgment |
|---|---|---|---|---|
| 1 | 家庭(family) | 0.233441499 | Abstract | Abstract |
| | 家人(family member) | 0.05937788 | Abstract | Abstract |
| | 父亲(father) | -0.053736733 | Concrete | Concrete |
| | 女儿(daughter) | -0.080239075 | Concrete | Concrete |
| | 宝贝(dotey) | -0.098889289 | Concrete | Abstract |
| | 爸爸(dad) | -0.13881706 | Concrete | Concrete |
| 2 | 医生(doctor) | -0.021514565 | Concrete | Concrete |
| | 职业(profession) | 0.361153675 | Abstract | Abstract |
| | 人物(personage) | 0.223654111 | Abstract | Abstract |
| | 疾病(disease) | 0.310301803 | Abstract | Abstract |
| | 感冒(cold) | 0.213516761 | Abstract | Abstract |
| | 咳嗽(cough) | -0.03520224 | Concrete | Concrete |
| 3 | 音乐(music) | 0.148178103 | Abstract | Abstract |
| | 歌曲(song) | 0.067723099 | Abstract | Concrete |
| | 歌声(singing) | 0.06083328 | Abstract | Concrete |
| | 音符(note) | -0.007958526 | Concrete | Concrete |
| | 钢琴(piano) | -0.18220036 | Concrete | Concrete |
| | 乐谱(notation) | -0.283403307 | Concrete | Concrete |
| 4 | 宠物(pet) | 0.205892004 | Abstract | Abstract |
| | 金鱼(goldfish) | -0.265222701 | Concrete | Concrete |
| | 猫咪(kitty) | -0.225567875 | Concrete | Concrete |
| | 鸟(bird) | -0.217372049 | Concrete | Concrete |
| | 老鼠(mouse) | -0.240931875 | Concrete | Concrete |
| | 母鸡(hen) | -0.265222701 | Concrete | Concrete |

As Example 4 shows, "宠物(pet)", which is regarded as a category concept, has the highest abstractness. The other five concepts are homonyms and their ratings are close. Their categories are in line with our cognition, indicting that our method is feasible.

Consequently, when calculating the degree of abstractness, our method performs well. When classifying concepts into concrete or the abstract categories, our method can achieve the coarse-grained division.

## 4    Conclusion

Because of the small amount of research analyzing the abstractness of Chinese words by computational ways, we proposed an easy-realized method for rating the abstractness of Chinese concepts. Our method identifies the abstract (or concrete) concepts according to their values of abstractness. We measured the abstractness by analyzing the semantic similarity with two pre-built paradigms. Semantic similarity was calculated based on the model of word embedding. To some degree, our method is context-sensitive. Based on the corpus, this method learns well the abstractness of concepts and analyzes the semantic similarity at the context level. Experimental results show that our method achieves good performances in rating the abstractness and classifying Chinese concepts.

The strength of our method is that the degree of abstractness is easy acquired without much training. Our method conforms with Chinese semantic knowledge. This method can be used for the baselines for the other fields of natural language processing. In future work, we will consider more lexical resources to supervise the process of rating abstractness, as well as the process of classification. Also, fine-grained analysis of abstractness and concreteness will be performed.

# References

1. Spreen, O., Schulz, R.: Parameters of abstraction, meaningfulness, and pronunciability for 329 nouns. J. Verbal Learn. Verbal Behav. **5**, 459–468 (1966). https://doi.org/10.1016/S0022-5371(66)80061-0

2. Lakoff, G., Johnson, M.: Metaphors we live by. University of Chicago Press, Chicago, Ill [u.a.] (2011). https://doi.org/10.3366/edinburgh/9780748643158.003.0003

3. Kammann, R., Streeter, L.: Two meanings of word abstractness. J. Verbal Learn. Verbal Behav. **10**, 303–306 (1971). https://doi.org/10.1016/s0022-5371(71)80058-0

4. Dunn, J.: Modeling abstractness and metaphoricity. Metaphor Symbol **30**, 259–289 (2015). https://doi.org/10.1080/10926488.2015.1074801

5. Altarriba, J., Bauer, L., Benvenuto, C.: Concreteness, context availability, and imageability ratings and word associations for abstract, concrete, and emotion words. Behav. Res. Methods Instrum. Comput. **31**, 578–602 (1999). https://doi.org/10.3758/bf03200738

6. McRae, K., Cree, G., Seidenberg, M., Mcnorgan, C.: Semantic feature production norms for a large set of living and nonliving things. Behav. Res. Methods **37**, 547–559 (2005). https://doi.org/10.3758/bf03192726

7. Tekiroglu, S.S., Özbal, G., Strapparava, C.: Sensicon: an automatically constructed sensorial lexicon. In: Proceedings of the 2014 Conference on Empirical Methods in Natural Language Processing (EMNLP), pp. 1511–1521 (2014). https://doi.org/10.3115/v1/d14-1160

8. Lynott, D., Connell, L.: Modality exclusivity norms for 400 nouns: the relationship between perceptual experience and surface word form. Behav. Res. Methods **45**, 516–526 (2012). https://doi.org/10.3758/s13428-012-0267-0

9. Brysbaert, M., Warriner, A., Kuperman, V.: Concreteness ratings for 40 thousand generally known English word lemmas. Behav. Res. Methods **46**, 904–911 (2013). https://doi.org/10.3758/s13428-013-0403-5

10. Changizi, M.: Economically organized hierarchies in WordNet and the Oxford English Dictionary. Cogn. Syst. Res. **9**, 214–228 (2008). https://doi.org/10.1016/j.cogsys.2008.02.001

11. Xing, X., Zhang, Y., Han, M.: Query difficulty prediction for contextual image retrieval. In: European Conference on Advances in Information Retrieval, pp. 581–585. Springer (2010). https://doi.org/10.1007/978-3-642-12275-0_52

12. Shutova, E., Teufel, S., Korhonen, A.: Statistical metaphor processing. Comput. Linguist. **39**, 301–353 (2013). https://doi.org/10.1162/coli_a_00124

13. Hill, F., Reichart, R., Korhonen, A.: Multi-modal models for concrete and abstract concept meaning. Trans. Assoc. Comput. Linguist. **2**(1), 285–296 (2014)
14. Huang, X., Zhang, H., Lu, B., et al.: An approach to Chinese metaphor identification based on word abstractness. New Technol. Libr. Inf. Serv. (2015). (in Chinese)
15. Turney, P., Neuman, Y., Cohen, Y.: Literal and metaphorical sense identification through concrete and abstract context. In: Conference on Empirical Methods in Natural Language Processing. Association for Computational Linguistics, pp. 680–690 (2011)
16. Coltheart, M.: The MRC psycholinguistic database. Q. J. Exp. Psychol. **33**(4), 497–505 (1981). https://doi.org/10.1080/14640748108400805
17. Zhao, H., Qu, W., Zhang F., et al.: Chinese verb metaphor recognition based on machine learning and semantic knowledge. J. Nanjing Normal Univ. (2011). (in Chinese)
18. Mikolov, T., Sutskever, I., Chen, K., et al.: Distributed representations of words and phrases and their compositionality, vol. 26, pp. 3111–3119 (2013)
19. Mikolov, T., Chen, K., Corrado, G.: Efficient estimation of word representations in vector space. Comput. Sci. (2013)
20. Keller, J., Gray, M., Givens, J.: A fuzzy K-nearest neighbor algorithm. IEEE Trans. Syst. Man Cybern. **SMC-15**, 580–585 (1985). https://doi.org/10.1109/TSMC.1985.6313426
21. Siegel, S.: Nonparametric Statistics for the Behavioral Sciences. McGraw-Hill, New York (1988). https://doi.org/10.4135/9781412961288.n273

# Effective Diagnosis of Diabetes with a Decision Tree-Initialised Neuro-fuzzy Approach

Tianhua Chen[1]($\boxtimes$), Changjing Shang[2], Pan Su[3], Grigoris Antoniou[1], and Qiang Shen[2]

[1] Department of Computer Science, School of Computing and Engineering,
University of Huddersfield, Huddersfield, UK
T.Chen@hud.ac.uk
[2] Department of Computer Science,
Institute of Mathematics, Physics and Computer Science,
Aberystwyth University, Aberystwyth, UK
[3] School of Control and Computer Engineering,
North China Electric Power University, Baoding, China

**Abstract.** Diabetes mellitus is a serious hazard to human health that can result in a number of severe complications. Early diagnosis and treatment is of significant importance to patients for the acquisition of a better quality life and precaution against subsequent complications. This paper proposes an approach by learning a fuzzy rule base for the effective diagnosis of diabetes mellitus. In particular, the proposed approach starts with the generation of a crisp rule base through a decision tree learning mechanism, which is data-driven and able to learn simple rule structures. The crisp rule base is then transformed into a fuzzy rule base, which forms the input to the powerful neuro-fuzzy framework of ANFIS, further optimising the parameters of both rule antecedents and consequents. Experimental study on the well-known Pima Indian diabetes data set is provided to demonstrate the promising potential of the proposed approach.

## 1 Introduction

Diabetes mellitus is a complex metabolic disorder characterised by persistent hyperglycemia, resulting from defects in insulin secretion, insulin action or both [1]. Diabetes is associated with a number of complications and it can increase the risk of developing blindness, blood pressure, heart disease, kidney disease and nerve damage [2]. Type 1 and type 2 are the most commonly seen forms of diabetes. In particular, type 2/adult-onset diabetes is associated with obesity and can be delayed or controlled with proper medication, healthy diet and exercise. It is therefore of significant importance to have an early detection of such disease to help patients suffering from it obtain a better quality life and subsequently prevail over the complications diabetes may bring.

© Springer Nature Switzerland AG 2019
A. Lotfi et al. (Eds.): UKCI 2018, AISC 840, pp. 227–239, 2019.
https://doi.org/10.1007/978-3-319-97982-3_19

The early diagnosis of type 2 diabetes is not an easy work, as diabetes patients show many symptoms in common with those that also appear in other types of disease. Despite recent medical progress, it has been reported that about half of the patients with type 2 diabetes are unaware of their disease and may take more than ten years as the delay from disease onset to diagnosis while early diagnosis and treatment is vital [3]. With the staggering development of computer technology and the rapid computerisation of business nowadays, huge volume of medical data taken from patients and evaluations of medical experts is being accumulated at a dramatic pace. However, raw data is barely of direct interest unless potentially useful information is extracted, which can then be utilised for future diagnosis.

Knowledge discovery in databases refers to the overall process of extracting useful high-level knowledge from low-level data with data mining methods. Such knowledge is typically required to be human interpretable as well as computationally useful. In particular, techniques developed on the basis of fuzzy set theory (FST) generally facilitate the tolerance of imprecision, uncertainty and approximation, where many problems in real-life cannot be handled with binary encoding. Amongst those, fuzzy rule-based systems (FRBSs) are one of the most important applications of FST in data mining. A fuzzy rule-based model consists of a set of fuzzy rules in the form of if-then statement, specifying what actions or behaviour should be taken under given circumstances, which allows such a system to reason about how it reaches a conclusion and provide an explanation of its reasoning to human users. Many approaches [4–7] have been proposed for generating and learning FRBSs to represent the input-output behaviour with applications to a number of domain areas [8,9], including applications to medical diagnosis problems [10,11].

In working towards providing assistance for medical doctors to conduct effective diagnosis of diabetes, this paper proposes a neuro-fuzzy approach with a dedicated application to the well-known Pima Indian diabetes benchmark data. The proposed approach works by first discretising each of the continuous attributes into a certain number of categorical ones, such that the original continuous data set can be mapped onto a new data set with only nominal values, facilitating the rapid generation of a set of crisp rules through the exploitation of advanced decision tree learning. The generated crisp rules are able to reveal the basic relationship between attribute-value pairs, whereas the attribute-value pairs which do not appear in the rules can be removed. The generated crisp rules are then transformed to corresponding fuzzy rules with categorical values replaced by Gaussian membership functions. Finally, the set of fuzzy rules are adapted in the neuro-fuzzy framework of ANFIS with gradient descent and least square estimation, for the acquisition of an optimal set of accurate fuzzy rules. Experiments are provided to illustrate the working mechanism of the proposed approach and its performance.

The reminder of this paper is organised as follows. Section 2 introduces the background of fuzzy rules and ANFIS. Section 3 describes the proposed methodology. Section 4 presents and discusses a comparative experimental study. Section 5 concludes the paper and outlines ideas for further development.

## 2    Background

The task of learning an FRBS for the detection of diabetes is to find a finite set of fuzzy production or if-then rules capable of classifying a given input. Without losing generality, the classification system to be modelled is herein assumed to be multiple-input-single-output, receiving $n$-dimensional input patterns and producing one output which is determined to be of one of the $M$ classes. The fuzzy rule set to be induced is required to perform the mapping $\varphi : X^n \to Y$, where $X^n = X_1 \times X_2 \times \cdots \times X_n$, $X_1, X_2, \ldots, X_n$ are the domains of discourse of the input variables and $Y$ represents the set of possible output classes of a cardinality of $M$. The information about the behaviour of the system is described by a set of input-output example pairs $E$, where for each instantiation of the input variables $\bar{x}^p = (x_1^p, x_2^p, \ldots, x_n^p)^T, x_i^p \in X_i, i = 1, 2, \ldots, n$, an associated class $y^p \in Y$ is indicated.

Owing to its capability to approximate nonlinear functions to any degree of accuracy in any convex compact region and being computationally efficient [12], a TSK fuzzy if-then rule is adopted in this paper. In general, a TSK fuzzy if-then rule $F_j$ can be represented as follows:

$$\text{If } x_1 \text{ is } A_{j1} \text{ and ... and } x_n \text{ is } A_{jn}, \text{Then } z = f(x_1, \ldots, x_n) \qquad (1)$$

where $j = 1, 2, .., N$, with $N$ denoting the number of all such fuzzy rules within the system; $x_i, i = 1, \ldots, n$ are the underlying domain variables, jointly defining the $n$-dimensional pattern space and respectively taking values from $X_i$; $A_{ji} \in X_i$ denotes a fuzzy set that the variable $x_i$ may take; and $z_j$ is the consequent of the fuzzy rule describing the output of the model within the fuzzy region specified by the antecedent part of the rule, which is defined as the first order polynomial in this paper, i.e., $z_j = \sum_{i=1}^{n} x_i p_i + q$.

ANFIS [13] is a popular TSK fuzzy inference system built under the framework of artificial neural network, capturing the benefits of both neural networks and fuzzy logic principles. The general ANFIS architecture containing five layers can be simply illustrated with a two-input and one-output system. Suppose that there are only two fuzzy if-then rules in the rule base as follows:

$$\begin{aligned} \text{Rule 1: If } x \text{ is } A_1 \text{ and } y \text{ is } B_1, \text{Then } z_1 = p_1 x + q_1 y + r_1 \\ \text{Rule 2: If } x \text{ is } A_2 \text{ and } y \text{ is } B_2, \text{Then } z_2 = p_2 x + q_2 y + r_2 \end{aligned} \qquad (2)$$

The structure of the neuro-fuzzy ANFIS that is equivalent to the flat fuzzy TSK rule base is shown in Fig. 1, where square nodes stand for the network nodes with parameters to be adapted, and circle nodes represent fixed ones without modifiable parameters. For completeness, details of individual layers within the ANFIS are briefly summarised below.

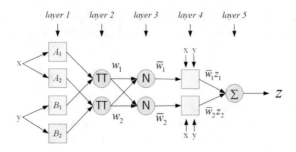

**Fig. 1.** Illustrative ANFIS structure

**Layer 1:** Every node $i$ in this layer is a square node with the following function:

$$O_i^1 = \mu_{A_i}(x) \tag{3}$$

where $x$ denotes the input variable to node $i$, and $A_i$ is the fuzzy set associated with this node. Usually the membership function $\mu_{A_i}(x)$ is chosen to be bell-shaped, typically defined by the Gaussian membership function:

$$\mu_{A_i}(x) = e^{-(\frac{x-c_i}{\sigma_i})^2} \tag{4}$$

where $\{c_i, \sigma_i\}$ are the parameters associated with the corresponding variable, representing the mean value and standard deviation of the Gaussian membership function. These parameters are named premise parameters hereafter. Note that other continuous and piecewise differentiable functions, such as trapezoid or triangular functions may also be utilised if desired.

**Layer 2:** Every node in this layer is a circle node which accumulates the incoming values through multiplication and outputs the product. The output $w_i$ in this layer acts as the firing strength of a certain rule (Rule $i, i = 1, 2$ for the present example). That is,

$$w_i = \mu_{A_i}(x) \times \mu_{B_i}(y) \tag{5}$$

**Layer 3:** Each node in this layer is a circle node, computing the ratio of the $i$th rule's firing strength to the sum of all rules' firing strengths:

$$\overline{w_i} = \frac{w_i}{\sum_{j=1}^{N} w_i} \tag{6}$$

where again, $i = 1, 2$ and the number of rules $N = 2$, for this particular example. The outputs of this layer are normalised firing strengths from previous layer.

**Layer 4:** Each node $i$ in this layer is a square node with the following function:

$$O_i^4 = \overline{w_i} z_i = \overline{w_i}(p_i x + q_i y + r_i) \tag{7}$$

where $\overline{w_i}$ is the output of layer 3, and $p_i, q_i, r_i$ are the parameters which appear in the rules and are referred to as consequent parameters hereafter.

*Layer 5:* The single node in this layer, the output layer, is a circle node that computes the overall output in response to all current inputs, defined as the summation of all incoming values, i.e.,

$$O_1^5 = \sum_i \overline{w_i}z_i = z \tag{8}$$

The parameters, including both premise and consequent ones, are trained using a hybrid learning method combining gradient descent and least square estimation.

# 3  Proposed Methodology

**Data Discretisation.** The Pima Indian Diabetes dataset is obtained from the National Institute of Diabetes and Digestive and Kidney Diseases, which made studies on Pima Indian women of at least 21 years old and living at Phoenix, Arizona, USA [14]. The objective of the dataset is to predict whether or not a patient has diabetes, based on certain diagnostic measurements included in the dataset. Several constraints were placed on the selection of these instances from a larger database. Among 768 cases, there are 268 (34.9%) positive tests for diabetes and 500 (65.1%) cases with tests being negative. There are eight continuous clinical measurements: (1) the number of times pregnant; (2) Plasma glucose concentration over 2 h in an oral glucose tolerance test (mg/dl); (3) Diastolic blood pressure (mm Hg); (4) Triceps skin fold thickness (mm); (5) 2-h serum insulin (mu U/ml); (6) Body mass index (kg/m$^2$); (7) Diabetes pedigree function; and (8) Age. A brief statistics of these eight features is given in Table 1.

**Table 1.** Statistics on Pima Indian diabetes data set

| Attribute | Abbreviation | Mean | Standard deviation | Min/max |
|-----------|--------------|------|--------------------|---------|
| 1 | Pregnant | 3.8 | 3.4 | 1/17 |
| 2 | Glucose | 120.9 | 32 | 56/197 |
| 3 | DBP | 69.1 | 19.4 | 24/110 |
| 4 | TSFT | 20.5 | 16 | 7/52 |
| 5 | INS | 79.8 | 115.2 | 15/846 |
| 6 | BMI | 32 | 7.9 | 18.2/57.3 |
| 7 | DPF | 0.5 | 0.3 | 0.0850/2.3290 |
| 8 | Age | 33.2 | 11.8 | 21/81 |

Each of the eight continuous variables is then discretised into a corresponding categorical one by simply partitioning the universe of discourse into a certain number of equal intervals. Each interval length is set to: $intl = \frac{max(X_i) - min(X_i)}{L}$ where $X_i$ is the domain of attribute $i$ with $max(X_i)$ and $min(X_i)$ being its maximum and minimal value, respectively; $L$ is the user-defined number of partitions. Any original value $x_i$ is mapped onto the integer $k$, if $intl_i^k \leqslant x_i < intl_i^{k+1}, k \in [1, \ldots, L+1]$, where $intl_i^k$ is the $k$-th interval point for attribute $i$. In so doing, the original continuous attributes are transformed into ordinal integers.

Although such heuristic partitioning may not be optimal at this stage, which is to be optimised later with an adaptive method, such a simple empirical data discretisation comes with two advantages: First, each interval can now be attached with a linguistic label with interpretable meanings over pure numerical numbers that hardly make any sense especially to novices in the domain. Second, from computational perspective, the small number of categorical values help expedite the construction of a decision tree for the acquisition of an initial crisp rule base as to be introduced in the subsequent section.

**Generating Crisp Rules with Decision Tree Learning.** Once the discretisation of the original data set is carried out, a set of crisp rules can be generated using a decision tree learning mechanism such as the Classification and Regression Tree (CART) [15]. The basic working of this learning method starts with the full data set at the root node and iteratively applies the Gini index as defined by

$$Gini(S) = \sum_{i=1}^{M} p_i \sum_{k \neq i}^{M} p_k = \sum_{i=1}^{M} p_i(1 - p_i) = \sum_{i=1}^{M} p_i - \sum_{i=1}^{M} p_i^2 = 1 - \sum_{i=1}^{M} p_i^2 \quad (9)$$

where $S$ denotes the current data set for which this index is calculated; $M$ is the number of class labels; $p_i, i \in \{1, \ldots, M\}$ is the probability of an object with the label $i$ being randomly chosen; and $\sum_{k \neq i} p_k = 1 - p_i$ represents the probability of a mistake in categorising an object. It can be seen that the Gini index reaches its minimum when all cases fall into a single category, and maximum when all items are equally distributed among all classes. As such, this index can be used to capture the amount of uncertainty in a dataset, measuring how often a randomly chosen object from the dataset may be incorrectly labelled, if it is randomly labelled according to the distribution of all the labels in the data.

At each split, a decision tree node is generated with the attribute for which the resulting Gini index is minimum. The same procedure is then recursed on each of its subsets using the remaining attributes. When there are no more attributes to be selected for further split or every element in the subset belongs to the same class, a complete decision tree is generated, which can be easily transformed into a set of crisp rules by retrieving paths from each leaf node backwards through its parent to the root node. Without losing generality, suppose that a crisp rule $C_j, j = 1, 2, \ldots, N$ (with $N$ denoting the number of all crisp rules available) is given as follows:

$$\text{If } x_1 \text{ is } I_{j1} \text{ and } \ldots \text{ and } x_n \text{ is } I_{jn}, \text{ Then class is } y^{C_j} \quad (10)$$

where $x_1, x_2, \ldots, x_n$ represent the underlying domain attributes; $I_{ji}, i \in \{1, 2, \ldots, n\}$, is the crisp interval of the antecedent attribute $x_i$ that may be attached with a meaningful label for linguistic interpretation; and $y^{C_j}$ is a class label, acting as the rule consequent.

**Converting Crisp Rules into Fuzzy Rules.** The above data-driven set of crisp rules can then be converted into a set of corresponding fuzzy rules for further optimisation. From the viewpoint of rule structures, a rule is made up of an antecedent and a consequent part, be it fuzzy or crisp. Both the fuzzy and crisp rule antecedent are of a conditional statement form, describing the values that the antecedent attributes should take in order to derive the corresponding consequent, which are connected by logical operators. The only difference is that the attributes in a crisp rule are described with a crisp intervals, whereas an attribute in a fuzzy rule is depicted by a fuzzy set. Therefore, a straightforward approach is to replace each crisp interval (represented by an integer) with a fuzzy membership function. In this work, a crisp interval $I_i$, $i \in \{1, 2, \ldots, n\}$ as shown in Eq. 10 is replaced with a Gaussian membership function $\mu_{A_i}(x_i) = e^{-(\frac{x-c_i}{\sigma_i})^2}$, where $c_i$ and $\sigma_i$ are the mean value and standard deviation of the Gaussian membership function. Here, the mean value is set to the average of those values belonging to the corresponding crisp interval $I_i$, such that

$$c_i = \frac{\sum_{x_i \in I_i} x_i}{|\{x_i \in I_i\}|} \tag{11}$$

Similarly, the standard deviation is set to:

$$\sigma_i = \sqrt{\frac{\sum_{x_i \in I_i} (x_i - c_i)^2}{|\{x_i \in I_i\}|}} \tag{12}$$

Once the process of replacing crisp intervals with the above Gaussian membership functions is complete, the transformation of the entire crisp rule antecedent finishes with the logical 'AND' connector in the original crisp rules replaced with a T-norm operator that performs fuzzy 'AND' operation such as product in ANFIS (or minimum in Mamdani models). The consequent of a crisp rule with a decision class is then directly mapped onto that of the corresponding fuzzy rule. As the first order polynomial TSK rule shown in Eq. 2 is utilised in this paper, the integer that represents the decision class in the crisp rule is taken as the bias term in the fuzzy rule. The resulting mapped fuzzy rule from an original crisp rule can thus, be generally represented as

$$\text{If } x_1 \text{ is } e^{-(\frac{x-c_1}{\sigma_1})^2} \text{ and } \ldots \text{ and } x_n \text{ is } e^{-(\frac{x-c_n}{\sigma_n})^2}, \text{Then } z = 0x_1 + \cdots + 0x_n + r \tag{13}$$

where $e^{-(\frac{x-c_i}{\sigma_i})^2}$ is the fuzzy membership function for attribute $x_i$, $i = 1, \ldots, n$ with $c_i$ and $\sigma_i$ calculated as above, and $r$ is the integer that represents the decision class of the corresponding crisp rule.

Note that running the conventional method of grid partitioning [16] of each and every input space may suffer from the curse of dimensionality as the number of inputs increases. Therefore, instead of considering all of the possible combinations of the input and class attributes, it is herein proposed to utilise the existing crisp rules, which have been generated by decision tree learning and which are able to efficiently and sufficiently generalise the given data to guide the transformation, without resorting to pure and brute force search. Being fundamentally data-driven, such a rule generation method will omit the empty parts of the input space, substantially expediting the subsequent optimisation process.

**Optimising Transformed Fuzzy Rules with ANFIS.** As introduced in Sect. 2, ANFIS uses the framework of neural networks to represent an existing set of fuzzy rules, which are then optimised by tuning premise and consequent parameters. The set of fuzzy rules converted from the generated decision tree rules are herein employed as the initial fuzzy rule set to specify the ANFIS architecture. The ANFIS parameters are optimised with a hybrid learning algorithm [13] that combines the gradient descent method and least squares estimator. In particular, each epoch of the hybrid learning procedure is composed of a forward pass and a backward pass. In running the forward pass, the antecedent parameters are fixed and a vector of input values is presented, and then the error between the actual output and the target output is calculated. In the backward pass, the error computed at the last forward pass is propagated backwards, from the output end towards the input end while fixing the consequent parameters, by the gradient method. The details of such an iteration of forward and backward computation is beyond the scope of this paper, but can be found in [13].

# 4    Experimental Study

To demonstrate the proposed approach at work for effectively aiding in diagnosis of diabetes, the experimental study is performed on the popular Pima Indian diabetes data set obtained from UCI machine learning repository [17], whose statistical information has already been listed in Table 1. Due to space limit only investigations into this single dataset are reported here.

As the range of different attributes vary significantly, a preprocessing step is to normalise each attribute so that their normalised values fall within the range of $[0, 1]$. This helps reduce the potential adverse effect caused by certain attributes being dominating, especially those undesired ones just because of their bigger value ranges. In the absence of testing data for the performance evaluation of the proposed approach, tenfold cross-validation (10-CV) is employed for result validation. In 10-CV, the given data set is partitioned into ten subsets. Of the ten, nine subsets are used to perform training, where the proposed approach is used to generate the desired fuzzy rule base, and the remaining single subset is retained as the testing data for assessing the learned classifier's performance. This cross-validation process is then repeated ten times in order to lessen the impact of any random factors.

The decision tree algorithm utilised in this paper is the popular Classification and Regression Trees (CART), which is characterised by its construction of a

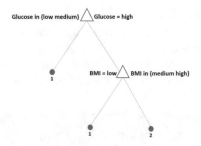

**Fig. 2.** Generated decision tree

binary tree with each internal node having exactly two outgoing branches. In particular, each original continuous attribute is empirically discretised into categorical one with 3 equally spaced bins. For instance, the resulting decision tree structure, which is taken from a single fold out of the complete 10-CV, is shown in Fig. 2, where '1' and '2' in the leaf nodes stand for negative and positive test respectively; and 'low', 'medium' and 'high' are the labels used to denote the corresponding discretised crisp intervals. Note that certain attributes can take more than one interval as its value such as that Glucose can take either low or medium in its left branch, which is attributed to the mechanism that CART grows the trees. However, this can easily be transformed into a rule base with each attribute only taking a single value as follows:

- Rule 1: If Glucose is *low*, Then test *negative*;
- Rule 2: If Glucose is *medium*, Then test *negative*;
- Rule 3: If Glucose is *high* and BMI is *low*, Then test *negative*;
- Rule 4: If Glucose is *high* and BMI is *medium*, Then test *positive*;
- Rule 5: If Glucose is *high* and BMI is *high*, Then test *positive*.

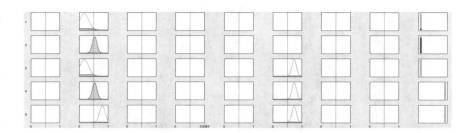

**Fig. 3.** Transformed fuzzy rule base

In this example, only two attributes out of the original eight are utilised to initiate the transformation process, with five rules generated, which could have been as many as $3^8 \times 2$ if grid partitioning were used for rule generation. The above crisp rule base is then transformed into a fuzzy one with crisp intervals replaced by Gaussian membership functions as specified in Sect. 3. The resultant fuzzy rule base is shown in Fig. 3. Each numbered row represents a converted fuzzy rule, where the last rectangle represents the output and the first eight rectangles are the input fuzzy membership functions. Obviously, only the second and the sixth rectangle come with Gaussian membership functions given that just the two corresponding attributes are utilised in the original crisp rules.

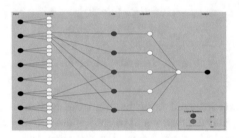

**Fig. 4.** ANFIS structure

The transformed rule base serves as the input to the neuro-fuzzy ANFIS structure as shown in Fig. 4. ANFIS then fine-tunes both the antecedent and consequent parameters based on the existing rule base structure by running the hybrid learning method. With regard to the two attributes used, Fig. 5 shows the definitions of the initial membership functions, whose optimised definitions are shown in Fig. 6. In order to better visualise the significant performance improvement after optimisation with ANFIS, Table 2 lists the accuracy of each 10-CV fold on training data with initial rule base (Trn_Initial_ANFIS), training data with ANFIS optimised rule base (Trn_Optimised_ANFIS) and testing data with optimised rule base (Tst_Optimised_ANFIS). Clearly, the performance of each individual fold on training data is significantly improved along with ANFIS optimisation.

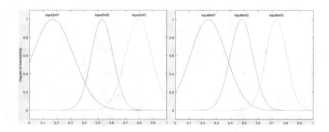

**Fig. 5.** Initial membership functions of 2nd attribute (Glucose) and 6th attribute (BMI)

To further compare the performance on testing data with alternative approaches, Table 3 lists the accuracy with another five popular learning classifiers. These include PTTD [5], a state-of-the-art fuzzy pattern-tree learning

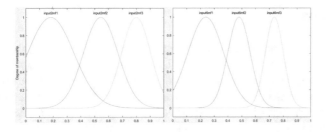

**Fig. 6.** Optimised membership functions of 2nd attribute (Glucose) and 6th attribute (BMI)

**Table 2.** Accuracy (%) over data classified with different rule sets

| Fold | Trn_Initial_ANFIS | Trn_Optimised_ANFIS | Tst_Optimised_ANFIS |
|---|---|---|---|
| 1 | 75.11 | 79.45 | 77.92 |
| 2 | 73.52 | 76.85 | 80.52 |
| 3 | 63.24 | 81.19 | 74.03 |
| 4 | 74.13 | 78.32 | 81.58 |
| 5 | 67.73 | 81.62 | 75.32 |
| 6 | 76.12 | 79.16 | 76.62 |
| 7 | 72.50 | 83.50 | 61.04 |
| 8 | 73.95 | 81.62 | 75.32 |
| 9 | 75.58 | 79.34 | 82.89 |
| 10 | 70.04 | 83.65 | 71.43 |
| Average | 72.19 | 80.47 | 75.67 |

classifier that is composed of an ensemble of pattern trees; GP-COACH [7], a genetic programming-based learning approach that learns rules of a disjunctive normal form; SMO, a sequential optimisation algorithm for building support vector machines with the polynomial kernel adopted as kernel function; IBk, the classical $k$-nearest neighbour approach, and RIPPER which is a crisp rule induction algorithm following a separate-and-conquer strategy. Apart from the powerful support vector machines, the proposed approach beats all the other popular learning classifiers, demonstrating its promising performance.

**Table 3.** Accuracy (%) comparison with alternative methods

| Proposed approach | PTTD | GP-COACH | SMO | IBk | RIPPER |
|---|---|---|---|---|---|
| 75.67 | 74.13 | 75.13 | 76.80 | 70.62 | 66.88 |

## 5   Conclusion

This paper has proposed an effective approach to learn a TSK fuzzy rule base with application to the diagnosis of diabetes mellitus. The proposed approach starts with the generation of a crisp rule base with a simple decision tree learning mechanism, which is data-driven and able to learn basic rule structures that reflect the characteristics between domain input and output attributes, with minor computational overheads. The crisp rule base is then transformed into a fuzzy rule base with crisp intervals replaced by Gaussian membership functions. This forms the input to the powerful neuro-fuzzy framework of ANFIS, which further optimises the rules' antecedent and consequent parameters. The resultant optimised fuzzy rules that are transformed from simple crisp rules show promising results with better or comparable performance to those derived from popular machine learning algorithms.

Whilst promising, interesting work remains. This includes: examining the influence of crisp rule bases of a different complexity upon the performance of the final fuzzy rules and the effect of any subsequent fine-tuning with ANFIS; investigating the use of more powerful data discretisation techniques (e.g., [18]); and comparing the proposed approach with alternative methods (e.g., [19]) that employ fuzzy systems for diabetes detection.

## References

1. Holt, R.I., Hanley, N.A.: Essential Endocrinology and Diabetes, vol. 41. Wiley, Chichester (2012)
2. Temurtas, H., Yumusak, N., Temurtas, F.: A comparative study on diabetes disease diagnosis using neural networks. Expert Syst. Appl. **36**(4), 8610–8615 (2009)
3. Polat, K., Güneş, S., Arslan, A.: A cascade learning system for classification of diabetes disease: generalized discriminant analysis and least square support vector machine. Expert Syst. Appl. **34**(1), 482–487 (2008)
4. Chen, T., Shang, C., Su, P., Shen, Q.: Induction of accurate and interpretable fuzzy rules from preliminary crisp representation. Knowl. Based Syst. **146**, 152–166 (2018)
5. Senge, R., Hüllermeier, E.: Fast fuzzy pattern tree learning for classification. IEEE Trans. Fuzzy Syst. **23**(6), 2024–2033 (2015)
6. Chen, T., Shen, Q., Su, P., Shang, C.: Fuzzy rule weight modification with particle swarm optimisation. Soft Comput. **20**(8), 2923–2937 (2016)
7. Berlanga, F.J., Rivera, A., del Jesús, M.J., Herrera, F.: GP-COACH: genetic programming-based learning of compact and accurate fuzzy rule-based classification systems for high-dimensional problems. Inf. Sci. **180**(8), 1183–1200 (2010)
8. Su, P., Shen, Q., Chen, T., Shang, C.: Ordered weighted aggregation of fuzzy similarity relations and its application to detecting water treatment plant malfunction. Eng. Appl. Artif. Intell. **66**, 17–29 (2017)
9. Su, P., Shang, C., Chen, T., Shen, Q.: Exploiting data reliability and fuzzy clustering for journal ranking. IEEE Trans. Fuzzy Syst. **25**(5), 1306–1319 (2017)
10. Zou, C., Deng, H.: Using fuzzy concept lattice for intelligent disease diagnosis. IEEE Access **5**, 236–242 (2017)

11. Wang, J., Hu, Y., Xiao, F., Deng, X., Deng, Y.: A novel method to use fuzzy soft sets in decision making based on ambiguity measure and Dempster-Shafer theory of evidence: an application in medical diagnosis. Artif. Intell. Med. **69**, 1–11 (2016)
12. Feng, G.: A survey on analysis and design of model-based fuzzy control systems. IEEE Trans. Fuzzy Syst. **14**(5), 676–697 (2006)
13. Jang, J.-S.: ANFIS: adaptive-network-based fuzzy inference system. IEEE Trans. Syst. Man Cybern. **23**(3), 665–685 (1993)
14. Knowler, W.C., Bennett, P.H., Hamman, R.F., Miller, M.: Diabetes incidence and prevalence in Pima Indians: a 19-fold greater incidence than in Rochester, Minnesota. Am. J. Epidem. **108**(6), 497–505 (1978)
15. Breiman, L.: Classification and Regression Trees. Routledge, New York (2017)
16. Wang, L.-X., Mendel, J.M.: Generating fuzzy rules by learning from examples. IEEE Trans. Syst. Man Cybern. **22**(6), 1414–1427 (1992)
17. Bache, K., Lichman, M.: UCI Machine Learning Repository (2013). http://archive.ics.uci.edu/ml
18. Boongoen, T., Shang, C., Iam-On, N., Shen, Q.: Extending data reliability measure to a filter approach for soft subspace clustering. IEEE Trans. Syst. Man Cybern Part B (Cybern.) **41**(6), 1705–1714 (2011)
19. Lukmanto, R., Irwansyah, E.: The early detection of diabetes mellitus (DM) using fuzzy hierarchical model. Proc. Comput. Sci. **59**, 312–319 (2015)

# A Comparison of Re-sampling Techniques for Pattern Classification in Imbalanced Data-Sets

Marcia Amstelvina Saul$^{(\boxtimes)}$ and Shahin Rostami

Faculty of Science & Technology, Bournemouth University,
Bournemouth BH12 5BB, UK
{msaul,srostami}@bournemouth.ac.uk
https://research.bournemouth.ac.uk/project/ciri/

**Abstract.** Class imbalance is a common challenge when dealing with pattern classification of real-world medical data-sets. An effective counter-measure typically used is a method known as re-sampling. In this paper we implement an ANN with different re-sampling techniques to subsequently compare and evaluate the performances. Re-sampling strategies included a control, under-sampling, over-sampling, and a combination of the two. We found that over-sampling and the combination of under- and over-sampling both led to a significantly superior classifier performance compared to under-sampling only in correctly predicting labelled classes.

**Keywords:** Machine learning · Imbalanced data · Over-sampling
Under-sampling

## 1 Introduction

There is an increasing interest in the application of machine learning in providing assistance to diagnosticians whom may otherwise be uncertain of a prognosis [26]. Previous research into predictive measures have found that pattern detections can be extracted from medical tracings and medical imaging, such as; identification of diabetic retinopathy [23], cancerous cells in dermatology [11] and identifying brain tumours in MRI scans [29]. Classification models which take into account previous medical cases could reduce the time taken to arrive at a prognosis, and even suggest possible onset of a disease to treat it before the harmful symptoms manifest [26]. Artificial neural networks (ANNs) have previously improved the performance of potentially out-of-date and ungeneralizable indexes or heuristics still used in the health-care industry [26] by allowing clinicians to make more informed decisions about their diagnosis. As such, machine learning is a vital tool in bridging the gaps of missing information within these tests to increase the validity and accuracy of the suggested prognosis. In addition, successful implementation of such an ANN will also reduce the risk of the disease

© Springer Nature Switzerland AG 2019
A. Lotfi et al. (Eds.): UKCI 2018, AISC 840, pp. 240–251, 2019.
https://doi.org/10.1007/978-3-319-97982-3_20

worsening and the corresponding financial implications. These workings ultimately lead towards one major outcome, that is overall patient satisfaction [26].

In the health-care industry, there is often the challenge of class imbalance within the data-set i.e. when classes are significantly over/under represented, particularly when concerning rare diseases or abnormalities. For example, consider a scenario where smallpox has become re-apparent and clinicians must quickly differentiate between spots symptomatic of chickenpox and those of smallpox to hasten eradication. The engineers may notice that out of the 1,000 case files, only 10 were reported of smallpox. If the engineers feed this data into a neural network, what they would find is that approximately 990 predicted output values would successfully estimate very close to the target output values. This suggests that the network model has 99% accuracy, however this is not necessarily a useful indication of its performance. Whilst it would indeed have 99% accuracy for correctly predicting chickenpox, they would also have 0.01% accuracy for correctly predicting smallpox (if the model could predict for smallpox at all). This means that the model overall has poor performance considering that they have built the model to specifically detect smallpox. This is an example of the class imbalance problem, which is typically addressed during pre-processing and manipulation of the data-set prior to ANN training.

The remainder of the paper is organised as follows: In Sect. 2 the data-set and approaches considered for comparison are discussed, followed by Sect. 3 which lists and discusses the numerical results complete with a statistical analysis. The paper is then concluded in Sects. 4 and 5 with recommendations for future research directions.

## 2   Methods

The data used in the experiments of this paper was obtained from the UCI Machine Learning Repository [7] and contains cardiotocography measurements. Data acquisition and analyses were carried out by [1], presenting a data-set with 23 attributes and 2,126 samples for each attribute (extracted from real-world consenting participants). These attributes include 21 input features and 2 possible output classification criteria to be utilised separately as 3-class or 10-class experiments. The study [1] authors state that the output classes in both criteria were labelled and substantiated by expert obstetricians. This paper uses the 3-class output criteria relating to the detection of foetal states: 'Normal', 'Suspect', and 'Pathologic' from the cardiotocograms. Thus, the preliminary network architecture incorporated 21 input nodes and 3 output nodes.

In order to solve this pattern classification problem, the first step was to address the data-set itself and its suitability for building classification models upon. The data-set source study on SisPorto 2.0, an automated cardiotocogram analysis system, is a performance test on a wide-scale evaluation. The system was tested on over 6,000 pregnancies across 14 centres in Europe and Australia producing an extent of generalisation, or domain representation, provided analysis is conducted on the tested demographics. In addition, tracings from the foetal heart rate (FHR), including baseline, accelerations, deceleration, and variability, were subject to Cohen's kappa coefficient testing. Whereby the clinicians over-all proportions gave "fair-to-good" agreements of the results [4]. The authors also found a 100% sensitivity and 99% specificity rating for their predictions in neonatal abnormalities [4]. These findings therefore propose a healthy quality of data for building a classification model.

The subsequent and focal issue to be addressed prior to classification was the imbalance of class frequencies. Whilst the experiment will implement different techniques to balancing the classes, the goal across all techniques is to obtain a 1:1:1 ratio. Four different experiments were conducted for balancing classes: over-sampling towards the majority class frequency, under-sampling towards the minority class frequency, a combination of over- and under-sampling towards a sufficiently representative sample size and a benchmark model with no re-sampling to allow for a control experiment.

The rule of thumb when choosing the number of hidden neurons is typically between the number of inputs and the number of outputs [13]. Similarly, for low-scale data-sets, one or two hidden layers are sufficient. The ANN classifier was defined by employing optimiser and loss function algorithms for the learning phase. The Adaptive Subgradient Method, or *adagrad*, is an optimisation function which has shown to accommodate for different measurement types across the input features [9]. Accompanied by the loss function often used for handling multi-class problems, Categorical Cross-Entropy. Following the implementation and training of the classifiers, performance evaluation measures were applied to test the quality of the predictions made. The evaluation techniques used were the receiver operator characteristic with area under curve (ROC-AUC) as literature suggests its suitability with medical data [2, 14, 22], and the f1-score which represents a harmonic mean between precision and recall measures. Each performance evaluation was computed from a confusion matrix and macro-averaged. The final stage of the experiment was to apply statistical testing to derive the presence of a significant difference between the performances of the trained classifiers across re-sampling techniques.

# 3 Numerical Results

## 3.1 Experimental Set-Up

The algorithms employed by this experiment were implemented using the Python 2.7 programming language, which leveraged the Keras (https://keras.io/) neural networks API running on top of the TensorFlow (https://www.tensorflow.org/) machine learning framework. All experiments were conducted within identical Docker containers to facilitate an isolated and reproducible research environment.

**Pre-processing of Data-Set.** On extraction from the UCI Machine Learning Repository, the first noticeable property of the raw data-set was that each input attribute had been measured on different scales during analysis on SisPorto 2.0.

Figure 1 illustrates the distribution of 4 input attributes and the extremities in scaling variance. Accordingly, a standardisation procedure was applied to the data-set as a whole. The data-set was divided into training and testing subsets, using the conventional division of 70% training and 30% testing [13]. Administering this division prior to re-sampling ensures that the testing data is untouched to provide pure values when testing the classifier model.

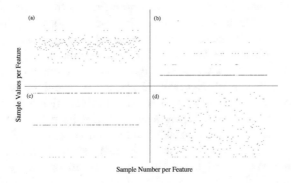

**Fig. 1.** Four features represented as scatter-plot; (a) Feature: UC, (b) Feature: Nzeros, (c) Feature: Variance, (d) Feature: Width. Extracted from the 'Suspect' data-set prior to pre-processing. Illustrates the complexity of feature measurements and necessity for standardisation of data.

The output nodes of 'Normal', 'Suspect', and 'Pathologic' foetal states comprise of 1,655, 295, and 176 samples per feature respectively. The over-sampling only experiment obtained the 1:1:1 ratio by re-sampling so each class was of 1,655 elements each. Conversely, the under-sampling only experiment was re-sampled so each class was of 176 elements each. The combination of over- and under-sampling techniques was where provisions were made in order to determine a representative class size. To derive the class sizes, ANOVA fixed-effects tests were used with a pre-determined power $(1-\beta$, where the $\beta$-value represents Type II error), alpha ($\alpha$, Type I error), and effect size ($f$). Clinically based studies suggest $\alpha$-values and $(1-\beta)$-values of at least 0.05 and 0.80 respectively as standard for optimal testing [16]. Subsequently, $f$ was determined using a type of effect size known as the risk ratio [12], recommended for binary-type classes [27]. Typically, effect sizes are calculated as a statistical measure between two classes and the number of classes in this experiment exceeded this limitation. Previous studies [3, 20] suggest that to overcome this problem, it is necessary to transfer the statistical inference by testing between the two outermost classes, i.e. 'Normal' (C1) and 'Pathologic' (C3). Therefore, the effect size (1) was calculated, where $n$ is the number of total elements across the classes. The G*Power 3.1 [8] software application was used to execute the ANOVA fixed-effects test and it was calculated that 861 elements per class were required as an optimal class size given the prior parameters.

$$RiskRatio = \frac{C1/n}{C3/n} = \frac{176/2126}{1655/2126} = 0.106 \tag{1}$$

Re-sampling of the data-set followed the determination of target class sizes. Over-sampling was applied by incorporating a variant of the synthetic data generation technique, SMOTE [5], known as SVM (Support Vector Machine) SMOTE. This variant of SMOTE generates new synthetic data using SVMs to predict new unknown elements at the borderline of the minority classes [17]. This method is capable of substantiating the decision boundary for the classifier and also expand minority classes which occur in the data space of majority class elements [17]. The intuition for the SMOTE technique is to generate new data elements in feature space as opposed to data space to reduce over-fitting. Under-sampling was applied using NearMiss-2 [28,30], a technique which employs a clustering $k$-NN (nearest neighbour) algorithm to select elements from the majority class which has the smallest averaged distance from the $k$ furthest minority elements. NearMiss is a controlled under-sampling method, which enabled the experiment to define the number of elements for each under-sampled class. In addition, NearMiss-2 has been found to perform optimally out of all the NearMiss variants [28]. The combination of over- and under-sampling employed SVM SMOTE on the 'Suspect' and 'Pathologic' classes whilst NearMiss-2 was be employed for the 'Normal' class.

**Network Architecture and Parameter Configuration.** The ANN topology consisted of 21 input nodes and 3 output nodes across all re-sampling experiments. There were 2 hidden layers implemented for the classifier model, with 21 hidden neurons integrated in each layer. Each hidden layer applied the Rectified Linear Unit (ReLU) activation function. The data fed into the ANN for training corresponds to the class sizes which had been established previously. The dimensions of the input data matrix consisted of $4,965$ by 21 for over-sampling only, 528 by 21 for under-sampling only, and 2,583 by 21 for over- and under-sampling. The control experiment had no re-sampling, and therefore consisted of the original 2,126 by 21 matrix. Each sample was fed into the ANN individually and the entire data-set was used for training the ANN iteratively for a maximum of 500 epochs.

To ensure accurate model predictions, it was essential that the output layer was a binary representation of the foetal state classes (Table 1). This is due to the fact that the original raw data-set classifies the foetal states by assigning '1', '2', and '3' to 'Normal', 'Suspect', and 'Pathologic' respectively. It would be difficult to confirm which class the ANN was attempting to predict if the values obtained were continuous (i.e. verifying a predicted output of 2.5 towards either 2 or 3). In addition, a typical output layer activation function (such as the sigmoid activation function which were used in these experiments) squashes retrieving values between 0 and 1, then activates if the value exceeded 0.5. Therefore, less dubious translations of model predictions to class predictions could be made when binary representations of the output were implemented.

**Table 1.** Binary output illustrated the activation of only one neuron in the output layer at a time.

| Foetal state | Normal | Suspect | Pathologic |
|---|---|---|---|
| Classification | 1 | 2 | 3 |
| Binary output | [ 1 0 0 ] | [ 0 1 0 ] | [ 0 0 1 ] |

During training, over-fitting preventative strategies were incorporated by setting a neuron dropout of 0.25 between layers (sets a percentage of neuronal output to 0) and neutralised the high epoch of 500 with an early stopping criterion [21]. Early stopping allowed the iterations of training to stop if the loss function does not improve for a specified amount of consecutive iterations, increasing robustness of the network.

**Statistical Analysis.** Once the classifier model was built and trained, the ROC-AUC and f1-score performance evaluations were calculated. In order to determine the presence of a significant difference between the re-sampling strategies, statistical analysis was conducted on these performance evaluations.

Firstly, the number of times each experiment was ran to determine the sample size for analysis was obtained by extracting the standard error of the mean (SEM) per number of trials. The number of times an experiment was trialled was grouped into an overall sample size. The SEM of the performance evaluation values were extracted from a starting point of 3 samples (i.e. running the experiment 3 times) and continued towards 80 samples. Figure 2 illustrates how the SEM changes with sample size. It could be observed that past 50 samples the SEM curve began to plateau and a larger sample size would not entail a notably greater effect worthy of additional trials when it came to the analysis.

Having established the sample size, the distribution normality of the performance evaluation data for each experiment was tested. The p-values from D'Agostino and Pearson's [6,18] test for normality was extracted for each class from measures of precision, recall and ROC-AUC (a macro-average of the f1-score as the harmonic mean of precision and recall was computed and therefore these distributions were tested for normality). Using an $\alpha$-value of 0.001, a collection of both normally and non-normally distributed data was found. The overall consensus, however, was that a non-parametric test was required to measure the statistical significance between the four experiments. This is because the macro-average of both the f1-score and the ROC-AUC were used as the components of the samples under statistical analysis. There were no two performance evaluation classes which contained purely normally or non-normally distributed data and a parametric test is bound in its ability to account for the non-parametric details of certain classes. Therefore, the non-parametric Wilcoxon signed-rank test (tests the medians of "two paired measurements made on identifiable population" [10]) was implemented to test for a significant difference between the performance evaluation outcomes of each re-sampling experiment.

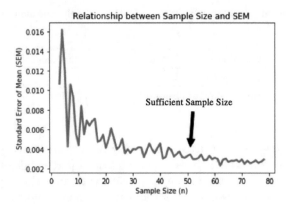

**Fig. 2.** The standard error of mean was measured for each number of times the experiment was run to determine an appropriate sample size for statistical analysis. The SEM data was extracted from the control group experiment.

## 3.2   Experimental Results

Performance evaluation measures were obtained by computing the confusion matrix for each experiment (Fig. 3). Hereafter, 'combi-sampling' refers to the over- and under-sampling experiment.

Correctly labelled samples (true positives) are represented diagonally in the boxes where the actual output class meets its corresponding predicted output class [24], i.e. 'Normal'-'Normal'. Whilst the off-diagonal values in the matrix represent mislabelled samples [24] (true negatives, false positives and false negatives). Throughout the experimental trials, over-sampling and combi-sampling exhibited consistent performance in the outcomes of the true positive values of the confusion matrix. For each class, both experiments held between 0.70 and 0.99 for each class true positive rate and low values of mislabelled samples over the 50 trials. On the other hand, there were fluctuating outcomes of correctly labelled samples for the control and under-sampling. Generally, both experiments had higher levels of mislabelled samples and were unable to predict all classes to the level exhibited by over-sampling and combi-sampling. Under-sampling either predicted one class well and poorly for the remaining two, or poorly for all classes. Control, as expected with no re-sampling, predicted only class 'Normal' well, with occasional good prediction for either one of the remaining classes but never all three classes.

From the confusion matrix, the ROC-AUC values for each class and its corresponding macro-average were computed for testing of the classifier performance. The ROC curve is defined as illustrating an excellent performance of a model when its area (AUC) is 1.00 [25]. In other words, the curve peaks at the beginning of the plot and maintains the inflated y-values (true positive rate) along x (false positive rate). Figure 4 portrays the ROC curves and AUC values for each re-sampling experiment.

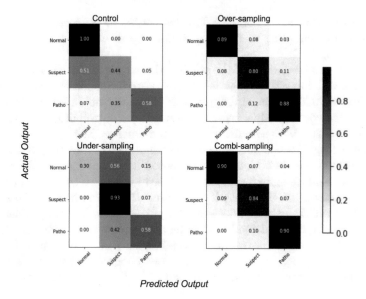

**Fig. 3.** Confusion matrices extracted from the last trial of each re-sampling experiment.

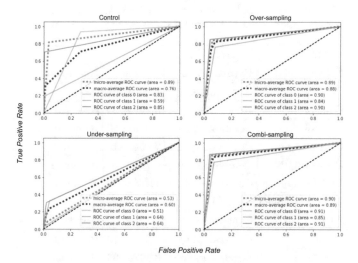

**Fig. 4.** Receiver operator characteristic (ROC) curves extracted from the last trial of each re-sampling experiment. Area under curve (AUC) values for each class and averaged values are indicated in the legend box of each plot.

All re-sampling techniques reduced the distances of the ROC curves for each class from each other, illustrating the 1:1:1 ratio across 'Normal', 'Suspect', and 'Pathologic' classes. Similarly to the results of the confusion matrix, the ROC-AUC metrics for over-sampling and combi-sampling were alike in shape and showed indications of correctly labelled classes by the classifier model. Also similarly, under-sampling was unable to perform as well as over-sampling and combi-sampling.

Table 2 shows the results of the statistical analysis made using the Wilcoxon signed-rank test for a difference between the performance evaluations of each re-sampling technique. The null hypothesis (there was no difference between the medians of the corresponding two measurements) was tested. Each test statistic was calculated at a 95% confidence level i.e. α-value = 0.05.

Statistical significance between each re-sampling experiment was determined by whether the null hypothesis was accepted or rejected for both evaluation metrics. All re-sampling experiments exhibited a significantly different classifier model performance to the control. This result drives the need for re-sampling when dealing with imbalanced data-sets for pattern classification problems using ANNs. Over-sampling and combi-sampling both presented significant differences in performance with under-sampling, whilst having no significant difference in performance with each other. This finding was congruous with the illustrations in Figs. 3 and 4 of the confusion matrices and ROC curves respectively.

**Table 2.** Results of the Wilcoxon signed-rank test between every re-sampling experiment from the macro-averaged f1 score and ROC-AUC. The null hypothesis of no difference between experiments is rejected if the p-value is less than α-value = 0.05.

| Macro-Averaged F1 Score | | | |
|---|---|---|---|
| Experiment 1 | Experiment 2 | p-value | Null hypothesis |
| Control | Over-sampling | 1.383e−09 | Rejected |
| Control | Under-sampling | 5.851e−09 | Rejected |
| Control | Combi-sampling | 2.349e−07 | Rejected |
| Over-sampling | Under-sampling | 7.554e−10 | Rejected |
| Over-sampling | Combi-sampling | 0.798 | Accepted |
| Under-sampling | Combi-sampling | 1.978e−09 | Rejected |
| Macro-Averaged ROC-AUC | | | |
| Experiment 1 | Experiment 2 | p-value | Null hypothesis |
| Control | Over-sampling | 7.550e−10 | Rejected |
| Control | Under-sampling | 3.197e−06 | Rejected |
| Control | Combi-sampling | 2.820e−09 | Rejected |
| Over-sampling | Under-sampling | 7.554e−10 | Rejected |
| Over-sampling | Combi-sampling | 0.263 | Accepted |
| Under-sampling | Combi-sampling | 8.271e−10 | Rejected |

# 4    Discussion

The experimental design of combi-sampling was incorporated as a proposition to address the distinctive frequency discrepancy between classes. By solely incorporating under-sampling, there is a risk of losing essential data in more than one feature of the data-set. On the other hand, pure over-sampling poses a risk of over-fitting the classifier model. Due to the use of early stopping and neuron dropout methods, there could be some exoneration of over-sampling from over-fitting the classifier model. However, it appeared from the statistical analyses that the under-sampling methods did indeed manage to lose essential data points in attempting to represent a complex data-set. By observing the methodology used in this paper for extracting a sufficient sample size for the combi-sampling experiment, it could be seen that the statistically minimum sample size required to be representative of the data-set was 861 elements per class. This meant that in the under-sampling experiment, the classes had a critically insufficient size of 176 elements per class. In addition, considering the existing limitation of obtaining a comprehensively generalizable cardiotocography data-set that includes the statistics of every pregnant woman, it would be favourable to build upon existing data and maintain the existing essential elements with over-sampling rather than lessen the data with under-sampling.

Although over-sampling and combi-sampling produced satisfactory performance evaluation scores, there is still room for improvement (i.e. bringing the AUC value closer to 1.00). Real-life applications of machine learning in healthcare would focus on this optimisation of state predictions to enhance preventative techniques. Therefore, some further work on the extension of this paper is required. Firstly, with the matter of classifier performance in general, the 'Suspect' data-set had perhaps been difficult to class because it contained aspects of data which could fall in either the 'Normal' data-set or the 'Pathologic' data-set. In other words, there were overlapping data-instances either already existent in the original data-set or introduced via the re-sampling strategies [19]. By comparing this paper to the results of an alternative machine learning technique (such as fuzzy logic systems which have been shown to work well with data-sets containing overlapping categories [15]), we may be able to determine if an alternative technique is able to provide a superior predictive performance. In addition, some further work could be done in establishing that over-fitting was indeed avoided within the over-sampling and combi-sampling experiments. Lastly, a larger data-set could be implemented to establish a more detailed view by introducing a larger set of unseen testing data, or an experiment incorporating and evaluating the effect of different ANN classifiers (i.e. variant network topologies, optimisers or loss-functions).

# 5    Conclusion

In this paper, the performance evaluation of an ANN classifier in predicting foetal states from cardiotocography measurements using different re-sampling techniques was presented. The main findings concluded that over-sampling and combi-sampling were able to reasonably predict all classes 'Normal', 'Suspect' and 'Pathologic'. Over-sampling and combi-sampling maintained a level of performance with each other (no significant differences) and accomplished significantly more accurate results than under-sampling, whose performance allowed the classifier to only sufficiently predict only one class at a time, if at all.

Furthermore, a significant difference between all re-sampling techniques and the control was found, reiterating the pivotal role of re-sampling of data-sets when dealing with class imbalances. The adverse effects of no re-sampling is additionally illustrated by the control ROC-AUC curve (Fig. 4), whereby an inconsistency in performance exists across classes. By balancing class frequencies, it eradicated means of false accuracy exhibited in the control and significantly improved the performance of an ANN classifier model in order to assess the differences in re-sampling techniques for a pattern recognition problems.

# References

1. Ayres-DeCampos, D., Bernardes, J., Garrido, A., MarquesDeS, J., PereiraLeite, L.: SisPorto 2.0: a program for automated analysis of cardiotocograms. J. Matern. Fetal Med. **9**, 311–318 (2000)
2. Bradley, A.P.: The use of the area under the ROC curve in the evaluation of machine learning algorithms. Pattern Recogn. **30**(7), 1145–1159 (1997). https://doi.org/10.1016/s0031-3203(96)00142-2
3. Brooks, G.P., Johanson, G.A.: Sample size considerations for multiple comparison procedures in ANOVA. J. Mod. Appl. Stat. Methods **10**(1), 97–109 (2011). https://doi.org/10.22237/jmasm/1304222940
4. de Campos, D.A.: The SisPorto automated analysis
5. Chawla, N., Bowyer, K., Hall, L., Kegelmeyer, W.P.: SMOTE: synthetic minority over-sampling technique. J. Artif. Intell. Res. **16**, 321–357 (2002)
6. Dagostino, R.B.: An omnibus test of normality for moderate and large size samples. Biometrika **58**(2), 341 (1971). https://doi.org/10.2307/2334522
7. UCI Machine Learning Repository Database: Cardiotocography Data Set (2010). https://archive.ics.uci.edu/ml/datasets/cardiotocography
8. HHU Düsseldorf: G*Power. http://www.gpower.hhu.de/en.html
9. Duchi, J., Hazan, E., Singer, Y.: Adaptive subgradient methods for online learning and stochastic optimization. J. Mach. Learn. Res. **12**, 2121–2159 (2011)
10. Ennos, A.R., Johnson, M.: Statistical and Data Handling Skills in Biology. Pearson Education, New York (2017)
11. Esteva, A., Kuprel, B., Novoa, R., Ko, J., Swetter, S., Blau, H., Thrun, S.: Dermatologist-level classification of skin cancer with deep neural networks. Nature **542**(7639), 115–118 (2017)
12. Gigerenzer, G.: Helping doctors and patients make sense of health statistics. In: Simply Rational, p. 2193 (2015). https://doi.org/10.1093/acprof:oso/9780199390076.003.0005
13. Heaton, J.: Introduction to Neural Networks for Java, p. 440. Heaton Research, Inc. (2008). https://dl.acm.org/citation.cfm?id=1502373. ISBN 1604390085 9781604390087

14. Huang, J., Ling, C.: Using AUC and accuracy in evaluating learning algorithms. IEEE Trans. Knowl. Data Eng. **17**(3), 299–310 (2005). https://doi.org/10.1109/tkde.2005.50
15. Ishibuchi, H., Nakaskima, T.: Improving the performance of fuzzy classifier systems for pattern classification problems with continuous attributes. IEEE Trans. Ind. Electron. **46**(6), 1057–1068 (1999). https://doi.org/10.1109/41.807986
16. Kim, H.Y.: Statistical notes for clinical researchers: type I and type II errors in statistical decision. Restor. Dentist. Endod. **40**(3), 249 (2015). https://doi.org/10.5395/rde.2015.40.3.249
17. Nguyen, H.M., Cooper, E.W., Kamei, K.: Borderline over-sampling for imbalanced data classification. Int. J. Knowl. Eng. Soft Data Paradigms **3**(1), 4 (2011). https://doi.org/10.1504/ijkesdp.2011.039875
18. Pearson, E.S., Dagostino, R.B., Bowman, K.O.: Tests for departure from normality: comparison of powers. Biometrika **64**(2), 231–246 (1977). https://doi.org/10.1093/biomet/64.2.231
19. Prati, R.C., Batista, G.E.A.P.A., Monard, M.C.: Class imbalances versus class overlapping: an analysis of a learning system behavior. In: MICAI 2004: Advances in Artificial Intelligence Lecture Notes in Computer Science, pp. 312–321 (2004). https://doi.org/10.1007/978-3-540-24694-7-32
20. Preacher, K.J., Rucker, D.D., Maccallum, R.C., Nicewander, W.A.: Use of the extreme groups approach: a critical reexamination and new recommendations. Psychol. Methods **10**(2), 178–192 (2005). https://doi.org/10.1037/1082-989x.10.2.178
21. Prechelt, L.: Early stopping but when? In: Neural Networks: Tricks of the Trade, vol. 7700 (2012). https://doi.org/10.1007/978-3-642-35289-8-5
22. Provost, F., Fawcett, T., Kohavi, R.: The case against accuracy estimation for comparing induction algorithms. In: Proceedings of the Fifteenth International Conference on Machine Learning (1998)
23. Saha, R., Chowdhury, A.R., Banerjee, S.: Diabetic retinopathy related lesions detection and classification using machine learning technology. Artificial Intelligence and Soft Computing Lecture Notes in Computer Science, pp. 734–745 (2016). https://doi.org/10.1007/978-3-319-39384-1-65
24. Scikit-Learn: Confusion Matrix. http://scikit-learn.org/stable/auto_examples/model_selection/plot_confusion_matrix.html
25. Tape, T.: The Area Under an ROC Curve. http://gim.unmc.edu/dxtests/roc3.htm
26. Thatcher, L.: The Benefits of Machine Learning in Healthcare (2017). https://healthcare.ai/the-benefits-of-machine-learning-in-healthcare
27. Penn State University: Power and Sample Size Determination for Testing a Population Mean. https://onlinecourses.science.psu.edu/stat500/node/46
28. Yen, S.J., Lee, Y.S.: Cluster-based under-sampling approaches for imbalanced data distributions. Expert Syst. Appl. **36**(3), 5718–5727 (2009). https://doi.org/10.1016/j.eswa.2008.06.108
29. Zacharaki, E.I., Wang, S., Chawla, S., Yoo, D.S., Wolf, R., Melhem, E.R., Davatzikos, C.: Classification of brain tumor type and grade using MRI texture and shape in a machine learning scheme. Magn. Reson. Med. **62**(6), 1609–1618 (2009). https://doi.org/10.1002/mrm.22147
30. Zhang, J., Mani, I.: KNN approach to unbalanced data distributions: a case study involving information extraction. In: Workshop on Learning from Imbalanced Datasets II (2003)

# Classification of Heterogeneous Data Based on Data Type Impact on Similarity

Najat Ali$^{(\boxtimes)}$, Daniel Neagu, and Paul Trundle

Artificial Intelligence Research (AIRe) Group, Faculty of Engineering and
Informatics, University of Bradford, Bradford, UK
{N.Ali50,D.Neagu,P.R.Trundle}@Bradford.ac.uk

**Abstract.** Real-world datasets are increasingly heterogeneous, showing a mixture of numerical, categorical and other feature types. The main challenge for mining heterogeneous datasets is how to deal with heterogeneity present in the dataset records. Although some existing classifiers (such as decision trees) can handle heterogeneous data in specific circumstances, the performance of such models may be still improved, because heterogeneity involves specific adjustments to similarity measurements and calculations. Moreover, heterogeneous data is still treated inconsistently and in ad-hoc manner. In this paper, we study the problem of heterogeneous data classification: our purpose is to use heterogeneity as a positive feature of the data classification effort by using consistently the similarity between data objects. We address the heterogeneity issue by studying the impact of mixing data types in the calculation of data objects' similarity. To reach our goal, we propose an algorithm to divide the initial data records based on pairwise similarity for classification subtasks with the aim to increase the quality of the data subsets and apply specialized classifier models on them. The performance of the proposed approach is evaluated on 10 publicly available heterogeneous data sets. The results show that the models achieve better performance for heterogeneous datasets when using the proposed similarity process.

**Keywords:** Heterogeneous datasets · Similarity measures
Two-dimensional similarity space · Classification algorithms

## 1 Introduction

Data classification is an important topic in data mining. Plenty of classifiers have been proposed for classifying data objects according to some constraints and requirements [1]. In the real world, data is heterogeneous: a mixture of numerical and categorical features; classifying such data using existing methods may lead to possible misclassifications and open-ended issues, due to the nature of heterogeneous data. Practically heterogeneity is seen in the process of classification as a special type of contamination, making it difficult to build credible and consistent classification model(s). The main challenge for classifying heterogeneous datasets is how to deal with a mixture of data types present in the dataset. We attempt to solve this issue by studying the impact of data similarity by their types on classifying instances from heterogeneous data sets.

© Springer Nature Switzerland AG 2019
A. Lotfi et al. (Eds.): UKCI 2018, AISC 840, pp. 252–263, 2019.
https://doi.org/10.1007/978-3-319-97982-3_21

The purpose of this paper is to utilize the influence of similarity measures on classification accuracy for heterogeneous data sets by generating a two-dimensional similarity space and classifying the data based on its similarity data values. Our motivation is to reduce the initial noisy data collection to more consistent subdomains that have all their data as similar as possible. Therefore, we first review the main notions of dissimilarity/similarity measures and present some of currently most known classification methods, and then we propose a new method to classify heterogeneous data set based on the newly introduced concept of the two-dimensional similarity space.

The rest of this paper is organized as follows: the next section provides the concepts, background and literature review relevant for the paper topic. Section 3 introduces the idea of the proposed similarity-based modeling. Section 4 reports experimental work and analysis of the results. Finally, Sect. 5 presents conclusions and future work.

## 2  Background

### 2.1  Distances and Similarity Measures

Many data mining algorithms use distance measures to determine and apply the similarity/dissimilarity (i.e. distance) between data objects. Similarity (and complementarily distance) functions are used to measure the degree to which data objects are comparably close (or not) to another [1].

**Definition 1:** Let $A$ be a set of $d$-dimensional observations (e.g. data objects). A mapping $d : A \times A \rightarrow R$ is called a **distance metric** on $A$ [2] if, for any $x, y, z \in A$, it satisfies requirements on:

1. $d(x, y) \geq 0$     $(non - negativity)$;
2. $d(x, y) = 0$   $if$   $x = y$   $(identity)$;
3. $d(x, y) = d(y, x)$     $(symmetry)$;
4. $d(x, z) \leq d(x, y) + d(y, z)(triangle\ inequality)$.

**Definition 2:** Let $A$ be a set of $d$-dimensional observations. A mapping $s : A \times A \rightarrow R$ is called a **similarity** on $A$ if it satisfies the following properties [2]:

5. $0 \leq s(x, y) \leq 1$     $(non - negativity)$;
6. $s(x, y) = 1$   $if$   $x = y$   $(identity)$;
7. $s(x, y) = s(y, x)$     $(symmetry)$.

A dissimilarity is generally a complementary mapping to the similarity definition. Plenty measures have been proposed for comparing data objects of same type in data mining applications. Some most popular distances for numerical data include Minkowski, Euclidean, Manhattan, and Chebyshev distances. The most common distances for categorical data types include Simple matching, Eskin, Tanimoto, Cosine and Goodall distances; more information about these distances can be found for example in

[1, 3, 4]. For comparing objects described by a mixture of features using a specific distance or similarity measures the area is not that rich; a general similarity coefficient measure proposed by Gower in [5] is the most common measure for comparing such data [3]. However, Ottaway in [6] highlighted some of the problems involved. Because of the additional challenges representation, the similarity for heterogeneous data is more complicated; researchers in different data mining studies have used a combination approach for computing the distance by combining different distances for different data types.

In our study, we define heterogeneous data as a combination of a mixture of features, some are numerical, and some are categorical at least; there may be examples using other data types, but we did not consider them hereby. This paper tackles the classification problem of heterogeneous data as a mixture of numerical and categorical records with variations in either or both types. For the sake of simplicity, we apply Minkowski distance for comparing numerical features and simple matching distance for comparing categorical features; both of them satisfy distance Definition 1 above. Minkowski and simple matching distances deal with the measurement of divergence between data objects; their similarity is calculated using relevant conversion methods.

## 2.2    Background: Similarity in Classification Algorithms

The classification problem in data mining is a supervised machine learning task that approaches the recognition of a given set of entries by a label based on previously presented samples. Many different algorithms have been proposed for solving the classification problem based on a variety of techniques and concepts, for example most commonly used methods for data classification tasks include decision trees such as ID3 [7], CART [8], C4.5 [9], K-nearest neighbour (KNN) [10], Artificial Neural Networks (ANN) [11], Support Vector Machines (SVMs) [12], and Naïve Bayes [13].

In many different studies, researchers have used the above-mentioned methods for classifying data described by a mixture of numerical and categorical features by initially transforming data (pre-process step) before or during the classifier training steps; an example of these studies include [14, 15] and relevant examples are described below.

Some authors studied the problem of heterogeneous data classification by improving existing classifiers to handle heterogeneous data. In [16] Pereira et al. have proposed a new distance for heterogeneous data which is used with a KNN classifier. This distance, called Heterogeneous Centered Distance Measure (HCDM), is based on a combination of two techniques; the proposed method relies on dividing the data set into pure numerical and pure categorical features, then applies Nearest Neighbor Classifier CNND distance to numerical features and Value Difference Metric to categorical features, and the result of the two distances is assembled in one single distance to form the HCDM value.

In [17] Jin et al. proposed a novel method for heterogeneous data classification called Homogeneous data In Similar Size (HISS); their method is based on dividing heterogeneous data into a number of homogeneous partitions of similar sizes. Although the method showed a good performance for heterogeneous data classification, the authors pointed out that they did not consider the effects of homogeneous subsets on all the relevant subspaces during training stage. Hsu et al. in [18] studied a mixed data classification problem by proposing a method called Extended Naïve Bayes (ENB) for mixed data with numerical and categorical features. The method uses the original Naive Bayes algorithm for computing the probabilities of categorical features: numerical features are statistically adapted to discrete symbols taking into consideration both the average and variance of numeric values. In [19] Li et al. proposed a new technique for mining large data with mixed numerical and nominal features. The technique is based on supervised clustering to learn data patterns and use these patterns for classifying a given data set. For calculating the distance between clusters. The authors have used two different methods; the first method was based on using specific distance measure for each type of features, and then combined them in one distance. The second method was based on converting nominal features into numeric features, and then numeric distance is used for all features. In [20] Sun et al. presented a soft computing technique called neuro-fuzzy based classification (NEF-CLASS) for heterogeneous medical data sets; the motivation at that time was based on the fact that most conventional classification techniques are able to handle homogeneous data sets but not heterogeneous ones. Their method has been tested on both pure numerical and mixed numerical and categorical medical datasets.

To summarise, the most commonly used approaches for handling data described by a mixture of numerical and categorical features use two approaches: (1) conversion methods of initial data components to a consistent standard data type for which relevant, specialized machine learning techniques are applied. For example, k-NN works naturally with numerical data, for heterogeneous data, the non-numerical data subset is converted into numerical data and sometimes calibrate/normalize or project that numerical data to reduce effects of disparate ranges. Alternatively, decision trees can be applied to heterogeneous data by converting numerical data into categorical data, and Naive Bayes is applied to learn discrete numeric attributes data converted into symbols. However, converting categorical features into numerical features (for example for SVM applications), may lead to loses of some useful information, a possible source of biased, or misclassification outcomes. (2) the hybrid ensemble development of classifiers by application of machine learning techniques to same data type component subsets, followed by a weighted average of all classifiers similar to computing the overall similarity value as a weighted average of same type data components. Each approach comes with added computational complexities and the need of data understanding and expertise to convert consistently either a priori or a posteriori the classification output.

## 3   Two-Dimensional Similarity Space Feature Selection-Based Classification Filter

In the proposed method, we intend to study the noise added by the numerical attributes and categorical attributes respectively, to the pairwise similarity of data records. This is approached by separating numerical features on one side, and categorical ones on the other side, and exploring when one becomes noisy for the other one, to leave just the case that they can still stay together when indeed full records are extremely similar.

Let $A = \{A_1, A_2, A_3, \ldots, A_N\}$ denote a set of $d$-dimensional objects of cardinality $N$, where each data object $A_i$, $i = 1, 2, 3, \ldots .N$, has $d$ mixed features: $d_1$ numerical features $\{x_1, x_2, \ldots x_{d_1}\}$, and $d_2$ categorical features $\{y_1, y_2, \ldots y_{d_2}\}$, where $d = d_1 + d_2$ (for sake of presentation clarity the indexes of the above-named features are ordered).

For each feature type, one relevant distance mapping is applied, to create the two-dimensional similarity space 2DSS. Each point $Z_{ij}$ in 2DSS is a pair of numerical and categorical similarity values $Z_{ij} = \left(s_{N_{ij}}, s_{C_{ij}}\right)$, where $0 \leq s_{N_{ij}} \leq 1$ and $0 \leq s_{C_{ij}} \leq 1$.

We define our similarity matrix (SM) as follows:

$$
SM = \begin{bmatrix}
\left(s_{N_{11}}, s_{C_{11}}\right) & \left(s_{N_{12}}, s_{C_{12}}\right) & \cdots & \left(s_{N_{1n}}, s_{C_{1n}}\right) \\
\left(s_{N_{21}}, s_{C_{21}}\right) & \left(s_{N_{22}}, s_{C_{22}}\right) & \cdots & \left(s_{N_{2n}}, s_{C_{2n}}\right) \\
\cdot & \cdot & \cdots & \cdot \\
\cdot & \cdot & \cdots & \cdot \\
\cdot & \cdot & \cdots & \cdot \\
\left(s_{N_{n1}}, s_{C_{n1}}\right) & \left(s_{N_{n2}}, s_{C_{n2}}\right) & \cdots & \left(s_{N_{nn}}, s_{C_{nn}}\right)
\end{bmatrix} \tag{1}
$$

SM is a symmetric matrix, the total number T of pairwise relevant points in the similarity space can be computed as $T = \dfrac{N(N-1)}{2}$.

The two-dimensional similarity space 2DSS is divided into four subspaces (see Fig. 1). Subspace A contains all points $Z_{ij} \in 2DSS$ with high similarity values for both numerical and categorical features $S_{NC}$. Subspace B contains all $Z_{ij} \in 2DSS$ that have a (relatively) high similarity value for numeric features, and low similarity values for categorical features $S_{N\overline{C}}$. Subspace C contains all points $Z_{ij} \in 2DSS$ with low similarity values for both numerical and categorical features $S_{\overline{NC}}$, and Subspace D contains all points $Z_{ij} \in 2DSS$ that have low similarity values for numerical features, and high similarity values for categorical features $S_{\overline{N}C}$. Figure 1 shows the division of the bi-dimensional similarity space in four relevant subspaces.

The proposed approach defines four directions for the original heterogeneous dataset to address the initial issues discussed in Sect. 2: data in subspace A is more homogeneous and requests a hybrid ensemble classifier or similar conversion methods that should learn data of high similarity; subspace C has noisiest samples that can be treated as outliers, subspaces B and D allow development of consistent, single-type machine learning models since data features of either numerical (subspace B) or categorical (subspace D) type are highly similar.

The successful selection of the subspaces based on the proposed filter depends on the boundaries Lower_SN, Upper_SN that can be moved right or left across the continuous numerical similarity values domain (e.g. $v1, v2, \ldots \ldots .n$) on the $S_N$ axis, and the boundaries Lower_SC and Upper_SC that can be moved up and down across the discrete categorical similarity values $c1, c2, \ldots \ldots .cn$ on the $S_N$ axis as shown in Fig. 1. The choice of these values can be a further optimization exercise. If these boundary values reach the maximum position then there will be just outliers, in the opposite case there will be no outlier cases. In the experiments reported in this paper these boundaries values are chosen by the rule of thumb.

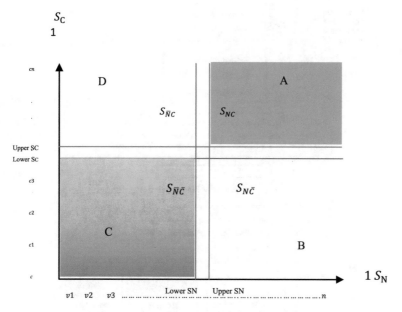

**Fig. 1.** The division of the two-dimensional similarity space 2DSS.

Figure 2 shows the proposed methodology that allows filtering (extraction) of highly similar records without dependence of the classification outputs (classes or labels).

## 4 Experimental Work

We performed our experiments on ten relevant heterogeneous datasets: three data sets from UCI Machine Learning Repository [21], and seven data sets from R packages datasets available in [22]. Each dataset contains different numbers of instances, attributes, and classes. A summary of properties of each data set is given in Table 1.

All datasets are pre-processed before we ran the experiments, erroneous, inconsistent, and missing entries being removed. Data columns with more than 10% missing values are removed; the ordinal features are also removed from the data sets. Numeric features are normalized. K-fold cross validation method has been used for model. For validation, the original data is randomly partitioned into k equal size subsamples, where k = 10.

Each data set has generally a limited number of categorical similarity values and a large number (practically any value in the numerical similarity domain0 of numerical similarity values, due to the intrinsic definition of these similarities. For example, Saratoga Houses data from Table 1, has categorical similarity values (0, 0.2, 0.4, 0.6, 0.8, 1), and any numerical similarity value between 0 and 1s. We defined the categorical similarity boundary value where the performance of the model starts increasing.

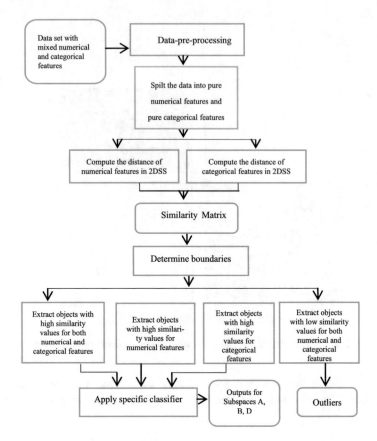

**Fig. 2.** The proposed classification filter stages.

**Table 1.** Summary of data sets properties.

| Dataset | Observations | Numerical features | Categorical features |
|---|---|---|---|
| Student alcohol consumption | 1044 | 9 | 22 |
| Credit approval | 690 | 6 | 9 |
| German credit risk | 1000 | 7 | 13 |
| Structure of demand for medical care | 5574 | 9 | 5 |
| Teatment | 2675 | 5 | 4 |
| Visits to Physician Office | 4405 | 10 | 8 |
| Saratoga Houses | 1728 | 10 | 5 |
| Job train | 2675 | 10 | 9 |
| Labour training Evaluation1 | 15992 | 5 | 4 |
| Wages and schooling | 2944 | 10 | 16 |

## 4.1    Results

Classifiers' performances are compared based on their accuracy. Decision Tree C5.0 classification was first applied to all datasets. The results are shown in Fig. 3:

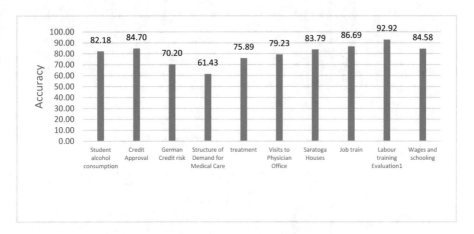

**Fig. 3.** Accuracy obtained by the classifier

As mentioned earlier, for each data set, the similarity values of numerical and categorical features are represented as coordinate pairs in the 2DSS space. The performance of the filtering technique applied on the similarity space to define the four subspaces A, B, C and D for the next action (such as classification or outlier identification) depends on the boundaries selection. This is exemplified within the experiments with Decision Tree C5.0 algorithm applied to each data records extracted from each subspace. For pure numerical and pure categorical subspaces cases (i.e. subspaces B, and D, respectively) feature selection has been also applied to the filtered attributes. Data objects with low similarity values for both numerical and categorical values (subspace C) $S_{\overline{NC}}$ are considered outliers and separated from the main classification exercise.

Figures 4, 5, 6, 7 and 8 show the results for the proposed method to the benchmark data sets. The models examined in the current experimental work perform well on the subsets of mixed numerical and categorical records with high similarity (subspace A) where data is homogeneous and similarity values $S_{NC}$ are high. The improvement exceeded 4%, reaching a maximum of 14% for the Student alcohol consumption dataset. In some cases though such as MedExp, Visits to Physician Office, and Labour training Evaluation1 datasets, the classifier performance increased just for the pure numerical features; also the performance of the classifiers increased just for pure categorical features in the case of Crx, German credit data set dataset. One of the reasons for such limited increase in subspace classifier performance is related to the chosen similarity distance.

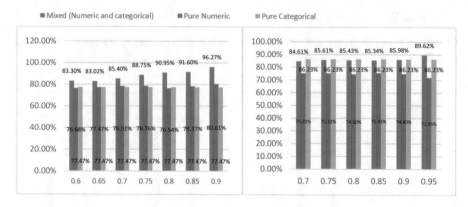

**Fig. 4.** Results for Student alcohol consumption and Crx datasets

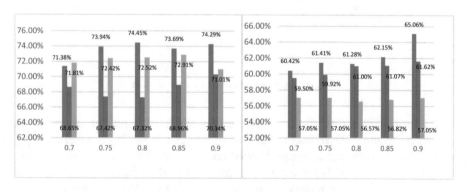

**Fig. 5.** Results for German credit card and MedExp datasets

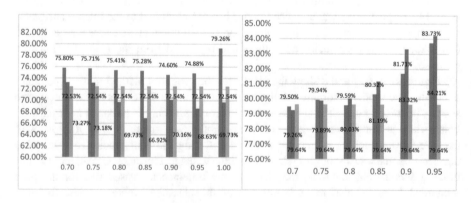

**Fig. 6.** Results for Treatment and OFP data sets

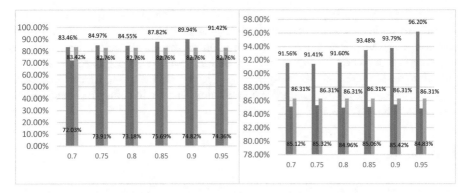

**Fig. 7.** Results for Saratoga Houses and Job train datasets

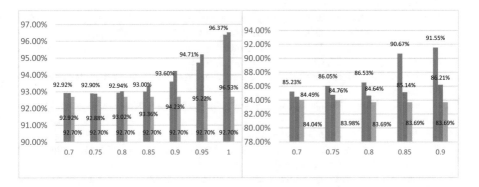

**Fig. 8.** Results for CPS1 and Schooling datasets

## 5  Conclusions and Further Work

A new approach to filter records for classifying heterogeneous datasets based on the impact of data types on their similarity is proposed and evaluated. The similarity space is built in the experiments using a Minkowski distance for numeric features and simple matching for categorical features. The influence of similarity measures on the performance of the classifiers was investigated by identifying and removing the outliers (data objects with overall low similarity values) from the initial data set. In the experiments, the influence of the similarity measures was investigated on classification accuracy.

It is important to point out that the proposed model may not handle efficiently small data sets because the subspaces become less relevant, and therefore we aim to investigate the performance and applicability of the proposed method for big heterogeneous data sets. There are also other wide areas for further work on this topic. We have tested currently the method on data samples with limited types of features (numerical and categorical) only. Future work may include other types of data features, such as ordinal, nominal, binary, and fuzzy, and extend the similarity space to multidimensional similarity space forms. We have used the most common distances (Minkowski and simple matching) for computing data objects similarities. More studies should be done to investigate the impact of the choice of similarities on the performance of the model. In addition, each subset may request optimisation of the specific classifier instead of applying just one classifier algorithm. The outliers can be also exploited for anomaly detection, data imputation and faulty records. Finally, an interesting future direction is related to the choice of the appropriate optimisation function to define the boundaries of subspaces automatically.

# References

1. Han, J., Pei, J., Kamber, M.: Data Mining: Concepts and Techniques. Elsevier, Waltham (2011)
2. Sarle, W.S.: Finding Groups in Data: An Introduction to Cluster Analysis. Wiley, New York (1991). JSTOR
3. Myatt, G.J., Johnson, W.P.: Making Sense of Data II: A Practical Guide to Data Visualization, Advanced Data Mining Methods, and Applications. Wiley, Cambridge (2009)
4. Deza, M.M., Deza, E.: Distances and similarities in data analysis. In: Encyclopedia of Distances, pp. 291–305. Springer, Heidelberg (2013)
5. Gower, J.C.: A general coefficient of similarity and some of its properties. Biometrics **27**, 857–871 (1971)
6. Ottaway, B.: Mixed data classification in archaeology. Revue d'Archéométrie **5**(1), 139–144 (1981)
7. Quinlan, J.R.: Induction of decision trees. Mach. Learn. **1**(1), 81–106 (1986)
8. Stone, C.J.: Classification and Regression Trees, vol. 8, pp. 452–456. Wadsworth International Group, Belmont (1984)
9. Salzberg, S.L.: C4. 5: Programs for machine learning by J. Ross Quinlan. Morgan Kaufmann Publishers, Inc., 1993. Mach. Learn. **16**(3), 235–240 (1994)
10. Cover, T., Hart, P.: Nearest neighbor pattern classification. IEEE Trans. Inf. Theory **13**(1), 21–27 (1967)
11. Hopfield, J.J.: Neural networks and physical systems with emergent collective computational abilities. Proc. Natl. Acad. Sci. **79**(8), 2554–2558 (1982)
12. Vapnik, V.: The Nature of Statistical Learning Theory. Springer, New York (2013)
13. John, G.H., Langley, P.: Estimating continuous distributions in Bayesian classifiers. In: Proceedings of the Eleventh Conference on Uncertainty in Artificial Intelligence. Morgan Kaufmann Publishers Inc. (1995)
14. Hu, L.-Y., et al.: The distance function effect on k-nearest neighbor classification for medical datasets. SpringerPlus **5**(1), 1304 (2016)
15. Chandrasekar, P., et al.: Improving the prediction accuracy of decision tree mining with data preprocessing. In: 2017 IEEE 41st Annual Computer Software and Applications Conference (COMPSAC). IEEE (2017)
16. Pereira, C.L., Cavalcanti, G.D., Ren, T.I.: A new heterogeneous dissimilarity measure for data classification. In: 2010 22nd IEEE International Conference on Tools with Artificial Intelligence (ICTAI). IEEE (2010)
17. Jin, R., Liu, H.: A novel approach to model generation for heterogeneous data classification
18. Hsu, C.-C., Huang, Y.-P., Chang, K.-W.: Extended Naive Bayes classifier for mixed data. Expert Syst. Appl. **35**(3), 1080–1083 (2008)
19. Li, X., Ye, N.: A supervised clustering and classification algorithm for mining data with mixed variables. IEEE Trans. Syst. Man Cybern. Part A Syst. Hum. **36**(2), 396–406 (2006)

20. Sun, Y., Karray, F., Al-Sharhan, S.: Hybrid soft computing techniques for heterogeneous data classification. In: Proceedings of the 2002 IEEE International Conference on Fuzzy Systems, FUZZ-IEEE 2002. IEEE (2002)
21. Frank, A., Asuncion, A.: UCI Machine Learning Repository. http://archive.ics.uci.edu/ml. University of California. School of Information and Computer Science, Irvine, p. 213 (2010)
22. R Data Sets. https://vincentarelbundock.github.io/Rdatasets/datasets.html

# Clustering-Based Fuzzy Finite State Machine for Human Activity Recognition

Gadelhag Mohmed$^{(\boxtimes)}$, Ahmad Lotfi, Caroline Langensiepen,
and Amir Pourabdollah

School of Science and Technology, Nottingham Trent University,
Nottingham NG11 8NS, UK
gadelhag.mohmed2016@my.ntu.ac.uk, ahmad.lotfi@ntu.ac.uk

**Abstract.** In this paper, a clustering-based fuzzy finite state machine approach for human activity modelling and recognition is proposed. It Incorporates the Fuzzy C-means (FCMs) clustering algorithm with a Fuzzy Finite State Machine (FuFSM) in order to generate the state transitions more effectively. This unsupervised approach will overcome the deficiency in identifying the knowledge-base required for FuFSM. To validate the proposed approach, experimental results are presented. The activities of two office workers are modelled/recognised using the proposed method. The approach taken for this research is based on ambient Intelligent sensory data rather than data coming from wearable sensors.

**Keywords:** Activity of daily working ADW · Human behaviour
Fuzzy finite state machine · Fuzzy C-means clustering

## 1 Introduction

To recognise human activities within a home or office smart environments, different recognition techniques have been proposed [2,4,9,10]. The Human Activities Recognition (HAR) area of research aims to understand the behaviour of a subject in an smart environment and provide appropriate actions to monitor and control the environment. For example in a home environment recognising activities such as walking, sleeping and eating representing Activities of Daily Living (ADL) would help independent living and provide assistance when required. In an office environment, recognising activities such as using a computer, meetings, taking breaks and some desk activities representing Activities of Daily Working (ADW) would help to optimising the energy consumption or overall office worker comfort. Two commonly used approaches to collect data representing human activities are based on wearable sensors or sensors integrated within an Ambient Intelligent (AmI) environment. Several studies have been conducted using either or both methods of data collection.

Most of the HAR systems are using a supervised learning approach to recognise human behaviour. In this paper, we propose the usage of a clustering-based

© Springer Nature Switzerland AG 2019
A. Lotfi et al. (Eds.): UKCI 2018, AISC 840, pp. 264–275, 2019.
https://doi.org/10.1007/978-3-319-97982-3_22

fuzzy finite state machine to model the human activities based on ambient sensory devices method. Fuzzy Finite State Machine (FuFSM) is a recommended tools for modelling and simulating uncertain temporal data such as human activities [2,3]. Considering the nature of human activities, it is possible to be in two separate states (activities) at the same time with a degree of belonging to either of these states. For example, an office worker could be answering a telephone call while still using a computer. This will be recognised in two states of computer and telephone activities with a certain degree of belonging to either of these states. Therefore, FuFSM is a suitable modelling technique to represent such system.

The most complex part of developing a FuFSM model for a certain task is to identify accurately the Knowledge Base (KB). Furthermore as the number of states increases, this will increase the complexity of the model and it would not be possible for an expert to define the KB. To overcome the deficiency, it is proposed to use an unsupervised learning method to enhance the performance of the FuFSM. Figure 1 illustrates the block diagram of the proposed model. In [2,3,10] it is shown that incorporating Genetic Algorithm (GA) with FuFSM will enhance the capability of fuzzy transition functions and improve its performance.

In the proposed approach in this paper, Fuzzy C-means (FCMs) clustering algorithm is incorporated with FuFSM to generate the states transition functions. This will replace the existing KB in the state model. FCM algorithm will be used primarily for automatically generating the fuzzy rules and membership functions that are responsible for states transitions.

To verify the proposed model, data representing the ADW for an office worker is used. Two datasets are used to represent the activities of two office workers.

This paper is organised as follows; related works about the human activity recognition using collected data from ambient smart environment and wearable sensory devices are presented in Sect. 2. Section 3 presents the details of clustering-based fuzzy finite state machine using FCMs algorithm and propose some improvements. The way of collecting the dataset and the office worker activities that used to validate the proposed approach are described in Sect. 4. In Sect. 5 the conducted experiments are explained. Results are discussed in Sect. 6. Eventually, the conclusion of this paper is drawn in Sect. 7.

## 2   Related Works

Several HAR techniques using the machine learning approaches were conducted in [8,10,11,19]. They extracted useful features from the collected sensory data and then made an attempt to model the human activities using computational intelligence techniques such as, Hidden Markov Model (HMM), Gaussian Mixture Models (GMM), Fuzzy Finite State Machine (FFSM) and Convolutional Neural Network (CNN) respectively. Recently, researchers have started to involve the deep learning methods in HAR, which obtained high performance for recognising and modelling the multimedia data that collected from real world environments (office or home) to understand and recognise the human behaviours.

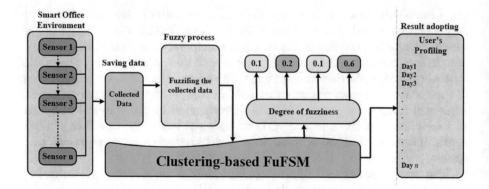

**Fig. 1.** A block diagram of the clustering-based FuFSM model.

The authors in [20] used deep learning methods for HAR, the advantage of their proposed approach is the system has its own data extraction process to extract the main feature needed to understand and recognise the human behaviours. A public dataset collected by smart phone that was located in a fixed position for example in a user's pocket or waist was used to evaluate and test the proposed systems [20].

For HAR within an indoor smart environment, Fuzzy Finite State Machine and Neural Network are widely used [5,6,10,15,18]. Evolutionary computing based techniques and other machine learning techniques based on Markov Modulated Poisson Process (MMPP) are also employed, thus enhance the human activity monitoring accuracy [1,5,7]. Hybrid computational techniques, such as data mining, pattern recognition and human activity profiling, are also used specially in the case of ADW to recognise activities and preferences of an office worker via smart office environment [10,13,16,17,21]. Authors in [2,3], have proposed an effective model-based Genetic Fuzzy Finite State Machine (GFFSM) that uses the GA for automatically generating the fuzzy KB instead of using the fuzzy rules.

In this proposed method, FCMs clustering algorithm is incorporated with FuFSM to generate the states transition functions. This will replace the existing KB in the state model. However, The input vectors would be still defined as fuzzy dataset before being expressed to the model.

## 3   Clustering-Based Fuzzy Finite State Machine for Human Activity Recognition

This section presents an unsupervised learning model using Clustering-based FuFSM for human activity recognition (HAR). As it has been explained in the earlier Sects. 1 and 2, the most difficult part in the FuFSM system is the definition of the fuzzy KB that representing the transitions between states. In order to overcome this deficiency, unsupervised learning FuFSM based on Fuzzy C-Means

(FCMs) clustering algorithm is proposed. The input data in this method will be expressed as a fuzzy set based on data observation to recognise the worker activities. The fuzzy states and the fuzzy rules for the transitions between states will be calculated automatically using FCMs algorithm as it is illustrated in Fig. 1.

There are three distinct steps included in this research; in the first step, a data collection system is used to record the office worker activities and to collect the office environment conditions. In the second step, the collected data is processed by applying a suitable data mining technique in order to distinguish the different activities for the office worker. In the third step, the processed data will be applied to the Cluster-based FuFSM system to recognise the worker activities. Eventually, the final results are used to create an individual profile for each worker. Once the individual profile being created, the environment (workspace) conditions can be automatically adjusted based on the workers' preferences and comfort factors.

The map structure of the office worker clustering-based FuFSM elements are explained in details as follows:

1. **Fuzzy State** (*Activities*). 5 states $\{S = s_1, s_2, ...., s_5\}$ that response to 5 different activities are created to recognise the office worker's activities during a normal working day, as it is shown in Fig. 2 and Table 1. The states will be defined numerically using state activation value $K(t)$ as:

$$K(t) = \{k_1(t), k_2(t), ..., k_n(t)\}$$

where $k_i(t) \in [0,1]$ and $\sum_{i=1}^{n} K_i(t) = 1$

2. **Input Vector** ($X$). The used data collection system, collects data form real environment using ambient sensory devices. The main data used in this section is a binary data that collected from PIR, Door, PC, Chair and Light sensors. Once the data being collected, it will be fuzzified to generate a fuzzy set of data $\{x_1, x_2, ..., x_n\}$ that will be used as an input to the proposed system. As the aim in this paper was to recognise the worker activities, the other data such as, inside and outside temperature, humidity and ambient light intensity are not considered in this experiment as they refer to the worker comfort and preferences.

3. **Transition Function.** Transitions between the different states in normal fuzzy state machine are expressed based on a set of fuzzy rules that require a source of KB, which is not an easy task for programmers and experts. Therefore, is proposed the unsupervised learning method for the FuFSM based on FCMs. In this method the transition between states will be done by a calculation that generate the rules automatically using FCMs algorithm.

A clustering-based FuFSM for HAR, can be expressed by:

$$\mathbf{IF}\{K(t) \text{ is } s_i\} \text{ } \mathbf{AND} \text{ } FC_{ij} \text{ } \mathbf{THEN} \text{ } K(t+1) \text{ is } s_j$$

where, $FC_{ij}$ is the value that responsible for representing the FCMs algorithm.

$K(t)$ is the state activation value.
$s_i$ is the system current state.
$s_j$ is the system next state.
$x_t$ is the input value

As is seen from the given expression, the antecedent part contains two terms; The first term is $\{K(t)$ is $s_i\}$, which uses the state activation value $K(t)$ to compute the states that can be in active mode at time $t$ based on the input value. The second term of the antecedent part is $FC_ij$, which will be responsible for representing the FCMs algorithm that is used to calculate the system's transitions and compute the states that will response to each calculation. The second process that will be done during this term is to determine the degree of membership (fuzziness) to each state (activities).

The consequent part of the given expression is $K(t + 1)$, represents the next state activation value. This value is calculated based on the current value of the state activation value $K(t)$ and the input value $x_t$.
In this method the next value of the state activation vector is characterised by Class Membership Function $(CMF)$ matrix $\overline{U} = [u_n^l]$, where $[u_n^l]$ is the grade of membership for $X_n$ object in the cluster [12].

By applying FCMs algorithm at the learning stage, cluster centres are determined and then classified. This process occurs by comparing the distance between the incoming features and each cluster centre. During this stage, the centre for each cluster is obtained by minimising a cost function. More details about FCMs algorithm can be found in an earlier publication [14].

At this point, as the CMF for each input value is calculated, the total output of the FCMs algorithm and therefore, the next state activation value $K(t+1)$ and a fuzziness degree of the individual input to each state (activity) will be computed as in formula 1.

$$K(t+1) = \begin{cases} \dfrac{\sum_{m=1}^{InputValue} f_m.(k_1, k_2, ...., k_n)}{\sum_{m=1}^{S}(s_1, s_2, ...., s_n)} & IF\ 1 < f_m < 0 \end{cases} \quad (1)$$

where $f_m$ is calculated using the FCMs algorithm based on the input value $x_n$ and some other features that extracted from the source data.

4. **Output Vector and Output Function.** $U(t)$, in this contribution, it simply can be considered as the degree of fuzziness for each sate activation value i.e. $U(t) = K(t)$.

# 4    Office Worker Activities

In order to record the office worker activities during the working day hours, an ambient conditions data is collected to verify the ADW. For this purpose, a data collection system is used to collect the data and save it into a database.

## 4.1   Sensory Devices

In order to collect real data from a smart office environment, the ambient office conditions and the office worker behaviour are recorded using several types of sensors. The majority of these sensors are low-cost sensors and commonly used in daily life. For example, PIR sensors, on/off switch sensors, mat pressure sensors and an agent application that installed in the office PC to record when the keyboard and/or mouse is in use. The used sensors are:

- Magnetic switch sensor, used for the office door to detect entry and leaving points and for window to measure their opening and closing states. These sensors give data in binary (on/off).
- Motion sensor, used to detect the movement within the office ambient and the output data from this sensor is in binary data (on/off).
- Pressure sensor to be used to detect chair occupancy, which gives binary data (on/off).
- PC usage, which is an application installed in the user's PC to record keyboard and mouse activities.
- Office temperature sensor.
- Office light sensor to measure light intensity.
- Office humidity sensor.

## 4.2   Challenges in Office Worker Activity Recognition

Each office worker has different characteristics that differentiate them from other workers in a given time. For example, coming to the workplace early, leaving it to late, taking a lunch break in a specific time everyday and having a regular visitors, all of these characteristics differentiate one worker from others. Based on this behaviour characterisation, the individual profiling for each worker can be generated. Recognising complex human activities based on an ambient sensory data is a challenging and active research area. In addition, the naturally uncertain human behaviour increases the complexity of this task. The challenges as they were reported in [8,10], in modelling and recognising office worker activities can be specified as follows:

- Recognising *"SimultaneousActivity"*, when more than one activity can take place at the same time. For instance, some worker can work on his/her computer and talking over the phone at the same time. Therefore, dealing with these kind of complex activities needs a different approach from what it is used in recognising a single activity.
- Recognising *"Interleaved Activities"*, which means when the current activity interrupted by another activity. For example, the current state of the office worker is recorded as "computer activity", while that a visitor comes to the office. In this case the first activity will be paused for the period of the second activity and then the first activity will be resumed.

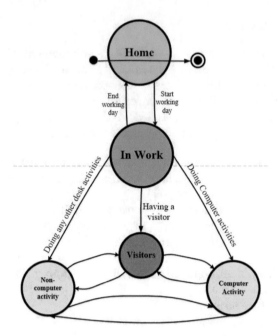

**Fig. 2.** A state diagram of the office worker daily activities.

This research focuses on recognition the activities that can occur at the same time. These activities will be mapped into two combinations of activities; The first combination will recognise the "*Simultaneous Activity*" e.g., (computer activity and Non-computer activity), where the office worker is using the computer and doing any other desk activities at the same time such as, reading or talking over the phone. The second combination will be for the "*Interleaved Activities*" e.g., to recognise the (visitor activity and in office activity), where in office activity here refers to presence of the office worker in his/her office and either using computer (computer activity) or doing any other desk activities (non-computer activity). The main five activities are displayed in Fig. 2.

## 5    Experiments

The data used in this experiment is taken from research conducted in [10], and it was collected from two intelligence office environments that were used as an academic staff offices for two users (user A and user B) at Nottingham Trent University. The main five different activities for the office workers have been recorded during the working days descried in Table 1, with the sensor being triggered for each activities. These activities are detailed (exclude the out office activity) as follows (Fig. 3):

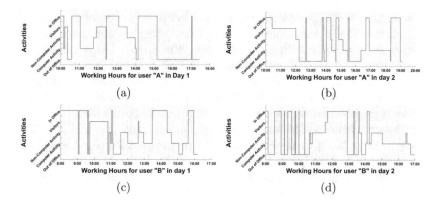

**Fig. 3.** Daily activities of a users (a) user A day 1 (b) user A day 2 (c) user B day 1 (d) user B day 2.

1. In the office, which refers to when the office worker comes to start his/her working day (in office activity). Later on, this activity will be divided into two sub-activities (computer and non-computer activities).
2. Visitors, it refers to when any visitor comes to the office worker during his/her working day (Visitor Activity).
3. Computer usage, which is one of the sub-activities that derived from the in office activity and it describes when the office worker uses computer during the working hours. It called as (computer-activity).
4. Doing different desk activities such as reading, talking over the phone or just seating on the chair. Also, this activity is the second sub-activity of the In Office activity, which is called (non-computer activity).

Once the data collection process is done, the collected data is processed using a suitable mining technique to first, pre-process it, and then extract main five activities that mentioned in Table 1, for users A and B in two different days. Figure 5 shows the office activities for two office workers and they are clearly not comparable. These plots show the activities for two different working days. Starting times for both users are not the same. User B comes earlier than user A, and user A leaves the office later than user B.

**Table 1.** Descriptions of the main five activities and related triggered sensors.

| Activity | Sensor being triggered |
|---|---|
| In Office | Door, PIR, Light and [Chair OR PC] → ON |
| Computer activity | Chair, PIR and Keyboard → ON |
| Non-computer activity | Chair, PIR → OFF |
| Visitors | Door, PIR, Light, and [Chair OR PC]→ ON |
| Out Office | PIR, Door, Chair, PC and Light → OFF |

Visualising the data has shown overlap between the recognised activities that occurred at the same time, for example, using the computer and doing any other desk activity such as talking over the phone. Other activities that will be recognised in this step are the interleaved activities, for example when the office worker is in the office and his/her state recognised as (computer or non-computer activities), a visitor comes to the office. In these both situations, recognising the actual activities are not an easy task to do. Therefore, we propose the clustering-based FuFSM. In this method, the collected data will pass through two steps before being recognised using the proposed clustering-based FuFSM. The first step is to fuzzify the collected data based on data observation and the daily annotations that are taken by the office worker to identify the office worker activities $X = \{x_1, x_2, ..., x_5\}$.

In this step, some features extracted from the data will be used such as, the activity's starting and ending time, its duration and its frequency. The second step, is to use the FCMs algorithm that already applied in the state machine to control the transitions using the state activation value $(K(t))$, together with the

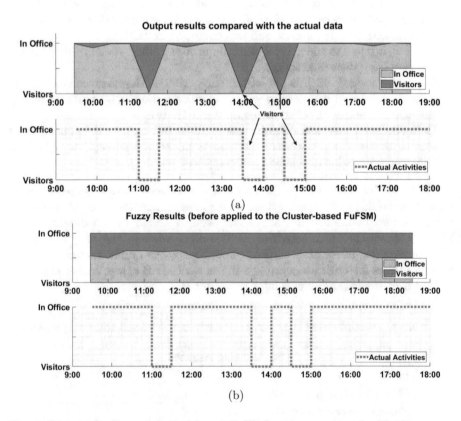

**Fig. 4.** The results from clustering-based FuFSM system compared with the output from normal FuFSM system for two interleaved activities for user A (a) clustering-based FuFSM (b) normal FuFSM.

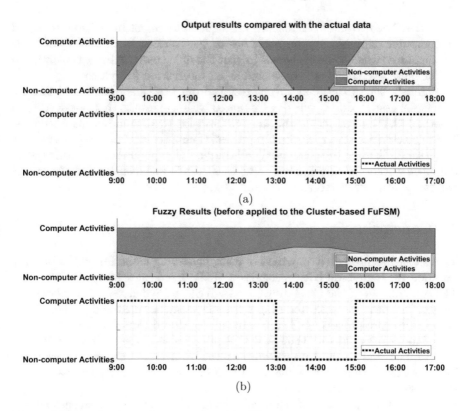

**Fig. 5.** The results from clustering-based FuFSM system compared with the output from normal FuFSM system for two simultaneous activities for user B (a) clustering-based FuFSM (b) normal FuFSM.

fuzzified data set. In this case the FCMs algorithm is in operation to give the fuzziness degree to each state $(S)$ for every single input data value $(x)$.

# 6 Results

The testbed data in this experiment was collected from academic offices environment at Nottingham Trent University by a researcher student over a period of 12 weeks as represented in [10].

Figure 5 demonstrates samples of two users activities (A and B) during their daily office activities. It can be seen that, the most complex part in these plots is to recognise the activities that are happening at the same time (Simultaneous Activities) or when the current activity is interrupted or paused by another activities (Interleaved Activities).

Figure 4, represents the final output results from the clustering-based FuFSM system for user A, during one day compared with the results that obtained from a normal FuFSM system for the same dataset and for two Interleaved

Activities. The two activities that were recognised represent In Office state (could be computer or non-computer activities or just seating in the office) and the Visitor state (when someone comes to visit the office user). As it was mentioned earlier those two states are considered as an Interleaved Activities.

Figure 5, shows the final output results from the Clustering-based FuFSM system for user B, during one day compared with the results that obtained from a normal FuFSM system for the same dataset for two Simultaneous Activities. The two activities that recognised represent Computer Activity state (when the office worker is using the computer) and the Non-Computer Activity state (when the office worker doing other desk activities). As it was mentioned earlier those two states are considered as an Simultaneous Activities.

## 7 Conclusions

This paper has presented a method of using clustering based Fuzzy Finite State Machine for human activity recognition HAR based on sensory devices data. The components of the Clustering-based FuFSM were defined and explained in details. The data was collected from a real office environment and applied to the system after pre-processing. The final results achieved from the system are displayed and compared with the result that are achieved from the normal FuFSM system. The final results have shown that, the behaviour of the office worker can be represented using Clustering-based FuFSM more accurately compared to a normal FuFSM systems.

some features in this research need to be improved and enhanced such adding a new computational technique with the FCMs clustering algorithm. Additionally, more sensory devices could be used in the office/house environment to capture much bigger set of data with more information and some extra features could be added to the transition function in the state machine.

## References

1. Aicha, A.N., Englebienne, G., Kröse, B.: Unsupervised visit detection in smart homes. Pervasive Mob. Comput. **34**, 157–167 (2017)
2. Alvarez-Alvarez, A., Trivino, G., Cordón, O.: Body posture recognition by means of a genetic fuzzy finite state machine. In: 2011 IEEE 5th International Workshop on Genetic and Evolutionary Fuzzy Systems (GEFS), pp. 60–65. IEEE (2011)
3. Alvarez-Alvarez, A., Trivino, G., Cordon, O.: Human gait modeling using a genetic fuzzy finite state machine. IEEE Trans. Fuzzy Syst. **20**(2), 205–223 (2012)
4. Atzmueller, M., Hayat, N., Trojahn, M., Kroll, D.: Explicative human activity recognition using adaptive association rule-based classification. In: 2018 IEEE International Conference onFuture IoT Technologies (Future IoT), pp. 1–6. IEEE (2018)
5. Barsocchi, P., Cimino, M.G., Ferro, E., Lazzeri, A., Palumbo, F., Vaglini, G.: Monitoring elderly behavior via indoor position-based stigmergy. Pervasive Mob Comput **23**, 26–42 (2015)

6. Cook, D.J., Augusto, J.C., Jakkula, V.R.: Ambient intelligence: technologies, applications, and opportunities. Pervasive Mob. Comput. **5**(4), 277–298 (2009)

7. Dawadi, P., Cook, D., Parsey, C., Schmitter-Edgecombe, M., Schneider, M.: An approach to cognitive assessment in smart home. In: Proceedings of the 2011 Workshop on Data Mining for Medicine and Healthcare, pp. 56–59. ACM (2011)

8. Garcia-Ceja, E., Brena, R.F.: An improved three-stage classifier for activity recognition. Int. J. Pattern Recogn. Artif. Intell. **32**(01), 1860003 (2018)

9. He, H., Tan, Y., Zhang, W.: A wavelet tensor fuzzy clustering scheme for multisensor human activity recognition. Eng. Appl. Artif. Intell. **70**, 109–122 (2018)

10. Langensiepen, C., Lotfi, A., Puteh, S.: Activities recognition and worker profiling in the intelligent office environment using a fuzzy finite state machine. In: 2014 IEEE International Conference on Fuzzy Systems (FUZZ-IEEE), pp. 873–880. IEEE (2014)

11. Lara, O.D., Labrador, M.A.: A survey on human activity recognition using wearable sensors. IEEE Commun. Surv. Tutor. **15**(3), 1192–1209 (2013)

12. Lotfi, A., Howarth, M.: Industrial application of fuzzy systems: adaptive fuzzy control of solder paste stencil printing. Inf. Sci. **107**(1–4), 273–285 (1998)

13. Lu-An, T., Jiawei, H., Guofei, J.: Mining sensor data in cyberphysical systems. Tsinghua Sci. Technol. **19**(1), 225–234 (2015)

14. Mohmed, G., Lotfi, A., Langensiepen, C., Pourabdollah, A.: Unsupervised learning fuzzy finite state machine for human activities recognition. In: The Pervasive Technologies Related to Assistive Environments (PETRA) Conference Proceedings (2018)

15. Palumbo, F., Ullberg, J., Štimec, A., Furfari, F., Karlsson, L., Coradeschi, S.: Sensor network infrastructure for a home care monitoring system. Sensors **14**(3), 3833–3860 (2014)

16. Panwar, M., Dyuthi, S.R., Prakash, K.C., Biswas, D., Acharyya, A., Maharatna, K., Gautam, A., Naik, G.R.: CNN based approach for activity recognition using a wrist-worn accelerometer. In: 2017 39th Annual International Conference of the IEEE Engineering in Medicine and Biology Society (EMBC), pp. 2438–2441. IEEE (2017)

17. Rashidi, P., Mihailidis, A.: A survey on ambient-assisted living tools for older adults. IEEE J. Biomed. Health Inform. **17**(3), 579–590 (2013)

18. Suryadevara, N.K., Mukhopadhyay, S.C., Wang, R., Rayudu, R.: Forecasting the behavior of an elderly using wireless sensors data in a smart home. Eng. Appl. Artif. Intell. **26**(10), 2641–2652 (2013)

19. Wang, Z., Jiang, M., Hu, Y., Li, H.: An incremental learning method based on probabilistic neural networks and adjustable fuzzy clustering for human activity recognition by using wearable sensors. IEEE Trans. Inf Technol. Biomed. **16**(4), 691–699 (2012)

20. Yang, J., Nguyen, M.N., San, P.P., Li, X., Krishnaswamy, S.: Deep convolutional neural networks on multichannel time series for human activity recognition. In: IJCAI, pp. 3995–4001 (2015)

21. Yin, J., Yang, Q., Pan, J.J.: Sensor-based abnormal human-activity detection. IEEE Trans. Knowl. Data Eng. **20**(8), 1082–1090 (2008)

# Deep Online Hierarchical Unsupervised Learning for Pattern Mining from Utility Usage Data

Saad Mohamad$^{(\boxtimes)}$, Damla Arifoglu, Chemseddine Mansouri,
and Abdelhamid Bouchachia

Department of Computing, Bournemouth University, Poole, UK
{smohamad,darifoglu,cmansouri,abouchachia}@bournemouth.ac.uk

**Abstract.** Machine learning approaches for non-intrusive load monitoring (NILM) have focused on supervised algorithms. Unsupervised
approaches can be more interesting and of more practical use in real
case scenarios. More specifically, they do not require labelled training
data to be collected from individual appliances and the algorithm can
be deployed to operate on the measured aggregate data directly. In this
paper, we propose a fully unsupervised NILM framework based on Deep
Belief network (DBN) and online Latent Dirichlet Allocation (LDA).
Firstly, the raw signals of the house utilities are fed into DBN to extract
low-level generic features in an unsupervised fashion, and then the hierarchical Bayesian model, LDA, learns high-level features that capture
the correlations between the low-level ones. Thus, the proposed method
(DBN-LDA) harnesses the DBN's ability of learning distributed hierarchies of features to extract sophisticated appliances-specific features
without the need of precise human-crafted input representations. The
clustering power of the hierarchical Bayesian models helps further summarise the input data by extracting higher-level information representing the residents' energy consumption patterns. Using Deep-Hierarchical
models reduces the computational complexity since LDA is not directly
applied to the raw data. The computational efficiency is crucial as our
application involves massive data from different types of utility usages.
Moreover, we develop a novel online inference algorithm to cope with
this big data. Another novelty of this work is that the data is a combination of different utilities (e.g., electricity, water and gas) and some
sensors measurements. Finally, we propose different methods to evaluate the results and preliminary experiments show that the DBN-LDA is
promising to extract useful patterns.

**Keywords:** Unsupervised non-intrusive load monitoring
Pattern recognition · Online Latent Dirichlet Allocation
Deep belief network

© Springer Nature Switzerland AG 2019
A. Lotfi et al. (Eds.): UKCI 2018, AISC 840, pp. 276–290, 2019.
https://doi.org/10.1007/978-3-319-97982-3_23

# 1 Introduction

The monitoring of human behaviour is highly relevant to many real-word domains such as safety, security, health and energy management. Research on human activity recognition (HAR) has been the key ingredient to extract patterns of human behaviour. There are three main types of HAR approaches, which are sensor-based, vision-based and radio-based. A common feature of these methods is that they all require equipping the living environment with embedded devices (sensors). On the other hand, non-intrusive load monitoring (NILM) requires only single meter per house or building that measures aggregated electrical signals at the entry point of the meter. Various techniques can then be used to disaggregate per-load power consumption from this composite signal providing energy consumption data at an appliance level granularity. In this sense, NILM's focus is not on extracting general human behaviour patterns but rather on identifying the appliances in use. This, however, can provide insight into the energy consumption behaviour of the residents and therefore can express users life style in their household. Traditional HAR methods introduce high costs of using various sensors, which makes NILM an attractive approach to exploit in general pattern recognition problems. On the other hand, taking the human behaviour into account can leverage the performance of NILM; thus, providing a better understanding of the resident's energy consumption behaviour. In this paper, we do not make a distinction between patterns and appliances recognition. The main goal of our approach (DBN-LDA) is to encode the regularities in a massive amount of utilities data into a reduced dimensionality representation. This is only possible if there are regular patterns in consumption behaviour of the residents. Working on an extra large amount of real world data makes this approach applicable to real-world scenario.

In contrast to supervised approaches for NILM, unsupervised algorithms can be deployed to operate directly from the measured aggregate data with no need for annotation. To the best of our knowledge, all existing unsupervised approaches to NILM [1] focus on disaggregating the whole house signal into its appliances' ones. In contrast, our approach, as mentioned earlier, does not focus on identifying per-appliance signal. We instead propose a novel approach that seeks to extract human behaviour patterns from home utility usage data. These patterns could be exploited for HAR as well as energy efficiency applications.

The proposed approach is a two-module architecture composed of DBN and a hierarchical Bayesian mixture model based on Latent Dirichlet Allocation (LDA). Hence, we call it DBN-LDA. It draws inspiration from the work in [2] where a hierarchical Dirichlet process (HDP) prior is plugged on top of a Deep Boltzmann Machine (DBM) network which allows learning multiple layer of abstractions. The low-level abstraction represents generic domain-specific features that are hierarchically clustered and shared to yield high-level abstraction representing patterns. Moreover, the success of LDA in text modelling domain is also an inspiration for DBN-LDA.

Recently, deep learning (DL) methods have achieved remarkable results in computer vision, natural language processing, and speech recognition while it has

278 S. Mohamad et al.

not been exploited in the field of NILM [3]. DL methods are good at extracting multiple layers of distributed feature representations from high-dimensional data. Each layer of the deep architecture performs a non-linear transformation on the outputs of the previous layer. Thus, high-dimensional data is represented in a structure of hierarchical feature representations, from low-level to high-level [4,5]. Instead of relying on heuristic hand-crafted features, DL learns to extract fruitful features. Relying on the big size of data and its high sampling rate (205 KHZ) which results in a very high-dimensional data, to the best of our knowledge, our study is the first to exploit unsupervised DL model in NILM. In contrast to existing electrical engineering and signal processing approaches, our method relies fully on the data to construct informative features. In this paper, first we pre-train DBN [6] to learn generic features from unlabelled raw electrical signal with 1 s granularity. The extracted features are then fed to the online LDA with 30 min granularity. Although, the bag-of-words assumption adopted here is a major simplification, it breaks down the unnecessary low-level hard-to-model complexity leading to computationally efficient inference with no much loss as shown in LDA [7].

In this work, we demonstrate that, similar to LDA in the domain of text mining, this approach can capture significant statistical structures in a specified window of data over a period of time. This structure provides understanding of regular patterns in the human behaviour that can be harnessed to provide various services including services to improve energy efficiency. For example, understanding of the usage and energy consumption patterns could be used to predict the power demand (load forecasting), to apply management policies and to avoid overloading the energy network. Moreover, providing consumers with information about their consumption behaviour and making them aware of any abnormal consumption patterns can influence their behaviour to moderate energy consumption [8].

In contrast to other studies, except [9–11], DBN-LDA is trained on a very huge amount of data with a high sampling rate of around 205 kHz of the electricity signal. High sampling rate allows extraction of rich features while this is not much possible with a low sampling rate. Online LDA helps deal with such a big size of data. This can be done by defining particular distributions for the exponential family in the class of models described in [12]. More details can be found in Sect. 3. Our method not only works on a big data, but also it is the only one including water and gas usage data, except [13,14] whose sampling rate is very low. Moreover, measurements provided by additional sensors are also exploited to refine the performance of the pattern recognition algorithm. More details on the data can be found in the Appendix A. The diversity of the data is another motivation for adopting a pattern recognition approach rather than a traditional disaggregation approach.

The rest of the paper is organised as follows. Section 2 presents a summary of the related work, while Sect. 3 explains the details of the proposed approach. Section 4 discusses the experimental results. Finally, Sect. 5 concludes the paper and hints to future work.

## 2 Related Work

We divide the related work into two parts: (i) Machine learning approaches and (ii) NILM datasets. There has been limited work on applying Machine learning in the area of NILM. Very few machine learning methods applied to NILM and those proposed are supervised methods [15–22]. These methods use labelled data involving an expensive and laborious task. In fact, the practicality of NILM stems from the fact that it comes with almost no setup cost. Recently, researchers have started exploring unsupervised machine learning algorithms to NILM. These methods have mainly focused on performing energy disaggregation to discern appliances from the aggregated load data directly without performing any sort of event detection. The most prominent of these methods are based on Dynamic Bayesian Network models, in particular different variants of Hidden Markov Model (HMM) [23–25].

The common characteristic of these approaches discussed so far is that the considered data is electricity data. In contrast, our data involves different utilities namely electricity, water and gas as well some sensors measurements that provide contextual features. To the best of our knowledge, the only data that considers water and gas usage is [13,14]. However, the sampling rate of this data is very low compared to that of our data. Authors in [26] exploit the correlation between appliances and side information, in particular temperature, in a convex optimisation problem for energy disaggregation. This algorithm is applied on low sampling rate electricity data with contextual supervision in the form of temperature information.

This work is a continuation of our previous work [27] where online Gaussian Latent Dirichlet Allocation (GLDA) is proposed to extract global components that summarise the energy signal. These components provide a representation of the consumption patterns. The algorithm is applied on the same data-set as in this paper. However, in contrast to [27], deep learning is employed in this paper to construct features rather than engineering them using signal processing techniques.

To wrap up this section, three traits distinguish our approach from existing ones. It bridges the gap between pattern recognition and NILM making it beneficial for a variety of different applications. Driven by massive amount of data, our method is computationally efficient and scalable, unlike the state-of-the-art probabilistic methods that posit detailed temporal relationships and involve complex inference steps. The approach is fully data-driven where DL is used to learn the features unlike existing feature engineering approaches. The available data has a high sampling rate electricity data allowing learning more informative features. It includes data from other utility usage and additional sensors measurements. Thus, our work also covers the research aspect of NILM concerned with the acquisition of data, prepossessing steps and evaluation of NILM algorithms.

## 3   The Proposed Method

Our proposed method has 2 main parts: (i) Features extraction and (ii) Pattern mining. As stated earlier, to achieve the first step, we use DBN. Note that we will not provide any background on the DL part of the model, which is the Deep Belief Network (DBN). Details about DBN can be found in [6]. Next, we will introduce stochastic variational inference for a family of graphical models [12] and derive online LDA [28]. Online LDA is an instance of the family of graphical models, operating online to accommodate high volume and speed data streams.

### 3.1   Stochastic Variational Inference

In the following, we describe the model family of LDA and review Stochastic Variational Inference (SVI).

**Model Family.** The family of models considered here consists of three random variables: observations $x = x_{1:D}$, local hidden variables $z = z_{1:D}$, global hidden variables $\beta$ and fixed parameters $\alpha$. The model assumes that the distribution of the $D$ pairs of $(x_i, z_i)$ is conditionally independent given $\beta$. Furthermore, their distribution and the prior distribution of $\beta$ belong to the exponential family as shown in the following:

$$p(\beta, x, z | \alpha) = p(\beta | \alpha) \prod_{i=1}^{D} p(z_i, x_i | \beta) \tag{1}$$

$$p(z_i, x_i | \beta) = h(x_i, z_i) \exp \left( \beta^T t(x_i, z_i) - a(\beta) \right) \tag{2}$$

$$p(\beta | \alpha) = h(\beta) \exp \left( \alpha^T t(\beta) - a(\alpha) \right) \tag{3}$$

Here, we overload the notation for the base measures $h(.)$, sufficient statistics $t(.)$ and log normalizer $a(.)$. While the soul of the proposed approach is generic, for simplicity we assume a conjugacy relationship between $(x_i, z_i)$ and $\beta$. That is, the distribution $p(\beta | x, z)$ is in the same family as the prior $p(\beta | \alpha)$.

   Note that this innocent looking family of models includes (but not limited to) Latent Dirichlet Allocation [7], Bayesian Gaussian Mixture, probabilistic matrix factorization, hidden Markov models, hierarchical linear and probit regression, and many Bayesian non-parametric models.

**Mean-Field Variational Inference.** Variational inference (VI) approximates intractable posterior $p(\beta, z | x)$ by positing a family of simple distributions $q(\beta, z)$ and find the member of the family that is closest to the posterior (closeness is measured with KL divergence). The resulting optimization problem is equivalent maximizing the evidence lower bound (ELBO):

$$\mathcal{L}(q) = E_q[\log p(x, z, \beta)] - E_q[\log p(z, \beta)] \leq \log p(x) \tag{4}$$

Mean-field is the simplest family of distributions, where the distribution over the hidden variables factorizes as follows:

$$q(\beta, z) = q(\beta | \lambda) \prod_{i=1}^{D} p(z_i | \phi_i) \tag{5}$$

where $\phi$ and $\lambda$ are the local and global variational parameters. Further, each variational distribution is assumed to come from the same family of the true one. Mean-field variational inference optimises ELBO with respect to the local and global variational parameters $\phi$ and $\lambda$.

$$\mathcal{L}(\lambda, \phi) = E_q \left[ \log \frac{p(\beta)}{q(\beta)} \right] + \sum_{i=1}^{D} E_q \left[ \log \frac{p(x_i, z_i | \beta)}{q(z_i)} \right] \tag{6}$$

It iteratively updates each variational parameter holding the other parameters fixed. With the assumptions taken so far, each update has a closed form solution. The local parameters are a function of the global parameters.

$$\phi(\lambda_j) = \arg \max_{\phi} \mathcal{L}(\lambda_j, \phi) \tag{7}$$

We are interested in the global parameters which summarise the whole dataset (clusters in Bayesian Gaussian mixture, topics in LDA).

$$\mathcal{L}(\lambda) = \max_{\phi} \mathcal{L}(\lambda, \phi) \tag{8}$$

To find the optimal value of $\lambda$ given that $\phi$ is fixed, we compute the natural gradient of $\mathcal{L}(\lambda)$ and set it to zero to obtain

$$\lambda^* = \alpha + \sum_{i=1}^{D} E_{\phi_i(\lambda_j)} [t(x_i, z_i)] \tag{9}$$

Thus, the new optimal global parameters are $\lambda_{j+1} = \lambda^*$. The algorithm works by iterating between computing the optimal local parameters given the global ones (Eq. 7) and computing the optimal global parameters given the local ones (Eq. 9).

**Stochastic Variational Inference.** Rather than analysing all the data to compute $\lambda^*$ at each iteration, stochastic optimization can be used. Assuming that the data samples are uniformly randomly selected from the dataset, an unbiased noisy estimator of $\mathcal{L}(\lambda, \phi)$ can be developed based on a single data point.

$$\mathcal{L}_i(\lambda, \phi_i) = E_q \left[ \log \frac{p(\beta)}{q(\beta)} \right] + D E_q \left[ \log \frac{p(x_i, z_i | \beta)}{q(z_i)} \right] \tag{10}$$

The unbiased stochastic approximation of the ELBO as a function of $\lambda$ can be written as follows

$$\mathcal{L}_i(\lambda) = \max_{\phi_i} \mathcal{L}_i(\lambda, \phi_i) \tag{11}$$

Following the same steps in the previous section, we end up with a noisy unbiased estimate of Eq. 8

$$\hat{\lambda} = \alpha + DE_{\phi_i(\lambda_j)}[t(x_i, z_i)] \tag{12}$$

At each iteration, we move the global parameters a step-size $\rho_j$ (learning rate) in the direction of the noisy natural gradient.

$$\lambda_{j+1} = (1 - \rho_j)\lambda_j + \rho_j\hat{\lambda} \tag{13}$$

With certain conditions on $\rho_j$, the algorithm converges ($\sum_{j=1}^{\infty} \rho_j = \infty$, $\sum_{j=1}^{\infty} \rho_j^2 < \infty$) [29].

## 3.2   Latent Dirichlet Allocation

Latent Dirichlet allocation (LDA) is an instance of the family of models described in Sect. 3.1 where the global, local, observed variables and their distributions are set as follows:

- the global variables $\{\beta\}_{k=1}^K$ are the topics in LDA. A topic is a distribution over the vocabulary, where the probability of a word $w$ in topic $k$ is denoted by $\beta_{k,w}$. Hence, the prior distribution of $\beta$ is a Dirichlet distribution $p(\beta) = \prod_k Dir(\beta_k; \eta)$
- the local variables are the topic proportions $\{\theta_d\}_{d=1}^D$ and the topic assignments $\{\{z_{d,w}\}_{d=1}^D\}_{w=1}^W$ which index the topic that generates the observations. Each document is associated with a topic proportion which is a distribution over topics, $p(\theta) = \prod_d Dir(\theta_d; \alpha)$. The assignments $\{\{z_{d,w}\}_{d=1}^D\}_{w=1}^W$ are indices, generated by $\theta_d$, that couple topics with words, $p(z_d|\theta) = \prod_w \theta_{d,z_{d,w}}$
- the observations $x_d$ are the words of the documents which are assumed to be drawn from topics $\beta$ selected by indices $z_d$, $p(x_d|z_d, \beta) = \prod_w \beta_{z_{d,w}, x_{d,w}}$

The basic idea of LDA is that documents are represented as random mixtures over latent topics, where each topic is characterized by a distribution over words [7]. LDA assumes the following generative process:

1. Draw topics $\beta_k \sim Dir(\eta, ..., \eta)$ for $k \in \{1, ..., K\}$
2. Draw topic proportions $\theta_d \sim Dir(\alpha, ..., \alpha)$ for $d \in \{1, ..., D\}$
   2.1. Draw topic assignments $z_{d,w} \sim Mult(\theta_d)$ for $w \in \{1, ..., W\}$
      2.1.1. Draw word $x_{d,w} \sim Mult(\beta_{z_{d,w}})$

According to Sect. 3.1, each variational distribution is assumed to come from the same family of the true one. Hence, $q(\beta_k|\lambda_k) = Dir(\lambda_k)$, $q(\theta_d|\gamma_d) = Dir(\gamma_d)$ and $q(z_{d,w}|\phi_{d,w}) = Mult(\phi_{d,w})$. To compute the stochastic natural gradient for LDA, we need to find the sufficient statistic $t(.)$ presented in Eq. (2). By writing the likelihood of LDA in the form of Eq. (2), we obtain $t(x_d, z_d) = \sum_{w=1}^W I_{z_{d,w}, x_{d,w}}$, where $I_{i,j}$ is equal to 1 for entry $(i, j)$ and 0 for all the rest. Hence, the stochastic natural gradient $g_i(\lambda_k)$ can be written as follows:

$$g_i(\lambda_k) = \eta + D \sum_{w=1}^W \phi_{i,w}^k I_{k, x_{i,w}} - \lambda_k \tag{14}$$

Details on how to compute the local variational parameters $\phi_i^*(\lambda^*)$ can be found in [12]. The analogy between LDA and the proposed approach can be explained as follows, the components are equivalent to topics in LDA. Because features extracted by DBN are in discrete space, the components represent categorical distributions over the input features like the LDA's categorical distributions over words. A pattern is a mixture of components generating the input features over a fixed period of time. In LDA, patterns are associated with documents that can be expressed by mixture of corpus-wide topics. One can clearly notice that this bag-of-words assumption, where temporal dependency in the data is neglected, is a major simplification. However, this simplification leads to methods that are computationally efficient. Such computational efficiency is essential in our case where massive amount of data (around 4 Tb) is used to train the model.

## 4    Experimental Settings

First, the data is pre-processed in 3 steps: (1) synchronisation of same utility data, (2) alignment of data coming from different utilities and (3) features extraction. Details about the pre-processing steps and data description are given in Appendix A. In this section, we focus on the experiments performed on the pre-processed data.

   In all experiments, we use the empirical Bayes method to online point estimate the hyper-parameters from the data. The idea is to maximise the log likelihood of the data with respect to the hyper-parameters. Since the computation of the log likelihood of the data is not tractable, the approximation based on the variational inference algorithm used in Sect. 3.2 is employed. The number of components is fixed to $K = 50$. We evaluated a range of settings of the learning parameters: $\kappa$ (learning factor), $\tau_0$ (learning delay) and batch size $BS$ on a testing set, where the parameters $\kappa$ and $\tau_0$, defined in [28], control the learning step-size $\rho_j$. We used the data collected during the last two weeks (see Appendix A) for testing. All experiments are run 30 times.

### 4.1    Evaluation and Analysis

In order to evaluate LDA, we use the perplexity measure which quantifies the fit of the model to the data. It is defined as the reciprocal geometric mean of the inverse marginal probability of the input in the held-out test set. Since perplexity cannot be computed directly, a lower bound on it is derived in a similar way to the one in [7]. This bound is used as a proxy for the perplexity.

   Moreover, to investigate the quality of the results, we study the regularity of the mined patterns by matching them across similar periods of time. For instance, it is expected that similar patterns will emerge in specific hours like breakfast in every morning, watching TV in the evening, etc. Hence, it is interesting to understand how such patterns occur as regular events.

   Finally, to provide a quantitative evaluation of the algorithm, we propose a mapping method that reveals the specific energy consumed for each pattern. By doing so, we can evaluate numerically the coherence of the extracted patterns by fitting a regression model to the energy consumption over components:

$$Aw = b \qquad (15)$$

where $w$ is a vector expressing energy consumption associated with components, $b$ is a vector representing per-pattern consumption and $A$ is the matrix of the per-pattern components proportions obtained by LDA. This technique will also allow us to numerically check the predicted consumption against the real consumption.

**A - Model Fitness:** Although online LDA converges for any valid $\kappa$, $\tau_0$ and $BS$, the quality and speed of the convergence may depend on how the learning parameters are set. We run online LDA on the training sets for $\kappa \in \{0.5, 0.6, 0.7, 0.8, 0.9\}$, $\tau_0 \in \{1, 64, 256, 1024\}$ and $BS \in \{1, 4, 8\}$. Table 1 summarises the best settings of each batch size along with the perplexity obtained on the test set.

**Table 1.** Parameter settings

| Batch size: $BS$ | 1 | 4 | 8 |
|---|---|---|---|
| Learning factor: $\kappa$ | 0.7 | 0.5 | 0.5 |
| Learning delay: $\tau_0$ | 256 | 64 | 64 |
| Perplexity | 334 | 333 | 350 |

The obtained results show that the perplexity for different parameters settings are similar. However, the computation complexity increases with the size of the batch. Hence, we set the batch size to 1, where the best learning parameters are $\kappa = 0.7$ and $\tau = 256$.

**B - Pattern Regularity:** Using the optimal parameters' setting, we examine in the following the regularity of the mined patterns. To do that, we use the data from 11-05-2017 10:10:10 to 25-05-2017 10:10:10 for testing. To study the regularity of the energy consumption behaviour of the residents, we compare the mined patterns across different days of the testing period. These patterns are represented by the proportions of the different components (topics) inferred from the training data. The dissimilarity of the patterns across the two weeks are computed as follows:

$$dissimilarity(day1, day2) = \frac{1}{K * F} \sum_{j=1}^{F} \sum_{i=1}^{K} |\gamma(day1)_{j,i} - \gamma(day2)_{j,i}| \qquad (16)$$

where $F = 48$ is the number of patterns within the day, $K$ is the number of components, $\gamma(day1)_{j,i}$ depends on component $i$ of pattern $j$ (see Sect. 3.2) of a day from the first week. Table 2 shows the per-day dissimilarity. It can be clearly seen from the table that there are regular patterns across the two weeks. That is, similar energy consumption patterns appear across different days. This similarity is a bit less (i.e., higher dissimilarity) for the weekend where more random activities could take place. Computing the dissimilarity measure between week and weekend days confirms this observation. For instance, the dissimilarity between Monday and Sunday of the first week is equal to 15.25.

**Table 2.** Patterns dissimilarity

| Day | Mon | Tue | Wed | Thu | Fri | Sat | Sun |
|---|---|---|---|---|---|---|---|
| Dissimilarity | 8.31 | 9.17 | 9.61 | 7.97 | 10.43 | 12.64 | 12.14 |

This regularity may be caused by regular user lifestyle leading to similar energy consumption behaviour within and across the weeks. Such regularity is violated in the weekend, as more irregular activities could take place. Having shown that there is some regularity in the mined patterns, it is more likely that specific energy consumption can be associated with each component. In the next section, we apply a regression method to map the patterns (i.e., components proportions) to energy consumption. Thus, the parameters of interest are the energy consumption associated with the components. By attaching an energy consumption with each component, we can help validate the coherence of the extracted patterns and do forecasting.

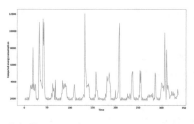

(a) Computed energy consumption

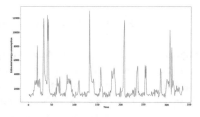

(b) Estimated energy consumption

**Fig. 1.** Evolution of the energy consumption over time

**C - Energy Mapping:** As shown in the previous section, LDA can express the energy consumption patterns by mixing global components summarising data. These global components can be thought of as a base in the space of patterns. Each component is a distribution over a high-dimensional feature space and understanding what it represents is not easy. Hence, we propose to associate consumption quantities to each component. Such association is motivated by the fact that an energy consumption pattern is normally governed by the usage of different appliances in the house. There should be a strong relation between components and appliances usage. Hence, a relation between components and energy consumption is plausible. Note that the best case scenario occurs if each component is associated with the usage of a specific appliance. Apart from the coherence study, associating energy consumption with each component can be used to forecast the energy consumption. This can be done through pattern forecasting which will be investigated in future work (see Sect. 5). We apply a simple least-square regression method to map patterns to energy consumption, expressed as follows:

$$\min ||A\boldsymbol{w} - \boldsymbol{b}||^2 \tag{17}$$

where $\boldsymbol{w}$ is the per-component energy consumption vector, $\boldsymbol{b}$ is the per-pattern consumption vector and $A$ is the matrix of the per-pattern components' proportions which is computed by LDA. We train the regression model on the week from 18-05-2017 23:45:22 to 25-05-2017 23:45:22 and run the model on the next one from 25-05-2017 23:45:22 to 01-06-2017 23:45:22. Figure 1 shows the energy consumption (in joules) along with the estimated consumption computed using the learned per-component consumption parameters.

The similarity between the estimated and computed energy consumption demonstrates that the LDA components express distinct usages of energy. Such distinction can be the result of the usage of different appliances likely having distinct energy consumption signatures. Thus, DBN-LDA produces coherent and regular patterns that reflect the energy consumption behaviour and human activities. Note that it is possible that different patterns (or appliance usages) may have the same energy consumption and that is why both estimated and computed energy consumption in Fig. 1 are not fully the same.

## 5    Conclusion and Future Work

In this paper, we presented a novel approach to extract patterns of the users' consumption behaviour from data involving different utilities (e.g., electricity, water and gas) as well as some sensors measurements. DBN-LDA is fully unsupervised and LDA works online which is suitable for dealing with big data. To analyse the performance, we proposed a three-steps evaluation that covers: model fitness, qualitative analysis and quantitative analysis. The experiments show that DBN-LDA is capable of extracting regular and coherent patterns that highlight energy consumption over time. In the future, we foresee three directions for research to improve the obtained results and provide more features: (i) developing online dynamic latent Dirichlet allocation (DLDA) to consider the temporal dependency in the data leading to better results and allowing forecasting, (ii) develop more scalable LDA by applying asynchronous distributed LDA which can be derived from [30] instead of SVI and (iii) involving active learning strategy to query users about ambiguous or unknown activities in order to guide the learning process when needed [31,32].

**Acknowledgment.** This work was supported by the Energy Technology Institute (UK) as part of the project: *High Frequency Appliance Disaggregation Analysis*

*(HFADA)*. A. Bouchachia was supported by the European Commission under the Horizon 2020 Grant 687691 related to the project: *PROTEUS: Scalable Online Machine Learning for Predictive Analytics and Real-Time Interactive Visualization.*

# A   Appendix

In this section, we will first introduce the experimental data, LDA will be tested on along with details about the data pre-processing stages.

## A.1   Datasets

The real-world multi-source utility usage data used here is provided by ETI[1]. The data includes electricity signals (voltage and current signals) sampled at high sampling rate around 205 kHz, water and gas consumption sampled at low sampling rate. The data also contains other sensors measurements collected from the Home Energy Monitoring System (HEMS). In this study we will use 4 Tb of utility usage data collected from one house over one month. This data has been recorded into three different formats. Water data is stored in text files with sampling rate of 10 s and is synchronised to Network Time Protocol (NTP) approximately once per month. Electricity data is stored in wave files with sampling rate of 4.88 s and is synchronised to NTP every 28 min 28 s. HEMS data is stored in a Mongo database with sampling rates differing according to the type of the data and sensors generating it (see Table 3).

**Data Pre-processing** In order to exploit raw utility data by LDA, a number of pre-processing steps are required. We read the data from the different sources, synchronise its time-stamps to NTP time-stamps, extract features and align the data samples to one time-stamp by measurement. For water data, the PC clock time-stamps of samples within each month are synchronised to NTP time-stamp. The synchronisation is done as follows:

$$timestampNTP(i) = timestampsclock(i) + i\frac{Total\_Time\_Shift}{Number\_of\_Samples} \qquad (18)$$

In this equation, we assume that the total shift (between NTP and PC clock) can be distributed over the samples in one month. Similarly, Electricity data samples' time-stamps are synchronised to NTP time-stamps. The shift is distributed over 28 min and 28 s.

$$timestampNTP(i) = timestampsclock(i) + i\frac{Total\_Time\_Shift}{Number\_of\_Samples} \qquad (19)$$

The time-stamps of HEMS data were collected using NTP and so no synchronisation is required. Having all data samples synchronised to the same reference

---

[1] Energy Technologies Institute: http://www.eti.co.uk/.

**Table 3.** Characteristics of the data

| Data | Range | Resolution | Measurement frequency | Total duration |
|---|---|---|---|---|
| Mains voltage | −500 V to +500 V | 62 mV | 4.88 s | 1 months |
| Mains current | −10 A to +10 A | 1.2 mA | 4.88 s | 1 months |
| Water flow volume | 0 to 100 L per min | 52.4 pulses per litre | 10 s | 1 months |
| Room air temperature | 0 to 40 DegC | 0.1 DegC | Once every minute | 1 months |
| Room relative humidity | 0 to 95% | 0.1 % | Once every 5 min | 1 months |
| Hot water | DegC | 0.1 DegC | Once every | 1 months |
| Feed temperature | | | 5 min | |
| Boiler: water | 0 to 85 DegC | 0.1 DegC | Once every | 1 months |
| Temperature (input) | | | 5 min | |
| Boiler: water | DegC | 0.1 DegC | Once every | 1 months |
| Temperature (output) | | | 5 min | |
| Household: mains cold | DegC | 0.1 DegC | Once every | 1 months |
| Water inlet temperature | | | 5 min | |
| Gas meter reading | Metric meter | 0.01 m$^3$ | Once every 15 min | 1 months |
| Radiator temperature | DegC | 0.1 DegC | Once every 5 min | 1 months |
| Radiator valve | 0 to 100% | 50% | Once every 5 min | 1 months |
| Boiler firing switch | Boolean | None | Once every 5 min | 1 months |

(NTP), we align the samples to the same time-stamps. The alignment strategy is shown in Fig. 2 where the union of all aligned data samples is stored in one matrix. Each row of this matrix includes a time-stamp and the corresponding values of the sensors. If for some sensors, there are no measurements taken at the time-stamp, the values measured at the previous time stamp are taken. The aligned data samples are the input of the feature extraction model (Deep Belief Network). Pushed by the complexity of the mining task and motivated by the informativeness and simplicity of the water and sensors data, at this stage, we apply DBN only on the electricity data over time windows of 1 s.

The employed DBN[2] consists of three Restricted Boltzmann Machine layers where the first layer reduces the input dimension from 204911 (1 s granularity) to 700. The second and third layers' outputs dimensions are 200 and 100 respectively. Note that the first layer's inputs are from continuous space while the rest is categorical data. The rest parameters are left to the default setting. The last

---

[2] https://github.com/lmjohns3/py-rbm.

| timestampNTP | Water | Electricity | | | HEMS | | | | | | |
|---|---|---|---|---|---|---|---|---|---|---|---|
| | | Real power | Reactive power | RMS Spectrum power over different frequency ranges | Gas meter | Temperature | | | humidity | Radiators' valve | Firing boiler |
| | | | | | | Rooms | radiators | water | | | |
| . | . | . | . | . | . | . | . | . | . | . | . |
| . | . | . | . | . | . | . | . | . | . | . | . |
| . | . | . | . | . | . | . | . | . | . | . | . |
| . | . | . | . | . | . | . | . | . | . | . | . |

**Fig. 2.** Alignment of the data

layer's outputs are aligned and concatenated with the other utility and sensors discretised data.

# References

1. Bonfigli, R., Squartini, S., Fagiani, M., Piazza, F.: Unsupervised algorithms for non-intrusive load monitoring: an up-to-date overview. In: 2015 IEEE 15th International Conference on Environment and Electrical Engineering (EEEIC). IEEE (2015)
2. Salakhutdinov, R., Tenenbaum, J.B., Torralba, A.: Learning with hierarchical-deep models. IEEE Trans. Pattern Anal. Mach. Intell. **35**(8), 1958–1971 (2013)
3. LeCun, Y., Bengio, Y., Hinton, G.: Deep learning. Nature **521**(7553), 436 (2015)
4. Deng, L.: A tutorial survey of architectures, algorithms, and applications for deep learning. APSIPA Trans. Sig. Inf. Process. (2014)
5. Bengio, Y., et al.: Learning deep architectures for AI. Found. Trends® Mach. Learn. **2**(1), 1–127 (2009)
6. Hinton, G.E., Osindero, S., Teh, Y.-W.: A fast learning algorithm for deep belief nets. Neural Comput. **18**(7), 1527–1554 (2006)
7. Blei, D.M., Ng, A.Y., Jordan, M.I.: Latent Dirichlet allocation. J. Mach. Learn. Res. **3**(Jan), 993–1022 (2003)
8. Fischer, C.: Feedback on household electricity consumption: a tool for saving energy? Energ. Effi. **1**(1), 79–104 (2008)
9. Kelly, J., Knottenbelt, W.: The UK-DALE dataset, domestic appliance-level electricity demand and whole-house demand from five UK homes. Sci. Data **2**, 150007 (2015)
10. Filip, A.: BLUED: a fully labeled public dataset for event-based non-intrusive load monitoring research. In: 2nd Workshop on Data Mining Applications in Sustainability (SustKDD) (2011)
11. Kolter, J.Z., Johnson, M.J.: REDD: a public data set for energy disaggregation research. In: Workshop on Data Mining Applications in Sustainability (SIGKDD). San Diego, CA (2011)
12. Hoffman, M.D., Blei, D.M., Wang, C., Paisley, J.: Stochastic variational inference. J. Mach. Learn. Res. **4**(1), 1303–1347 (2013)
13. Makonin, S., Popowich, F., Bartram, L., Gill, B., Bajic, I.V.: AMPds: a public dataset for load disaggregation and eco-feedback research. In: 2013 IEEE Electrical Power & Energy Conference (EPEC). IEEE (2013)
14. Makonin, S., Ellert, B., Bajić, I.V., Popowich, F.: Electricity, water, and natural gas consumption of a residential house in Canada from 2012 to 2014. Sci. Data **3**, 160037 (2016)

15. Hart, G.W.: Nonintrusive appliance load monitoring. In: Proceedings of the IEEE (1992)
16. Liang, J., Ng, S.K., Kendall, G., Cheng, J.W.: Load signature study part I: basic concept, structure, and methodology. IEEE Trans. Power Delivery **25**(2), 551–560 (2010)
17. Kolter, J.Z., Batra, S., Ng, A.Y.: Energy disaggregation via discriminative sparse coding. In: Advances in Neural Information Processing Systems (2010)
18. Srinivasan, D., Ng, W., Liew, A.: Neural-network-based signature recognition for harmonic source identification. IEEE Trans. Power Delivery **21**(1), 398–405 (2006)
19. Berges, M., Goldman, E., Matthews, H.S., Soibelman, L.: Learning systems for electric consumption of buildings. In: Computing in Civil Engineering (2009)
20. Ruzzelli, A.G., Nicolas, C., Schoofs, A., O'Hare, G.M.: Real-time recognition and profiling of appliances through a single electricity sensor. In: 2010 7th Annual IEEE Communications Society Conference on Sensor Mesh and Ad Hoc Communications and Networks (SECON). IEEE (2010)
21. Kelly, J., Knottenbelt, W.: Neural nilm: deep neural networks applied to energy disaggregation. In: Proceedings of the 2nd ACM International Conference on Embedded Systems for Energy-Efficient Built Environments. ACM (2015)
22. Lai, Y.-X., Lai, C.-F., Huang, Y.-M., Chao, H.-C.: Multi-appliance recognition system with hybrid SVM/GMM classifier in ubiquitous smart home. Inf. Sci. **230**, 39–55 (2013)
23. Kim, H., Marwah, M., Arlitt, M., Lyon, G., Han, J.: Unsupervised disaggregation of low frequency power measurements. In: Proceedings of the 2011 SIAM International Conference on Data Mining. SIAM (2011)
24. Kolter, J.Z., Jaakkola, T.: Approximate inference in additive factorial HMMs with application to energy disaggregation. In: Artificial Intelligence and Statistics (2012)
25. Johnson, M.J., Willsky, A.S.: Bayesian nonparametric hidden semi-Markov models. J. Mach. Learn. Res. **14**(Feb), 673–701 (2013)
26. Wytock, M., Kolter, J.Z.: Contextually supervised source separation with application to energy disaggregation. In: AAAI (2014)
27. Saad, M., Abdelhamid, B.: Online Gaussian LDA for unsupervised pattern mining from utility usage data. In: ECML-PKDD (2018, submitted)
28. Hoffman, M., Bach, F.R., Blei, D.M.: Online learning for latent Dirichlet allocation. In: Advances in Neural Information Processing Systems (2010)
29. Robbins, H., Monro, S.: A stochastic approximation method. Ann. Math. Stat. **22**, 400–407 (1951)
30. Mohamad, S., Bouchachia, A., Sayed-Mouchaweh, M.: Asynchronous stochastic variational inference. arXiv preprint. arXiv:1801.04289 (2018)
31. Mohamad, S., Sayed-Mouchaweh, M., Bouchachia, A.: Active learning for classifying data streams with unknown number of classes. Neural Netw. **98**, 1–5 (2018)
32. Mohamad, S., Bouchachia, A., Sayed-Mouchaweh, M.: A bi-criteria active learning algorithm for dynamic data streams. IEEE Trans. Neural Netw. Learn. Syst. (2016)

# Online Object Trajectory Classification Using FPGA-SoC Devices

Pranjali Shinde[1], Pedro Machado[2(✉)], Filipe N. Santos[1],
and T. M. McGinnity[2]

[1] INESC TEC Campus da Faculdade de Engenharia da Universidade do Porto,
Rua Dr. Roberto Frias, Porto, Portugal
{pranjali.shinde,filipe.n.santos}@inesctec.pt
[2] Computational Neuroscience and Cognitive Robotics Laboratory,
Nottingham Trent University, Nottingham, UK
{pedro.baptistamachado,martin.mcginnity}@ntu.ac.uk

**Abstract.** Real time classification of objects using computer vision techniques are becoming relevant with emergence of advanced perceptions systems required by, surveillance systems, industry 4.0 robotics and agricultural robots. Conventional video surveillance basically detects and tracks moving object whereas there is no indication of whether the object is approaching or receding the camera (looming). Looming detection and classification of object movements aids in knowing the position of the object and plays a crucial role in military, vehicle traffic management, robotics, etc. To accomplish real-time object trajectory classification, a contour tracking algorithm is necessary. In this paper, an application is made to perform looming detection and to detect imminent collision on a system-on-chip field-programmable gate array (SoC- FPGA) hardware. The work presented in this paper was designed for running in Robotic platforms, Unmanned Aerial Vehicles, Advanced Driver Assistance System, etc. Due to several advantages of SoC-FPGA the proposed work is performed on the hardware. The proposed work focusses on capturing images, processing, classifying the movements of the object and issues an imminent collision warning on-the-fly. This paper details the proposed software algorithm used for the classification of the movement of the object, simulation of the results and future work.

**Keywords:** SoC-FPGA · Computer vision · Colour detection
Contour tracking · Trajectory detection · Object tracking

## 1 Introduction

Image processing using computer vision techniques has gained wide importance in large scale systems and embedded platforms [1]. The conventional method of object tracking using standard computers consumes more power, requires more space for installing components than using highly integrated devices; such

© Springer Nature Switzerland AG 2019
A. Lotfi et al. (Eds.): UKCI 2018, AISC 840, pp. 291–302, 2019.
https://doi.org/10.1007/978-3-319-97982-3_24

as the Intel (formerly Altera) Cyclone V SoC-FPGA (System-on-Chip-Field-Programmable Gate Array). This paper presents contour tracking algorithm to detect and track the object in real time and detects imminent collision. The proposed work focusses on classifying the movement of the object and a novel idea of performing looming detection on the hard processor of Intel Cyclone V SoC-FPGA which gives an idea of the object approaching or receding the camera is executed. Imminent collision is also detected which aids in knowing if the object has occupied 80% of the frame. During imminent collision the object movement cannot be distinguished. Such feature is desirable in several applications like security purposes, controlling robotic operations, etc. Thus, this system can be used to alert and take the necessary actions when the movement of the object cannot be perceived. FPGAs are fully-reconfigurable devices employed with memory, programmable logics, Input/output (I/O), routing capabilities and a matrix of resources connected via the programmable interconnects [2]. Object detection and recognition has been appealing as it involves numerous applications. Classification of the objects is performed based on size, shape, colour, etc., which aids in tracking and detecting the movement of the object which is achieved by employing computer vision technique. This technique collects the pictorial information from the real-world, processes and analyses the information and later converts it to binary values. The hard processor (dual core ARM Cortex A9), an integrated circuit on the FPGA fabric is used to further process the output and to detect the movements appropriately. The purpose of selecting FPGA over other computing devices is, it provides a robust platform for running the operating system accelerated by the custom hardware. The programmable features of the FPGA get it into the limelight as it supports parallel processing, reconfiguration of the logic elements and provides cost effectiveness for each logic element. The traits of the FPGA which includes low power consumption, faster processing speed, portability, increased memory, Intellectual property, digital signal processing (DSP) block make it tremendously a powerful computational appliance [3]. Although FPGAs deliver such flexibility, programming FPGAs is a complex task and, in this work, only the hard-processor was used. The optimisation of the tracking algorithm using the FPGA will be discussed in the Future Work section. This paper is structured as follows. Section 2 presents the related work, Sect. 3 discusses the design methodology used, along with the hardware and software configuration and explains the detection and tracking algorithm as well. Section 4 portrays the simulation of the results and Sect. 5 discusses the results and the future work.

## 2    Related Work

Real time classification in computer vision has been widely studied over the last decades. The real time moving target detection on UAV using FPGA is presented in [4] which allows to perform the computer vision algorithms on-the-fly platform. The vehicle detection algorithm used in [4] is Motion Estimation and object segmentation process and supports complex embedded video processing.

The work of [5] uses camera to track the objects based on colours and the movement of the cursor is controlled by the tracked objects. The work of [5] is sceptical in the use of computer vision technique as it does not specify the method used for distinguishing the objects. The frame rate of the camera is not specified as well. Although, the work produced by [5] showcases the segmentation process but does not provide any evidence if a still image or video is captured and lacks the details about the processing. Besides, the work of [5] does not specify if the detection is performed in real-time and does not indicate the type of objects detected for controlling the cursor movement and lacks details about the operating system. Furthermore, the angle is not mentioned at which the object could track the cursor movement. A similar approach of multiple object detection in real time is described in [1] using computer vision on embedded system. This work employs the cascade classifiers used for object detection based on Haar feature and provides a prospect to cascade other complex classifiers thereby augmenting the speed of object detection. The application was developed on desktop and on the embedded system. It was observed from the results of [1] that the execution time on windows-based PC is lower than the embedded platform.

In [6] is described two cases which are considered while detecting objects in real time. They are: Detecting objects in complex background and Detecting objects with colour against a background. The second case is more challenging as it involves detecting an object of a colour same as the background. Differentiating the work of [6,7] uses Kalman filter for motion estimation and tracking of the objects. The work of [1,5,7] faced problems in implementation of the system in real time. Because of the constraints related to algorithms which dealt with white or plain background. However, the research produced by [6] has developed and implemented a system which detects and tracks the object in an unknown environment. The real time object classification is presented in [8] which aims in detecting the vehicle, classifying and tracking it in real time. Also, they have described computer vision algorithms are used to classify the objects based on the size, shape and movement of the vehicles. The work of [8] does not give a clear view of the test performed like the road and weather conditions, period of the day, etc.

Occlusion and Illumination are two big issues in object tracking. The work of [9] stresses on tracking of objects with partial or full occlusion in real time. [9] has proposed a colour-based algorithm which is used to track the objects through colours and contours. It is a challenging task to obtain real time performance on embedded vision as there is no perfect hardware which meets all the conditions of the levels of processing. Computer vision applications finds three different levels of processing: low, mid and high level. The iterative operations at pixel level is termed as the low level which includes filtering operations such as noise reduction, edge detection, etc. The Simple Instruction Multiple Data (SIMD) architectures better serve the low-level processing. The region of interest considered by the mid-level is the classification criteria such as segmentation, classification of objects, feature extraction, etc. Parallelism with mid-level can be achieved to a certain extent. The sequential processing is obtained at the high level which is responsible for decision making [10]. In addition to [10,11] presents the embedded vision challenges regarding the hardware and software issues such as power consumption, timing etc. From, the literature, it can be observed that object recognition, detection and tracking has been demanding in real time. Various object detection and tracking algorithms are illustrated in the works though not including looming detection. Different algorithms are used to detect the objects like the Binary histogram which is based on split or merge [12], Haar classifiers, multi-object detection using grid based Histograms of Oriented Gradient (HoG) is presented in [13], colour detection, segmentation, etc., which aids in detecting the objects. Another approach of small object detection using infrared camera is shown in [14] and makes use of FPGA and Digital Signal Processing (DSP) for the computations. The infrared camera an capture 22 frames per second leading to slower processing speed in comparison to the camera used in the proposed work. Like, the work of [9], the proposed research incorporates the colour and contour tracking algorithm to classify the object trajectory and detects imminent collision in real time considering the illumination and background changes.

In this work, it is presented an object tracking system for tracking object movements and emit a warning when the object gets too close to the camera. Such algorithm can be used as part of Advanced Driver Assisted Systems (ADAS) for tracking the movement of objects that may, or may not interfere, with autonomous system. This algorithm can also be used in defence application where the enemy could be identified, and the required action could be taken.

## 3 Methodology

The work described in this paper was carried out on an Intel Cyclone V SoC-FPGA. The goal of this work is to implement an embedded object tracking device which can be used for general-proposed applications where object tracking is required. Such a system requires a camera connected to a SoC-FPGA with a standard Operating System (OS) running on the Hard Processor (HP) from a SD-Card. A SoC-FPGA was chosen because of its flexibility feature. A standard OS runs on the SoC device, for providing all the services and applications (e.g. network services, OpenCV library, security, etc.) which is connected via a high-speed bus to the FPGA. This provides the flexibility for performing hardware acceleration (which will be discussed in the Conclusions and Future Work section). In this work, the Ubuntu Linux 16.04 LTS was chosen over other possible embedded Linux distributions. Robotic Operating System (ROS) was installed because it was desirable to (i) increase the compatibility of the system with most Robotic Platforms and (ii) ROS has the OpenCV library already integrated. The SoC-FPGA board is interfaced with camera, keyboard, mouse, monitor.

## 3.1   Overall System Architecture

The hardware and software configuration for the proposed system involves setting the hardware platform and developing the object tracking algorithm on the SoC-FPGA. The classification of the movement of the object and looming detection is observed on the monitor.

### 3.1.1   Hardware Layer

The Altera Cyclone V SoC FPGA is used as the embedded platform. The feature of reconfigurability, low power processor system leverages the user as it supports flexibility in the design. The SoC kit is an integration of hard processor Cortex A9, comprising of processor, memory, peripherals tangled impeccably with FPGA through the interconnects. Also, it possesses high speed DDR3 memory, video, audio, networking capabilities, etc.[1]

The Logitech camera provides a maximum resolution of 640X480 and the image sensor used is of type Complementary Metal Oxide Semiconductor (CMOS) and as well supports USB interface. This camera is made to operate on Ubuntu by running the guvcview program. A 64 GB micro secure digital (SD) card is used to set up the operating system, OpenCV on the SoC-FPGA to perform the classification. Due to its huge storage capacity of 64 GB it provides flexibility in setting the additional packages on the Hard Processor which include the editor, libraries, etc. The main feature of the SD card is that it reads the data at a speed of 80 MBps.

### 3.1.2   Software Layer

The Ubuntu 16.04 LTS Linux operating system and OpenCV 3.2.0 is installed on the hard processor as well on ThinkPad laptop. The laptop has Intel i5 processor 2nd generation and operates at a processing speed of 2.5 GHz. It supports a memory of 8 GB and hard drive of 1 TB with windows 10 operating system installed whereas Linux operating system is installed for the implementation of the work. The environment set up on the laptop is the same as on the FPGA.

The programming platform used for the proposed work is python, running on Ubuntu and uses the OpenCV library. Whereas, python does not require compilation instead needs an interpreter which generates the python executable file. The elf file is used to run on the hard processor through RS232 port [9]. The flowchart of the contour tracking algorithm used in the proposed system is as shown in Fig. 1.

The conversion of the grabbed frame (RGB) to HSV plays an important role in detecting the object. RGB colour space is used in all the systems like computer, TV, etc. as it is easy to implement as described in [15]. RGB is not suitable for processing the colour images. When considered the perceptual non-linearity, machine dependency and the integration of luminance and chrominance

---

[1] Retrieved from https://www.terasic.com.tw/cgi-bin/page/archive.pl?Language=English&CategoryNo=165&No=836&PartNo=2, Last accessed: 2018-06-06.

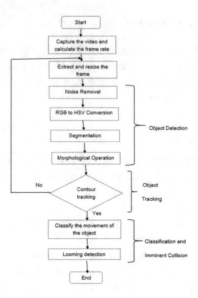

**Fig. 1.** Flowchart of the proposed system

the data obtained makes RGB incompatible for colour image processing. HSV is mainly used because it is capable of handling images influenced by illumination such as lighting, shade, contrast light [16].

The colour detection algorithm aids in recognizing and identifying the object. The video is captured, and the frame rate is calculated determining the processing speed of the frames. The extracted frame is scaled down in size because the frames with lower size can be processed at a faster rate. The raw input or the real-world frame captured is influenced by the presence of the noise and noise removal is important to process the frame. After removal of the noise the frame is converted from RGB to HSV. The segmentation technique used in the work is threshold segmentation. After performing segmentation, morphological operations are performed to further process the frame. This operation filters out the noise and is useful to get the segmented frame. The HSV based contour will be applied to the object based on the area of the frame. If at least one contour is found, then it is used to compute the minimum enclosing circle and centroid.

The classification is basically performed by storing the x-y coordinates of the previous frame and finding the difference with the current frame. The difference obtained is used to classify the movement of the object horizontally, vertically and diagonally. Next, the looming detection is performed by calculating the binary difference between the previous and the current frame which helps in determining the imminent collision, proceeding or receding of the object. Imminent collision is basically detecting the proximity of the object to the camera and this technique can be employed in automotive cars which would aid in avoiding accidents. Imminent collision is detected when number of 1s in the binary image occupy more than 80% of the frame size. The movement of the object cannot be classified when imminent collision is detected as the object has occupied the maximum frame size. Single object is being tracked and the distance is computed to perform the looming detection.

The algorithm used for tracking the object in the work is the contour tracking algorithm. Contours are basically the line joining the boundary of the object. When the object in the frame is detected a contour is formed. After finding the contours then the centroid of the object is found which is obtained from the image moments. The centroid coordinates search are ruled by Eq. 1.

$$c_x = int\Big(\frac{M\,[m_{10}]}{M\,[m_{00}]}\Big), c_y = int\Big(\frac{M\,[m_{01}]}{M\,[m_{00}]}\Big) \tag{1}$$

M is the image moments inside the contour, $M_{00}$ is the average of image pixel intensity, $M_{01}$ is the pixel value of y coordinate and $M_{10}$ is the pixel value of x coordinate.

The centroid points are used to put the bounding box (circular boundary) for the object. The dimensions of the box are calculated, and the object enclosed in the box aids in tracking the object. The difference between the x and y coordinate is used in classifying the movement of the object. If the object moves through the x coordinates, then the horizontal movement of the object is tracked. Likewise, the vertical movement of the object is tracked when the object moves over the y coordinates. The diagonal movement of the object is tracked based on the x and y coordinates. The looming detection is performed by comparing the binary matrix of the image. More number of 1s in the image determine the object is approaching. Less number of 1s in the image determine the object is receding and when the 1s occupy more than the maximum size of the image, imminent collision is determined. The algorithm for looming detection is as described in Algorithm 1. This differs from the existing algorithms as it detects the imminent collision which is not performed in any of the related works. This algorithm was designed for providing a good balance between low latency in an Embedded System and accuracy.

## 4   Results

The application was initially tested on a laptop and then ported to the SoC-FPGA. Classification of the movement of the object was performed by moving the ball in various directions. The object performs the trajectory in pink based on the movement of the ball. The contour tracking algorithm was producing an accuracy above 95% when the object had a white background and with no other objects moving in the background and the accuracy was dropped in more complex background with natural images. Also, the object tracking was not so robust when the background had colours similar to the object. The horizontal, vertical, diagonal and looming tests are reported in Sects. 4.1, 4.2, 4.3 and 4.4 respectively.

---

**Algorithm 1.** Classification of the object motion.

1: **if** $average > P_{MIN} \times MaxValue$ & $average < P_{MAX} \times MaxValue$ & $binSum[n] < BinSum[n-1] + Th$ **then**
2:     $out = Receding$
3: **else**
4:
5:     **if** $average > P_{MIN} \times MaxValue$ & $average < P_{MAX} \times MaxValue$ & $BinSum[n] > BinSum[n-1] + Th$ **then**
6:         $out = Approaching$
7:     **else**
8:
9:         **if** $average < P_{MIN} \times MaxValue$ **then**
10:             $out = FarAway$
11:         **else**
12:
13:             **if** $average \geq P_{MAX} \times MaxValue$ **then**
14:                 $out = ImminentCollision$
15:             **else**
16:                 $out = Inconclusive$
17:             **end if**
18:         **end if**
19:     **end if**
20: **end if**

---

### 4.1   Horizontal Movement Test

If the change in the x co-ordinate was positive provided there was no change in the y co-ordinate it was determined that the object was moving towards the right (east) direction else in the left (west) direction. The classification of the horizontal movement of the object is shown in Fig. 2.

**Fig. 2.** Object moving towards east (left image) and west (right image) directions

### 4.2   Vertical Movement Test

The movement of the object in the y co-ordinate would determine the vertical movement of the object. If the change in the y co-ordinate was positive provided there is no change in the x co-ordinate then it was determined the object is moving upwards in North direction else in South direction. The classification of the vertical movement of the object is shown in Fig. 3.

**Fig. 3.** Object moving towards east (left image) and west (right image) directions

### 4.3 Diagonal Movement Test

The movement of the object in the x, y coordinate would determine the diagonal movement. When the change in the x value and y value is greater than the pixel difference it is determined that the ball is moving diagonally. If the change in the x co-ordinate is positive/negative and change in the y co-ordinate is positive/negative, then the ball is termed to move either in north-east, north-west, south-east or south-west direction which is portrayed in Fig. 4.

**Fig. 4.** Object moving towards east (left image) and west (right image) directions

### 4.4 Looming Movement Test

This test was performed by comparing the previous and current segmented matrix to determine if the object was approaching or receding. The imminent collision was detected when the computed average of the matrix was greater than the maximum value of frame size. Imminent collision was determined when the object was at a distance approximately 15 cm from the camera. The approaching and receding of the object was determined when the object was between a distance approximately of 15 cm and 120 cm from the camera. After, a distance of 120 cm from the camera, the object to be detected is termed as far away. The approaching, receding and the imminent collision can be seen from Fig. 5.

**Fig. 5.** Object moving towards east (left image) and west (right image) directions

## 4.5   Performance Results

The proposed system was implemented on the laptop and SoC-FPGA which resulted in the different execution speed which is determined from Table 1 followed by the graphical representation.

**Table 1.** Execution time per processor type

| Platform | CPU clock | Camera frame rate | Execution time |
|---|---|---|---|
| Dual-core Intel i5 processor | 2.5 GHz | 30 fps | $0.4 \times 10^{-6}$ ms |
| Dual-core ARM Cortex A9 | 925 MHz | 30 fps | $1.8 \times 10^{-6}$ ms |

It can be observed from Table 1, the processing speed of the proposed system on the laptop is six times more than the execution of the proposed system on the hard processor of FPGA. The laptop here is used as ground truth and that the goal is to improve the SoC-FPGA for achieving the same performance with low power. When compared to the energy consumption of the laptop and SoC-FPGA, the energy consumed by the SoC-FPGA is less as it operates on 12 V supply. In addition, laptops are not suitable for robotics and real time object detection for many reasons: size, performance, robustness in comparison to SoC-FPGA. Also, it is not possible to use a laptop on a Drone or a Robotic Platform used in resilient environments.

## 5   Discussion and Future Work

In this paper, we have presented a system which detects the object trajectory and detects imminent collision in real time with the specified distance of the object from the camera. It can be observed from the results; the movement of the ball is displayed on the screen. When the object is at a distance of 15 cm from the camera it detects imminent collision. Similarly, when the ball moves away from the camera and is at a distance of 120 cm, the ball is identified, displaying far away on the monitor. The implementation of the proposed work makes it more representable in real time because of the execution of looming detection. The results show that the performance must be improved. However, the results demonstrate that the classification results were very similar to the results obtained with a powerful computer. The proposed algorithm can be improved by lowering the resolution, porting some of the filters to the SoC-FPGA and increasing the speed and performance of the system. The implementation of looming detection in the work presents new prospective which is not considered in any of the previous works. This makes the proposed system more reliable for security purposes- military, vehicle traffic system, UAVs, etc. The object tracking in the proposed system can be made more specific in terms of minute movement of the object. This can be achieved by knowing the angular movement of the object which would make the system more robust. In future, the system can be made to detect more than one object of different or same colours and shapes. The performance of the algorithm on the laptop is better than the FPGA. The variation in the processing time was not taken into consideration as the main motive of the work was to classify the movement of the object and to perform looming detection. In addition, the functionality of the present contour tracking algorithm can be increased by considering the colour of the object same as the background. The proposed system can be made to operate at a faster speed than the processor of laptop by forwarding the captured frames to the SoC-FPGA. This was not implemented due to shortage of time. By ignoring the difference of the execution time on both the platforms-laptop and DE1-SoC, the SoC makes the system portable and energy efficient which is recommended in real time robotics applications, UAVs, etc. The Terasic 8-megapixel daughter card can be employed in future to capture the video and process the video frame wise. The video from the camera can be forwarded to the FPGA through the high speed General Purpose Input Output (GPIO) interconnects. The camera consists of a Mobile Industry Processor Interface (MIPI) decoder and a module in addition to (High Definition Multimedia Interface (HDMI-TX). The captured video from the camera MIPI module is forwarded for further processing in video signal MIPI package. The MIPI decoder converts the packet from the module into 10-bit Bayer pattern. The voice coil motor (VCM) is employed in the camera to adjust the focus through the I2C interface. As the FPGA does not possess display capability, the HDMI-TX allows the video to be displayed on the monitor[2]. This interfacing would make the system to operate at a significantly higher speed and would provide drastic increase in the performance of the system.

---

[2] Retrieved from: http://www.terasic.com.tw/cgi-bin/page/archive.pl?Language= English&No=1051, last accessed: 2018-06-06.

# References

1. Guennouni, S., Ahaitouf, A., Mansouri, A.: Multiple object detection using openCV on an embedded platform. In: 2014 Third IEEE International Colloquium in Information Science and Technology (CIST), pp. 374–377, October 2014
2. Intel FPGA devices. https://www.intel.co.uk/content/www/uk/en/fpga/devices.html. Accessed 06 June 2018
3. Machado, P., Wade, J., McGinnity, T.M.:. Si elegans: modeling the C. elegans nematode nervous system using high performance FPGAS. In: Londral, R.A., Encarnação, P. (eds.) Advances in Neurotechnology, Electronics and Informatics, Chap. Si elegans, 12th edn., pp. 31–45. Springer, Heidelberg (2016)
4. Tang, J.W., Shaikh-Husin, N., Sheikh, U.U., Marsono, M.N.: FPGA-based real-time moving target detection system for unmanned aerial vehicle application. Int. J. Reconfigurable Comput. (2016)
5. Firmanda, D., Pramadihanto, D.: Computer vision based analysis for cursor control using object tracking and color detection. In: 2014 Seventh International Symposium on Computational Intelligence and Design (2014)
6. Prasad, S., Sinha, S.: Real-time object detection and tracking in an unknown environment. In: 2011 World Congress on Information and Communication Technologies, pp. 1056–1061, December 2011
7. Uke, N.J., Futane, P.R.: Efficient method for detecting and tracking moving objects in video. In: 2016 IEEE International Conference on Advances in Electronics, ICAECCT, Communication and Computer Technology, p. 2017 (2016)
8. Pea-Gonzlez, R.H., Nuo-Maganda, M.A.: Computer vision based real-time vehicle tracking and classification system. In: 2014 IEEE 57th International Midwest Symposium on Circuits and Systems (MWSCAS), pp. 679–682, August 2014
9. Gajbhiye, S.D., Gundewar, P.P.: A real-time color-based object tracking and occlusion handling using arm cortex-a7. In: 2015 Annual IEEE India Conference (INDICON), pp. 1–6, December 2015
10. Nieto, M., Otaegui, O., Vélez, G., Ortega, J.D., Cortés, A.: On creating vision-based advanced driver assistance systems. In: IET Intelligent Transport Systems (2015)
11. Stein, F.: The challenge of putting vision algorithms into a car. In: IEEE Computer Society Conference on Computer Vision and Pattern Recognition Workshops (2012)
12. Appiah, K., Meng, H., Hunter, A., Dickinson, P.: Binary histogram based split/merge object detection using FPGAs. In: 2010 IEEE Computer Society Conference on Computer Vision and Pattern Recognition - Workshops, pp. 45–52, June 2010
13. Chayeb, A., Ouadah, N., Tobal, Z., Lakrouf, M., Azouaoui,O.: HOG based multi-object detection for urban navigation. In: 2014 17th IEEE International Conference on Intelligent Transportation Systems, ITSC 2014 (2014)
14. Wang, Z., Song, H., Xiao, H., He, W., Gu, J., Yuan, K.: A real-time small moving object detection system based on infrared image (2014)
15. Saravanan, G., Yamuna, G., Nandhini, S.: Real time implementation of RGB to HSV/HSI/HSL and its reverse color space models. In: 2016 International Conference on Communication and Signal Processing (ICCSP), pp. 0462–0466, April 2016
16. Xue, T., Wang, Y., Qi, Y.: Multi-feature fusion based GMM for moving object and shadow detection. In: International Conference on Signal Processing Proceedings, ICSP (2012)

# Key Frame Extraction and Classification of Human Activities Using Motion Energy

David Ada Adama$^{(\boxtimes)}$, Ahmad Lotfi, and Caroline Langensiepen

School of Science and Technology, Nottingham Trent University,
Nottingham NG11 8NS, UK
david.adama2015@my.ntu.ac.uk

**Abstract.** One of the imminent challenges for assistive robots in learning human activities while observing a human perform a task is how to define movement representations (states). This has been recently explored for improved solutions. This paper proposes a method of extracting key frames (or poses) of human activities from skeleton joint coordinates information obtained using an RGB-D Camera (Depth Sensor). The motion energy (kinetic energy) of each pose in an activity sequence is computed and a novel approach is proposed for extracting key pose locations that define an activity using moving average crossovers of computed pose kinetic energy. This is important as not all frames of an activity sequence are key in defining the activity. In order to evaluate the reliability of extracted key poses, Long Short-Term Memory (LSTM) Recurrent Neural Network (RNN) which is capable to learn a sequence of transition from states in an activity is applied in classifying activities from identified key poses. This is important for assistive robots to identify key human poses and states transition in order to correctly carry out human activities. Some preliminary experimental results are presented to illustrate the proposed methodology.

**Keywords:** Human activity segmentation · Key pose extraction
Assistive robotics · Motion energy

## 1 Introduction

Humans have the ability to learn activities by observing while activities are executed by another human. One important aspect of this process is extracting segments of key aspects of activities and exploiting this information to be able to replicate the action executed. Similarly, with recent advancement in assistive technology, there is an increase in research on human-robot interaction related to assistive robots learning to execute human activities by extracting key information as they observe humans carry out activities [5,9]. In order to equip assistive robots with capabilities to perform certain activities associated with human movement, it is necessary to identify key aspects of human movement

© Springer Nature Switzerland AG 2019
A. Lotfi et al. (Eds.): UKCI 2018, AISC 840, pp. 303–311, 2019.
https://doi.org/10.1007/978-3-319-97982-3_25

which can be adapted to an assistive robot platform while learning to perform human activities.

Learning skill set for executing human activity can be divided into two processes. Firstly, creating a model used in identifying or recognizing various activities in order to differentiate one activity from another [1]. This can also be called learning or classification of activities. The second process involves generating activity representations required to understand sequential movements of different body parts towards actualizing the activity. The fusion of these two processes in a single platform would enhance the ability of an assistive robot to learn human activities and be able to carry out such activities even if executed by different people.

In this work, we propose a method of extracting key frames of human activities by analyzing moving average crossovers of kinetic energy of poses in an activity sequence. The pose information is obtained from $3D$ coordinates of human skeleton joints generated using an RGB-D camera (depth sensor) to observe while human activities are executed. The extracted key poses are used in a Long Short-Term Memory (LSTM) Recurrent Neural Networks (RNNs) [8] to learn the human activities from the sequence of key pose. LSTM are used in sequence-based learning applications such as time series prediction [8]. This provides information of relevant sequential movement in an activity.

The rest of this paper is divided into the following sections: In Sect. 2 works related to the proposed approach are discussed. Section 3 describes the proposed approach to key frame extraction of human activity with experiments and results presented in Sect. 4. Section 5 concludes the paper.

## 2    Related Work

The development of assistive technology such as assistive robots in human environments has increasingly received attention in the research community [5] with robots used as socially assistive platforms [3] for elderly care and other applications. Most applications of assistive robots require robots to execute fixed tasks which are already embedded in the robots knowledge base which makes such robot platforms dependent on users to provide instructions for tasks to execute. One imminent challenge encountered is how to make assistive robots execute tasks independently by extracting representations and sequences of movements from observation of activities.

As mentioned earlier in Sect. 1, recognizing human activities is also necessary in having an assistive robot execute such activities by understanding different segments of activities. The authors in [13] proposed a method for human action segmentation and recognition using atomic action templates of pose kinetic energy. In [4], the authors combined body pose estimation and 2D body shape to detect key poses used to recognize human action. The authors in [2,6,7] proposed different methods used in human activity learning from RGB-D sensor information. These works focused on recognizing human action from visual observation and are not self-sufficient in enabling assistive robots effectively execute activities as they focus on categorizing activities. However, if combined with a method

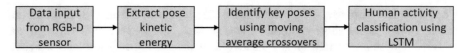

**Fig. 1.** Architecture of proposed key frame extraction and classification for human activity

of identifying key frames that define relevant joint movements in an activity, will enhance the ability of assistive robots learning human activities.

In learning transition states, Hidden Markov Models (HMM) and Neural Networks (NN) [11] have been applied to perform classification of human activities. The authors in [11] used LSTMs and Convolutional Neural Networks for activity and hand gesture recognition. LSTM networks have proven to maintain temporality in sequence-based learning applications compared with other base models [10]. With these benefits the proposed work shows how LSTMs can be used in classifying of human activities from sequence of extracted key poses.

## 3   Proposed Approach

The model proposed for human activity movement representations of body parts is illustrated in Fig. 1. The process begins with data input of an activity sequence from an RGB-D sensor. $3D$ coordinate positions of different joints in the body are obtained from the skeleton information provided by the sensor. The information is obtained for all joints provided by the sensor and pose kinetic energy is computed for all the frames of an activity sequence. Two simple moving averages of pose kinetic energies are computed and poses corresponding to the crossover points of the moving averages are selected as key poses that distinctively define an activity. The crossover points are important as they identify key points through an activity where there are distinct changes in the increase and decrease of motion energy associated with movement. These poses are the key points that define relevant movements of body joints associated with an activity through time. The selected poses are then passed into an LSTM recurrent neural network which learns the activity sequences to predict an activity being executed.

### 3.1   Key Pose Selection

Human activity consists of a sequence of movements of different body parts. It is worth noting that not all aspects of an activity sequence are necessary to define an activity. Certain aspects of the sequence can be executed in different forms and still result in a similar activity. In order to simplify an activity to the necessary pose configurations that constitute the sequence, key poses are selected. In this paper, a method of moving average crossovers of pose kinetic energy is applied as the process of selecting key poses used in segmenting an activity.

**Pose Kinetic Energy.** The pose kinetic energy adopted from [13] is based on the fact that joints show changes in acceleration and deceleration through an activity and this information is considered in identifying relevant poses. From the sequence of an activity obtained using an RGB-D sensor, given that a pose consists of $3D$ skeleton joints represented by $J$ and defined by;

$$J = [j_1, j_2, j_3, \ldots, j_i], \quad \text{for } J \in \mathbb{R}^{3 \times i} \tag{1}$$

where $i$ is the number of joints contained in a pose with $3D$ coordinates. An activity $A$ is a collection of a number of $J$ poses. Given kinetic energy $E_k = \frac{1}{2}mv^2$, where $m$ and $v$ are the mass and velocity respectively of the body part whose kinetic energy is computed. The kinetic energy for each pose is computed as the sum of kinetic energies for each joint in the pose;

$$E_k(J) = \sum_{n=1}^{i} E_k(j_n) \tag{2}$$

where $n$ is the $n^{th}$ joint in the pose. By assuming the mass $m$ of all joints to be unity and computing the joint velocities using the temporal change $\Delta T$ in the position $d$ of joints during an activity, the kinetic energy can be expressed as;

$$E_k(J) = \frac{1}{2} \sum_{n=1}^{i} (v_{j_n}), \tag{3}$$

$v_{j_n}$ represents the velocity of joint $j_n$ and is expressed as $v_{j_n} = \frac{d_c^n - d_p^n}{\Delta T}$, where $d_c^n$ is the current joint position and $d_p^n$ is the previous joint position. By substituting $v_{j_n}$ in Eq. 3, the kinetic energy of each joint is computed using the following equation:

$$E_k(J) = \frac{1}{2} \sum_{n=1}^{i} (\frac{d_c^n - d_p^n}{\Delta T}), \tag{4}$$

**Moving Average of Pose Kinetic Energy.** The Moving Average (MA) is a filter technique often applied to get overall trends in data. This technique is used to highlight long-term cycles in time series data by smoothing out short-term variations.

Moving averages of computed kinetic energy of poses are used in selecting the key poses in an activity sequence. Most of the works employing pose kinetic energy for key pose selection [12–14] select the key poses by using threshold values to select poses. The energy thresholds are selected by repeated experiments of different threshold values and the poses below the threshold value are selected as key poses. In this paper, we propose a different approach to use crossovers of two simple moving averages of pose kinetic energy in selecting key poses. Simple moving average is an un-weighted mean of a set of data points. This is taken from equal sets of data to ensure variations in the mean and data points are aligned and not shifted in time. Two simple moving averages of pose kinetic energy are selected - a short-term average (fast moving average) and a long-term moving average (slow moving average). The crossovers are obtained from points where the slow moving average crosses the fast moving average. These points indicate the change in kinetic energy of activity poses which are selected as key poses representing an activity. Given the computed kinetic energy for an activity pose as $E_k(J)$, the simple moving average $SMA_{E_k}$ for all frames in the activity is obtained using following equation;

$$SMA_{E_k} = \frac{E_k(J)_t + E_k(J)_{t-1} + \ldots + E_k(J)_{t-(m-1)}}{m} \tag{5}$$

where $m$ is the value of the period selected for either a slow or fast moving average.

### 3.2   Long Short-Term Memory

In the proposed approach, a LSTM RNN is used to evaluate the key poses identified from crossover points of the moving averages of kinetic energy poses. Traditional RNNs make predictions sequentially and the hidden layers from each prediction is fed into a succeeding prediction's hidden layer. This gives RNNs the characteristics of having a *memory* due to the fact that results from previous predictions can influence future predictions. LSTM is an extension to RNN and uses a gating mechanism and cell activation state in addition to the hidden layers to learn a sequence. It also has the characteristics of learning when to forget long-term information and when to incorporate new information.

In this work, an input vector of identified key poses is passed into the LSTM node and this is combined with a previous hidden state to obtain a new cell activation. More details of the LSTM configuration can be found in [8].

## 4   Experiments and Preliminary Results

In this section, the experiment conducted to verify the proposed approach is described and results obtained are presented. Human activity data is collected using an RGB-D sensor from which we obtain joint coordinates of 15 human skeleton joints as shown in Fig. 2. Data is collected from three volunteers which perform four activities namely; *Brushing teeth, Pick up object, sit on sofa* and *stand up*. The statistics for the activities performed is given in Table 1.

From the data obtained. The kinetic energy of poses are computed for each activity using the process described in Sect. 3.1. The plot of the computed kinetic energy of poses for *picking up object* activity for actor 2 is represented in Fig. 3.

From the computed pose kinetic energy as shown in Fig. 3, two moving averages are chosen and used in computing the simple moving average from the computed kinetic energy. In this work, 15 and 30 activity frames are selected for short-term and long-term moving averages respectively after experimentation. A result from the moving average computed for the pose kinetic energy represented in Fig. 3 is given in Fig. 4. The crossover points of the moving averages are used

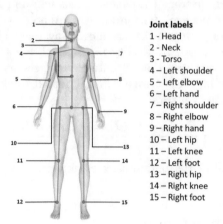

**Fig. 2.** Skeleton joint representation used in the proposed approach

**Table 1.** Summary of experimental human activity data collected from 3 actors using an RGB-D sensor. Activities performed comprise: *Brushing teeth, Pick up object, Sit on sofa, Stand up.*

| Activity | Number of frames | | |
|---|---|---|---|
| | Actor 1 | Actor 2 | Actor 3 |
| Brushing teeth | 2202 | 1876 | 1781 |
| Pick up object | 1804 | 1663 | 1355 |
| Sit on sofa | 1489 | 1672 | 2736 |
| Stand up | 2126 | 2059 | 2100 |
| Total | 7621 | 7270 | 7972 |

**Table 2.** Activity classification performance of LSTM model compared with Support Vector Machine (SVM) using a 4-fold cross validation test strategy.

| Activity | LSTM | | SVM | | Key poses |
|---|---|---|---|---|---|
| | Prec. | Recall | Prec. | Recall | |
| Brushing teeth | 1.00 | 1.00 | 0.86 | 0.92 | 286 |
| Pick up object | 1.00 | 0.71 | 0.81 | 0.44 | 202 |
| Sit on sofa | 0.90 | 1.00 | 1.00 | 1.00 | 527 |
| Stand up | 0.99 | 1.00 | 0.85 | 0.96 | 562 |
| Average/Total | 0.97 | 0.96 | 0.90 | 0.90 | 1577 |

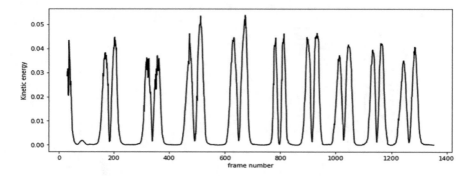

**Fig. 3.** Pose kinetic energy computed for *picking up object* activity for actor 2

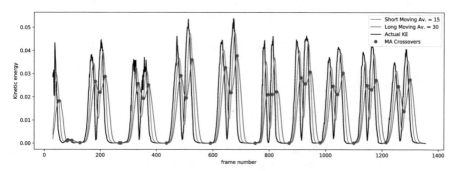

**Fig. 4.** Pose kinetic energy computed with simple moving averages of kinetic energy. Short-term moving average = 15 and Long-term moving average = 30. The moving average crossover points are represented by dotted points on the plot

as key pose locations from which activity poses corresponding to this locations are extracted and used as input to an LSTM model.

The LSTM model is trained over 150 epochs and tested after each training epoch. The model attains an optimal performance just after 100 epochs of training as shown in Fig. 5 which presents the training and testing performance of the model over the number of epochs. A K-Fold cross-validation test (with $k = 4$) is performed in order to extensively test the trained model on all data samples. The data is divided into 4-folds where 3-folds (75%) of the data used in training the model and the remainder fold (25%) used in validating the trained model. This is repeated for all folds and the result obtained is the average performance of all folds. The result from the LSTM model is shown in Table 2 with the precision and recall values obtained for each activity. This result is also compared with the performance achieved with a base classifier model - Support Vector Machine (SVM) classifier using a one-against one configuration - with the extracted key poses as the input features to the classifier. The LSTM model achieves an accuracy of 96.26% which outperforms that of 90.04% achieved with SVM. This shows that the key poses extracted are sufficient in identifying human activities.

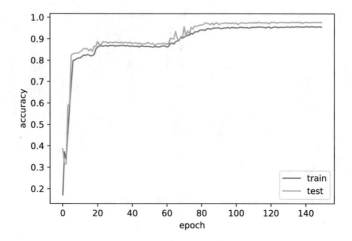

**Fig. 5.** Performance of LSTM model trained and tested for 150 epochs

## 5    Conclusion and Discussion

The result presented in Table 2, it can be observed that from the total number of 22863 frames of activities given in Table 1, the proposed method identifies 1577 key frames which are the activity poses that are most important in describing the activities. With the selected poses, the LSTM network is able to learn the activity sequence with a good performance on 3 out of the 4 activities tested. This is also validated by using the selected key poses with an SVM classifier.

This paper proposes a method of key pose extraction from human activity by identifying moving average crossovers of the kinetic energy of poses through an activity sequence. Activity poses corresponding to the identified crossover points are selected as key poses which define an activity. This approach is important due to the fact that not all poses within an activity are relevant in discriminating one activity from another. The preliminary results presented have shown the performance of identified key poses from an activity with an LSTM sequence based learning algorithm for human activity learning. For future work, extensive experiments will be conducted in order to generalize the performance across a wider range of subjects. This is aimed at increasing the robustness of the proposed method. In addition, the proposed method is planned to be incorporated in an assistive robot for identifying key poses used in learning human activity.

## References

1. Adama, D.A., Lotfi, A., Langensiepen, C., Lee, K.: Human activities transfer learning for assistive robotics, pp. 253–264. Springer, Cardiff (2018)
2. Adama, D.A., Lotfi, A., Langensiepen, C., Lee, K., Trindade, P.: Learning human activities for assisted living robotics. In: Proceedings of the 10th International Conference on PErvasive Technologies Related to Assistive Environments, Island of Rhodes, Greece, PETRA 2017, pp. 286–292. ACM (2017)

3. Bemelmans, R., Gelderblom, G.J., Jonker, P., de Witte, L.: Socially assistive robots in elderly care: a systematic review into effects and effectiveness. J. Am. Med. Dir. Assoc. **13**(2), 114–120.e1 (2012)
4. Chaaraoui, A.A., Padilla-López, J.R., Flórez-Revuelta, F.: Fusion of skeletal and silhouette-based features for human action recognition with RGB-D devices. In: Proceedings of the 2013 IEEE International Conference on Computer Vision Workshops, ICCVW 2013, pp. 91–97 (2013)
5. Espingardeiro, A.: Social assistive robots, reframing the human robotics interaction benchmark of social success. Int. J. Soc. Behav. Educ. Econ. Bus. Ind. Eng. **9**(1), 377–382 (2015)
6. Gaglio, S., Re, G.L., Morana, M.: Human activity recognition process using 3-D posture data. IEEE Trans. Hum. Mach. Syst. **45**(5), 586–597 (2015)
7. Gupta, R., Chia, A.Y.-S., Rajan, D.: Human activities recognition using depth images. In: Proceedings of the 21st ACM International Conference on Multimedia, pp. 283–292 (2013)
8. Hochreiter, S., Schmidhuber, J.: Long short-term memory. Neural Comput. **9**(8), 1735–1780 (1997)
9. Koskinopoulou, M., Piperakis, S., Trahanias, P.: Learning from demonstration facilitates human-robot collaborative task execution. In: 2016 11th ACM/IEEE International Conference on Human-Robot Interaction (HRI), pp. 59–66 (2016)
10. Lipton, Z.C., Kale, D.C., Elkan, C., Wetzel, R.C.: Learning to diagnose with LSTM recurrent neural networks. CoRR abs/1511.03677 (2015)
11. Nez, J.C., Cabido, R., Pantrigo, J.J., Montemayor, A.S., Vlez, J.F.: Convolutional neural networks and long short-term memory for skeleton-based human activity and hand gesture recognition. Pattern Recognit. **76**, 80–94 (2018)
12. Nunes, U.M., Faria, D.R., Peixoto, P.: A human activity recognition framework using max-min features and key poses with differential evolution random forests classifier. Pattern Recognit. Lett. **99**, 21–31 (2017)
13. Shan, J., Akella, S.: 3D human action segmentation and recognition using pose kinetic energy. In: 2014 IEEE International Workshop on Advanced Robotics and its Social Impacts, pp. 69–75, September 2014
14. Zhu, G., Zhang, L., Shen, P., Song, J., Zhi, L., Yi, K.: Human action recognition using key poses and atomic motions. In: 2015 IEEE International Conference on Robotics and Biomimetics (ROBIO), pp. 1209–1214 (2015)

# A Comprehensive Obstacle Avoidance System of Mobile Robots Using an Adaptive Threshold Clustering and the Morphin Algorithm

Meng Yuan Chen[1,2], Yong Jian Wu[1], and Hongmei He[3(✉)]

[1] Key Lab of Electric Drive and Control of Anhui Province,
Anhui Polytechnic University, Wuhu 241000, China
mychen@ahpu.edu.cn
[2] Department of Precision Machinery and Precision Instrumentation,
University of Science and Technology of China, Hefei 230027, China
[3] Manufacturing Informatics Centre, SATM, Cranfield University, Bedford, UK
h.he@cranfield.ac.uk

**Abstract.** To solve the problem of obstacle avoidance for a mobile robot in unknown environment, a comprehensive obstacle avoidance system (called ATCM system) is developed. It integrates obstacle detection, obstacle classification, collision prediction and obstacle avoidance. Especially, an Adaptive-Threshold Clustering algorithm is developed to detect obstacles, and the Morphin algorithm is applied for path planning when the robot predicts a collision ahead. A dynamic circular window is set to continuously scan the surrounding environment of the robot during the task period. The simulation results show that the obstacle avoidance system enables robot to avoid any static and dynamic obstacles effectively.

**Keywords:** Adaptive threshold clustering · Morphin algorithm
Obstacle detection · Obstacle classification · Collision prediction
Collision avoidance

## 1 Introduction

Obstacle avoidance in mobile robots' navigation is a key issue that has attracted much attention from researchers [1–3]. To make right decisions in response to the dynamics of surrounding environments of a robot, the robot should be able to collect data from sensors and do appropriate information processing. Currently, sensors commonly used for obstacle avoidance, include visual sensors [4], ultrasonic sensors [5], infrared sensors [6] and laser sensors [7, 8]. Laser sensors are popularly used due to their wide detection range and high measurement accuracy. The obstacle avoidance includes include such steps as obstacle detection, collision prediction, avoidance, and finally plan an appropriate path.

Yu et al. [9] used the confidence distance theory to process the Velodyne data from a 3D laser sensor and the motion state information from a 4-wire laser sensor Ibeo, and thus to derive the position of the moving obstacle in a grid map according to the fusion results. Huang et al. [10] proposed a dynamic obstacle detection, using a support vector

© Springer Nature Switzerland AG 2019
A. Lotfi et al. (Eds.): UKCI 2018, AISC 840, pp. 312–324, 2019.
https://doi.org/10.1007/978-3-319-97982-3_26

machine based on the space-time feature vector to recognize dynamic obstacles. Although these methods are well performed for the detection of obstacles, they didn't show the obstacle's trajectory prediction and collision avoidance strategy. Liu et al. [11] proposed an improved vector field histogram avoidance algorithm by adjusting the adaptive threshold using the positional relationship between the obstacle and the target point. However, the researchers only discussed the obstacle avoidance of static obstacle without the exploration of dynamic obstacle detection and collision prediction. Yang et al. [12] proposed an approach to detecting the speed and direction of an obstacle in terms of least Euclidean distance between the robot and the edge of the obstacle, thus to implement the plan of the robot's obstacle avoidance. But this approach is not appropriate for diversity of obstacles, as the shape of dynamic obstacles is limited to circular objects.

Most of research focused on partial process of obstacle avoidance. There is little exploration on the entire obstacle avoidance system. This does not benefit the performance assessment of the whole robot's navigation process. In this paper, we develop a comprehensive obstacle avoidance system, comprised of laser data acquisition, obstacle detection, collision prediction and avoidance by combining the Adaptive Threshold Clustering, based on the method in [13], and the Morphin algorithm [15, 16], named as ATCM.

The basic idea of ATCM is that a mobile robot first constructs a rolling window with the robot as a centre, classifies obstacles in the window using the adaptive-threshold clustering algorithm, predicts possible collision, and uses the Morphin algorithm to avoid obstacles when a collision in front of the robot is predicted; then, the robot updates the state and the dynamic circular window, and move toward the generated local sub-target, until it reaches the global target. Algorithm 1 shows the pseudocode of the ATCM System.

## 2    Obstacle Detection

The dynamic and static obstacles existing in the environment can be recognized based on the data from sensors. The data points within the window are clustered to produce a chain of obstacles, of which the types (static, dynamic or new) will be further classified. For dynamic obstacles, their speed and moving direction will be calculated.

### 2.1    Laser Sensor Data Acquisition

It is easy to build and maintain a Grid Map with a specified resolution [14]. In this research, a grid map is used to establish the environment model. Assume the robot itself adopts the same coordinate system as the world coordinate system. Namely, the starting point of the robot coincides with the origin of the global coordinate system. To obtain obstacle information, the German company SICK's two-dimensional laser sensor is used to scan the surrounding environment. The distance between a robot to an obstacle is calculated based on the time interval from the time when laser pulse is emitted to hit an object to the time when a laser pulse is received back from the object.

The laser scanner is configured to scan the front semicircle of the robot in the range [0°, 180°] with the angular resolution of 0.5°.

---

**Algorithm 1** ATCM($r_w$, $R$, $Target$)

---
1: Initialise($R$, $r_w$); /*set window with radius $r_w$*/;
2: t=0;
3: **while** ($R$ has not reached $Target$) **do**
4:    D=Read_Data(); /*from laser scanner*/;
5:    $Ob_{chain}(t)$=Clustering(D);
6:    **for** ($O_k(t) \in Ob_{chain}(t)$) **do**
7:       $O_{type}$=Classification($O_k(t),Ob_{chain}(t-1)$));
8:       $O_{info}$=Calculate_Obstacle_Info($O_k(t), O_{type}$);
9:       $Collision$=Collision_Prediction($R$, $O_{info}$);
10:      **if** ($Collision \neq$ NULL) **then**
11:         Morphin($R$, $Collision$)); /*collision avoiding*/
12:         Break;
13:      **end if**
14:   **end for**
15:   Update($R,r_w$);
16:   t=t+1;
17: **end while**

---

Figure 1 shows the positions of a robot and an obstacle. The location of an obstacle can be represented with the pair of ($\rho_i$, $\alpha_i$) in the local polar coordinates with the robot as an original point, $\rho_i$ is the length of a laser beam, indicating the distance between the obstacle and the robot, $\varphi_i \in$ [0°, 180°], i is the index of the laser beam, $i = 0...360$; $\alpha_i$ is the angle between laser beam and the robot's direction, θR is the angle of the robot. The position of the obstacle in the global coordinate system is expressed as Eq. (1). The location ($x'_o, y'_o$) in the grid map of an object with global coordinates ($x_o, y_o$) can be calculated with Eq. (2).

$$\begin{cases} x_o = x_R + \rho_i\cos(\theta_R + \alpha_i) \\ y_o = y_R + \rho_i\sin(\theta_R + \alpha_i). \end{cases} \tag{1}$$

$$\begin{cases} x'_o = \pi r^2 = floor\left(\frac{x_o}{r} + \frac{1}{2}\right) \\ y'_o = floor\left(\frac{y_o}{r} + \frac{1}{2}\right) \end{cases} \tag{2}$$

where, $r$ is the resolution of the grid map (see Fig. 2).

## 2.2    Barrier Point Clustering

When a robot is moving in a grid map, a set of data points is obtained by scanning obstacles with the laser sensor. To determine an obstacle (i.e. a cluster), an adaptive threshold nearest neighbor clustering method is developed to separate data points.

The data out of the circular window (i.e. $\rho > r_w$) will not be used for clustering, for instance, $O_3$ in Fig. 3, as $\rho_3 > r_w$. The distance between two data points is calculated with Euclidean distance. Two available consecutive data points (ρ2 and ρ4) belong to different obstacles, respectively, if the distance between them is larger than the distance between two neighbouring laser beams with the same length and the scanning resolution 0.5° (e.g. $\rho_1$ and $\rho_2$). Therefore, a threshold $\Theta$ is defined as Eq. (3). It is proportional to the value of [$\rho(t)$ *sin* (0.5)], which approximates the distance between two neighbouring data points, belonging to a cluster. Obviously, the threshold $\Theta$ is adaptive to the current laser beam $\rho(t)$. As an obstacle could have an irregular shape, the adaptive rate $\lambda$ ($\lambda > 1$) is introduced to represent the irregularity of obstacle shapes. The introduction of $\lambda$ could also allow putting two closed obstacles to one cluster. Hence, the adaptive threshold makes the clustering algorithm robust. We set $\lambda$ to a value close to 1 for rectangle of obstacles.

$$\Theta = \lambda\rho(t)\sin(0.5°) \qquad (3)$$

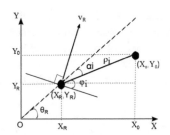

Fig. 1. The robot model

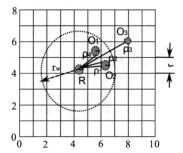

Fig. 2. A robot and obstacles on the grid map

In each step of data clustering, the distance between the current data point and the prior data point is calculated, and compared with the adaptive threshold. If it is larger than the threshold, then the current data point and the prior data point belong to the same cluster, otherwise, the current data point represents a new obstacle. After clustering, a chain of obstacles *Ob_chain* (Eq. 4) at time $t$ is obtained. Each obstacle $O_k(t)$ can be expressed with vector of four items as shown in Eq. (5).

$$Ob\_chain = \{O_1(t), O_2(t), \ldots, O_n(t)\} \qquad (4)$$

$$O_k(t) = (Z_k(t), S_k(t), \xi_k(t), V_k(t)) \qquad (5)$$

$Z_k(t)$ indicates the centroid of the obstacle; $S_k(t)$ indicates the area that the obstacle occupies in the grid map; $\xi_k(t)$ is the coincidence of data points in cluster $C_k$ to the data points in previous clusters. $V_k(t)$ indicates the speed of the dynamic obstacle. When $V_k(t) = 0$ indicates the obstacle is static. Assume cluster $C_k$ represents obstacle $O_k$. In the cluster, there are $n_k$ laser beams, $\{l_1, \ldots, l_{n(k)}\}$ in the dynamic window of the robot, each laser beam is represented with a pair of ($\rho$, $\alpha$).

The center $Z_k(t)$ of Obstacle $O_k$ can be calculated with Eq. (6), and using the Eq. (1), we can get the global coordinates of the center.

$$\bar{\alpha}_k = \frac{\sum_{i=1}^{n(k)} \alpha_i}{n(k)}, \quad \bar{\rho}_k = \frac{\sum_{i=1}^{n(k)} \rho_i}{n(k)}. \tag{6}$$

We can get the $\min(\rho)$, and $\max(\rho)$, $\min(\alpha)$, and $\max(\alpha)$ in the cluster $C_k$. Using Eq. (1), we can calculate $(x_{k,min}, y_{k,min})$, and $(x_{k,max}, y_{k,max})$; Further, using Eq. (2), we can get $(x'_{k,min}, y'_{k,min})$, and $(x'_{k,max}, y'_{k,max})$ in the grid map. Therefore, the area $S_k$ includes all grids within the ranges of coordinates in Eq. (7).

$$\begin{cases} x'_k \in \left[ x'_{k,min}, x'_{k,max} \right], \\ y'_k \in \left[ y'_{k,min}, y'_{k,max} \right]. \end{cases} \tag{7}$$

Using Eqs. (1) and (2), we can calculate the grid coordinates of all data points in a cluster $C_k$ at time $t$. A 3 * 3 grid template is used, where the data point is in the center of the template. The evaluation if the template of each data point at time $t$ is matched to the templates of data points at time $t - 1$ that fall into the area of $S_k(t)$ is done by comparing their coordinates in the grid map. For each data point at time $t$ in $C_k$, representing $O_k$, the coincidence is defined as:

$$\zeta_i = \frac{\tau}{9}, i = 1 \ldots n(k). \tag{8}$$

where, $\tau$ is the overlapped grid number between templates of two data points at times $t$ and $t - 1$, respectively. Hence, the coincidence of obstacle $O_k$ is defined as:

$$\xi_k(t) = \frac{\sum_{i=1}^{n(k)} \zeta_i}{n_k}. \tag{9}$$

To determine the type of an obstacle, we need to analyse the correlation between two obstacles in current clusters and previous clusters, respectively. Two parameters are used to represent the correlation between two obstacles: the distance between two cluster's centers, denoted as $\delta$, and the non-overlapping rate, denoted as $\eta$. The spatial correlation function is shown in Eq. (10), where, $\delta$ and $\eta$ can be calculated with Eq. (11), and $\gamma_\delta$, $\gamma_\eta$ represent the efficiencies.

$$\varsigma_{k_1,k_2} = \varsigma(O_{k_1}(t), O_{k_2}(t-1)) = \gamma_\delta \frac{1}{\delta+1} + \gamma_\eta \frac{1}{\eta+1} \tag{10}$$

$$\delta = \|Z_{k_1}(t), Z_{k_2}(t-1)\|, \eta = 1 - \frac{S_{O_{k_1}}(t) \cap S_{O_{k_2}}(t-1)}{S_{O_{k_1}}(t) \cup S_{O_{k_2}}(t-1)} \tag{11}$$

When two obstacles at times $t$ and $t - 1$ are the same obstacle, they should have the same center, namely $\delta = 0$. When two obstacles are complete overlap, $\eta = 0$; when two obstacles are complete separated, $\eta = 1$. Therefore, if we set $\gamma_\delta = 0.5$, and $\gamma_\eta = 0.5$, when the two obstacles completely overlap, $\varsigma = 1$. The maximum spatial correlation of

Obstacle $O_{k(t)}$ is the maximum value among all spatial correlations between $O_{kt} \in$ $Ob\_chain(t)$ and all obstacles in $Ob\_chain(t - 1)$, expressed as Eq. (12).

$$\varsigma_{k(t),max} = \max_{k_2=1..n_k(t-1)} \left( \varsigma_{k(t),k_2} \right). \tag{12}$$

## 2.3  Identify the Type of an Obstacle

An obstacle chain, $Ob\_chain = \{O_1(t), O_2(t),..., O_k(t)\}$, is produced after clustering. For each obstacle, we can calculate the center $Z_k(t)$, the grid area $S_k(t)$, and the coincidence $\xi_k(t)$, respectively. Then the spatial correlation can be calculated, and the obstacle with maximum spatial correlation $\varsigma_{k(t),max}$ can be obtained. Three possible obstacle types are static, new and dynamic, as shown in Fig. 3(a)–(c).

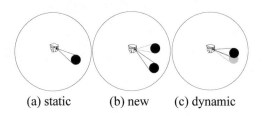

(a) static        (b) new        (c) dynamic

**Fig. 3.**  Different types of obstacles

Two thresholds $\theta_{\varsigma 1}$ and $\theta_{\varsigma 2}$ of the spatial correlation are set for distinguishing a new or a static obstacle. Figure 3(a) shows a static obstacle, when $\varsigma_{k,max} > \theta_{\varsigma 2}$; Fig. 3(b) shows a new obstacle, when $\varsigma_{k,max} < \theta_{\varsigma 1}$; Fig. 3(c) shows a dynamic obstacle. When the value $\varsigma_{k,max} \in [\theta_{\varsigma 1}, \theta_{\varsigma 2}]$, then the obstacle is possibly a dynamic obstacle. It will be further evaluated in terms of center distance $\delta$ ($0 \leq \delta < r_w$) between two obstacles and the coincidence $\xi_k(t)$. A threshold $\theta_\delta$ is set to distinguish static and dynamic obstacle. If $\delta < \theta_\delta$, the obstacle is static, otherwise, the obstacle is evaluated in terms of the coincidence $\xi_k(t)$. A threshold $\theta_\xi$ is set as well. If $\xi_k(t) \geq \theta_\xi$, then the obstacle is static, otherwise, it is dynamic. Figure 4 illustrates the process of the obstacle classification using a simple decision tree.

## 2.4  Movement of a Dynamic Obstacle

The dynamic obstacles are further analyzed to calculate the speed and angle of their movement (Fig. 5). Assume the time interval of two rounds of laser scanning is $T$. We can catch up the global coordinates of a moving robot at time $t$ and $t - T$, $R(x(t), y(t))$ and $R(x(t - T), y(t - T))$. Hence, we always can calculate the global coordinates of an obstacle, $O_k(x_k(t), y_k(t))$ and $O_k(x_k(t - T), y_k(t - T))$, given the values from laser beams at time $t$ and $t - T$, using Eq. (1).

The moving distance $d_o$, speed $v_o$ and the direction angle $\alpha_o$ of the dynamic obstacle from $t - T$ to $t$ can be calculated with Eqs. (13)–(15). Similarly, the distance $d_R$, speed

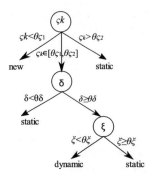

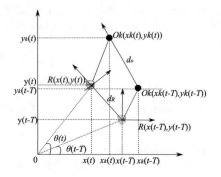

**Fig. 4.** The decision tree of obstacle classification

**Fig. 5.** Motion process of an obstacle

$v_R$ and the direction angle $\alpha_R$ of the moving robot from time $t - T$ to $t$ can be calculated as well.

$$d_o = \sqrt{[x_k(t) - x_k(t - T)]^2 + [y_k(t) - y_k(t - T)]^2} \tag{13}$$

$$v_o = \frac{d_o}{T} \tag{14}$$

$$\alpha_o = \arctan\left(\frac{y_k(t) - y_k(t - T)}{x_k(t) - x_k(t - T)}\right) \tag{15}$$

According to the states of the robot and the obstacle, the robot can predict the potential collisions ahead. There are six scenarios, when a robot is moving in its surrounding environment: (1) a static obstacle is in front of the robot, where the collision is just at the point of obstacle; (2) an obstacle at the probing area, moving at a faster speed than the robot, could pass the path before robot arrives the potential collision point; (3) an obstacle at the probing area, moving slower than the robot, has not arrived at the path, when robot arrives the potential collision point; (4) the robot and the obstacle may collide at the crossing point between robot's path and obstacle's path when the current speed, distance, and position of the robot and the obstacle make them arrive the point at the same time; (5) the robot and the obstacle are running in opposite direction, hence, the robot and the obstacle could collide at a point between the robot and the obstacle; (6) the robot is running in the same direction as an obstacle but is faster than the obstacle, then the robot may collide with the obstacle at some time.

## 2.5   Avoiding Obstacle Collision

We use the classic local path avoidance algorithm—Morphin algorithm [15, 16] to implement the path planning. As shown in Fig. 6, an obstacle that may collide with the robot is detected, a few alternative paths to avoid the obstacle are set in the front of the robot, and the optimal obstacle avoidance path is selected according to the current state of the robot and the evaluation function of the alternative paths.

In the Morphin algorithm, the robot is always assumed to face the obstacle (e.g. scenarios (1), (5), (6)). Hence, we can connect the robot's current position and the center of the obstacle to form a centerline, and draw several arcs on the left and right sides of the centerline and evaluate each arc by using Eq. (16).

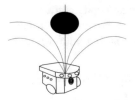

**Fig. 6.** Alternative paths in the Morphin algorithm

$$y = \begin{cases} \infty, \text{ the obstacle is above the arc,} \\ \varepsilon_1 P + \varepsilon_2 M + \varepsilon_3 \Delta L + \varepsilon_4 W, \text{ others.} \end{cases} \tag{16}$$

where, $P$ represents the length of each arc path; M represents the inflection point parameter of each arc path; $\Delta L$ represents the average distance from each grid point to the sub-target point through which the arc passes; $W$ represents the arc endpoint and the number of times the sub-target points intersect the obstacle grid; $\varepsilon_1$, $\varepsilon_2$, $\varepsilon_3$, $\varepsilon_4$ are the weights of the items, respectively. When the obstacle is above the arc, the value of the evaluation function $y$ is infinite, and the arc with the smallest $y$ value represents the local optimal path. For more details about the Morphin algorithm, please refer to the research in [15, 16]. In the scenarios (1), (5) and (6), a collision could occur, hence, the Morphin algorithm is called to update the robot's path with the optimal path for collision avoidance. In the scenarios (2) and (3), no collision could occur, hence, the robot will not take any measures and continue to run. In the scenario (4), the robot will stop running until the obstacle passes the predicted collision point.

# 3   Experiments

To validate the effectiveness of the proposed ATCM system, we conducted some experiments for robot moving from a specific starting point to a specific target. The experimental platform is MATLAB. All parameters in the experiments are set with trial and error method. The simulation environment is set to a $20 \times 20$ grid map with many obstacles. Grid resolution $r = 500$ mm. The radius $r_w$ of the dynamic window is set to 8 grids. As all obstacles added to the grid map have a regular shape, the adaptive rate $\lambda$ of the threshold $\Theta$ is set to 1.2; for simplicity, the parameters in the obstacle classification tree (Fig. 4) are set as: $\theta_{\varsigma 1} = 0.30$, $\theta_{\varsigma 2} = 0.7$, $\theta_{\delta} = 0.4$, $\theta_{\xi} = 0.5$, respectively. The parameters $(\varepsilon_1 - \varepsilon_4)$ of Morphin Algorithm are set to 1, 1, 1.3 and 0.6, as in [16]. Three experiments are conducted: (1) without dynamic obstacles; (2) adding some dynamic obstacles in the grid environment; (3) a mixed case with static and dynamic obstacles.

### 3.1   Without Dynamic Obstacles

Figure 7 shows the local path planning process of a mobile robot when there are no temporary obstacles in the environment. The mobile robot does not find any dynamic obstacles, except the fixed obstacles in the rolling window in the grid environment. Using the rolling window algorithm, each rolling step draws the rolling window centered on the current position and updates the window map. The mobile robot moves towards the next step. The local sub-targets are rolled forward step by step, and finally a trajectory is formed from the starting point to the target.

### 3.2   Adding Instantly Dynamic Obstacles

Dynamic obstacle avoidance increases the computing complexity, as the robot needs to predict potential collision point in terms of the dynamics of the obstacle. Figure 8 shows the simulation of the local path planning process by adding three types of dynamic obstacles to the grid map. The mobile robot starts from the starting point and relies on the sensor, and identifies the dynamic obstacle Ob1 in the current window. The robot calculates the motion trajectory of Ob1, and predicts that it will not collide with Ob1. Hence, it will keep the original speed and moving direction. When the

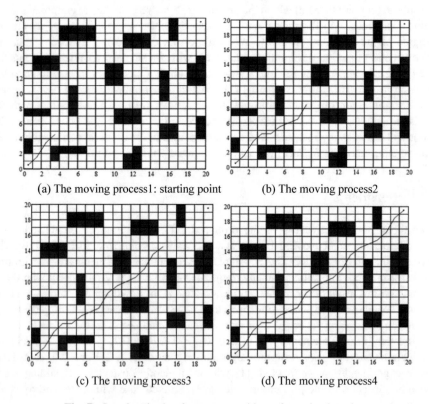

(a) The moving process1: starting point          (b) The moving process2

(c) The moving process3          (d) The moving process4

**Fig. 7.**   Local path planning process without dynamic obstacles

mobile robot moves and finds Ob2, a moving obstacle, and predicts that it could collide with Ob2. Hence, it stays in the place for a while until Ob2 passes the collision point. After the dynamic obstacle passes the collision point, the robot generates local sub-goal in the current window and moves. When the robot detects the dynamic obstacle Ob3 from the opposite side, the collision is unavoidable, and the collision point is predicted, then the robot calls the Morphin algorithm to get an optimal path, and moves following the path until reaching the specified target.

### 3.3   A Mixed Case

The simulation results are shown in Fig. 9. In a real case, the circular window is updated in real time, and the center trajectory of the dynamic circular window is the trajectory of the robot. For demonstration, $V_1$, $V_2$, $V_3$ and $V_4$ denote circular windows, $S_1$, $S_2$, $S_3$ and $S_4$ denote detected static obstacles, and $D_1$, $D_2$, $D_{3-1}$, $D_{3-2}$, $D_4$ represent the detected dynamic obstacles. At $C$, $E$, $F$, the robot detects dynamic obstacles, and at $A$ *and* $D$, the robot detects the static obstacles, respectively. Table 1 provides the speeds and angles of the four detected dynamic obstacles at different positions in the grid map. Figure 9 shows the obstacle avoidance process in the mixed case.

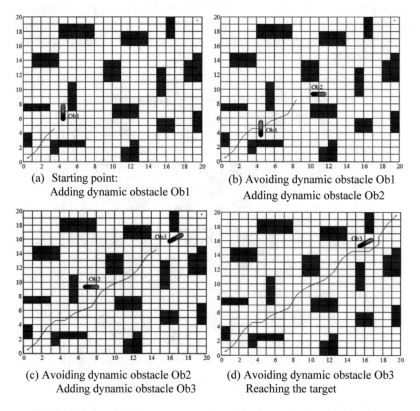

(a)  Starting point:
Adding dynamic obstacle Ob1

(b) Avoiding dynamic obstacle Ob1
Adding dynamic obstacle Ob2

(c) Avoiding dynamic obstacle Ob2
Adding dynamic obstacle Ob3

(d) Avoiding dynamic obstacle Ob3
Reaching the target

**Fig. 8.** Local path planning process for adding instantly dynamic obstacles

**Table 1.** Movement speed and angle of obstacle

|   | D1 | D2 | D3 | D4 |
|---|----|----|----|----|
| $v$ | 450 mm/s | 760 mm/s | 510 mm/s | 805 mm/s |
| $\alpha$ | 80.83° | 62.47° | 91.05° | 180.09 |

As shown in Fig. 9, the robot moves within the V1 window and builds a dynamic window map, where, the robot detects obstacle S1 at point A and calls Morphin algorithm to avoid the obstacle S1. In the phase from A to B, the moving obstacle D1 is recognised at the speed of 450 mm/s and the angle of 80.83°; once the robot moves beyond the effective area of V1, a circular window of V2 is constructed immediately, and the moving obstacle D2 is detected at the phase from C to D, with the speed of 760 mm/s and the angle of 62.47°; a static obstacle S3 is detected at D, hence, the Morphin algorithm is called immediately. Similarly, beyond the valid area of V2, a circular window of V3 is constructed. When reaching at E, the robot obtains the information that obstacle D3-1 is running at a speed of 510 mm/s and the angle of 91.05°, and predicts that a collision could occur at O1. In this case, the robot will stop, and wait for the detected obstacle D3-1 running away from the collision point to become D3-2, and then start running again. In the V4 window, at F, an opposite obstacle is detected, running at the speed of 805 mm/s and the angle of 180.05°, which would produce a collision at O2 if robot does not change its direction immediately. In this case, the robot calls the Morphin algorithm immediately to change the path, thus to avoid the collision.

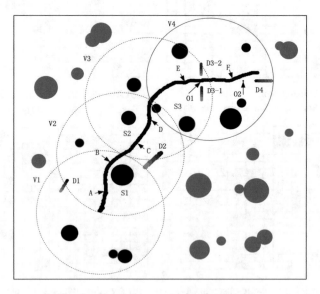

**Fig. 9.** Simulation results for the mix case

# 4 Conclusions

This research provided a comprehensive obstacle avoiding system, ATCM, for a mobile robot in unknown environment. A dynamic circular window is updated in real time during the navigation. The laser data in the circular window are clustered with the adaptive threshold nearest neighbor clustering algorithm. Three types of obstacles can be classified in terms of three parameters, spatial correlation, center distance, and coincidence of two obstacles. Then potential collision with detected static and dynamic obstacles is predicted. The Morphin algorithm is applied to find an alternative path when robot detects a potential collision ahead. The simulation results show that the mobile robot can effectively detect obstacles, predict potential collision, and avoid obstacles. The future work is to apply the ATCM system to laboratory robots, and improve the system.

**Acknowledgments.** This work was supported by 2018 Natural Science Foundation of Anhui, China (1808085QF215), 2018 Foundation for Distinguished Young Talents in Higher Education of Anhui, China (gxyqZD2018050) and Anhui Key Research and Development Programs (Foreign Scientific and Technological Cooperation, 1804b06020375).

# References

1. Wang, J.L., Zhou, J., Gao, H., et al.: Obstacle avoidance method for mobile robots based on the identification of local environment shape features. Inf. Control **44**(1), 91–98 (2015)
2. Luo, J., Liu, C., Liu, F.: Piloting-following formation and obstacle avoidance control of multiple mobile robots. CAAI Trans. Intell. Syst. **12**(02), 1–10 (2017)
3. Zhang, Q., Wang, P., Chen, Z.: Velocity space based concurrent obstacle avoidance and trajectory tracking for mobile robots. Control Decis. **32**(02), 358–362 (2017)
4. Zhang, Q., Yang, X., Liu, T., et al.: Design of a smart visual sensor based on fast template matching. Chin. J. Sens. Actuators **26**(8), 1039–1044 (2013)
5. Wang, Z., Cui, X., Hou, C.: Analysis and countermeasures to the problem of ultrasonic sensor receives the ultrasonic signal asymmetric. Chin. J. Sens. Actuators **28**(1), 81–85 (2015)
6. Wang, M., Fan, Y., Wang, X., et al.: Design of infrared FPA detector simulator. Laser Infrared **46**(12), 1481–1485 (2016)
7. Zhang, Y., Xu, J., Chen, L., et al.: Design of terrain recognition system based on laser distance sensor. Laser Infrared **46**(03), 265–270 (2016)
8. Zhang, D., Li, W., Wu, H., et al.: Mobile robot adaptive navigation in dynamic scenarios based on learning mechanism. Inf. Control **45**(05), 521–529 (2016)
9. Xin, Y., Liang, H., Mei, T., et al.: Dynamic obstacle detection and representation approach for unmanned vehicles based on laser sensor. Robot **36**(6), 654–661 (2014)
10. Huang, R., Liang, H., Chen, J., et al.: Lidar based dynamic obstacle detection, tracking and recognition method for driverless cars. Robot **38**(4), 437–443 (2016)
11. Liu, J., Yan, Q., Tang, Z.: Simulation research on obstacle avoidance planning for mobile robot based on laser radar. Comput. Eng. **41**(4), 306–310 (2015)
12. Yang, Y., Han, F., Cao, Z., et al.: Laser sensor based dynamic fitting strategy for obstacle avoidance control and simulation. J. Syst. Simul. **25**(4), 118–122 (2013)

13. Zhong, X., Peng, X., Zhou, J.: Detection of moving obstacles for mobile robot using laser sensor. In: The 20th Chinese Control Conference (CCC), Yantai, China, 22–24 July 2011
14. Zhu, J., Zhou, Y., Wang, C., et al.: Grid map merging approach based on image registration. Acta Automatica Sinica **41**(2), 285–294 (2015)
15. Zhu-Ge, C., Tang, Z., Shi, Z.: UGV local path planning algorithm based on multilayer Morphin search tree. Robot **04**, 491–497 (2014)
16. Wan, X., Hu, W., Zheng, B., et al.: Robot path planning method based on improved ant colony algorithm and Morphin algorithm. Sci. Technol. Rev. **33**(3), 84–89 (2015)

# Analysis and Detection

# Physarum Inspired Connectivity and Restoration for Wireless Sensor and Actor Networks

Abubakr Awad$^{(\boxtimes)}$, Wei Pang, and George M. Coghill

Department of Computing Science, University of Aberdeen, Aberdeen, UK
{abubakr.awad,pang.wei,g.coghill}@abdn.ac.uk

**Abstract.** Wireless sensor-actor networks (WSANs) are a core component of Internet of Things (IOT), and are useful for environments that are difficult and/or dangerous for sensors to be deployed deterministically. After random deployment, the sensors are required to disperse autonomously without central control to maximize the coverage and re-establish the connectivity of the network. In this paper, we propose a Physarum inspired self-healing autonomous network connectivity restoration algorithm that minimize movement overhead and keep load balance. The mechanism to select the alternative nodes only involves the one-hop information table, and depends on actor node location from base station (regions of k-influence), and residual energy. Our model achieved almost complete coverage, and fault repair in one or two rounds with minimal number of movement overhead.

**Keywords:** Physarum polycephalum · Hexagonal cellular automaton
Wireless sensor-actor networks · Connectivity · Fault repair

## 1 Introduction

### 1.1 Wireless Sensor-Actor Networks (WSANs)

Recently, wireless sensor-actor networks (WSANs) have attracted attention as one of the core technologies of the "Internet of Things" (IoT) [20,21], and has been also extended to the "Internet of Nano Things" (IoNT) [6]. WSANs are used to collect ground data for various purposes such as battle field monitoring, bio-environmental surveillance, earthquake observation, and wildlife reservoir [10,16]. WSANs are formed of a number of static sensors for monitoring and collecting data from specific area and actors are responsible for processing data from sensors to base station through one/multiple hop [20,30]. It is desired that all actors are connected at any time, however, if any actor fails to work due to low energy, hardware degradation, or harsh environment the network connectivity will be prone to be lost resulting in disjoint network components [30].

© Springer Nature Switzerland AG 2019
A. Lotfi et al. (Eds.): UKCI 2018, AISC 840, pp. 327–338, 2019.
https://doi.org/10.1007/978-3-319-97982-3_27

## 1.2 Physarum Intelligent Behaviour

Physarum polycephalum (a type of slime mould) has been shown to exhibit intelligent behavior while foraging for food, it senses gradients of chemo attractants and repellents to form a yellowish network connecting all food resources [25]. Physarum is capable of making complex foraging decisions based on trade-offs between risks, hunger level and food patch quality [17,24].

Physarum demonstrates an excellent ability in network construction without central consciousness in the process of foraging [23,32]. This has motivated many researchers to take inspiration from their biological phenomena to come up with a novel, biologically inspired models for unconventional computational methods capable of solving many NP-hard problems [5]. Many mathematical models have been proposed to simulate Physarum foraging behavior [4,11,27]. However, most of these models focus on modeling single Physarum and have not studied multiple Physarum foraging behavior. In this research, we think that competition among different Physarum individual may facilitate global exploration, and avoid falling in local minimum.

## 1.3 Motivation and Aim

Physarum protoplasmic flux is changing in a continuous way with the change of environment during evolution process. This characteristic allows Physarum to have great potentials in dealing with graph-optimization problems in dynamic environment as WSANs. In this research, we have taken advantage of the excellent characteristic heuristics similarity between Physarum and WSANs to deal with as a complicated graph problem and proposed a self-healing, and autonomous network connectivity restoration algorithm that minimize movement overhead and keep load balance.

## 2 Related Work

The connectivity restoration algorithms of WSANs were investigated thoroughly, and several fault-tolerant techniques were proposed with various capability and limitations [1,8,28,31]. They can be classified into: the provisioned mechanism; which involves pre-configuration of some backup nodes [14], and the reactive mechanisms; which will not perform the restoration until one of the nodes appears abnormal [26].

Active spare designation algorithm (NORAS) [28]; which belongs to provisioned mechanism, works by finding the spare nodes inside the network prior to the abnormity. The Distributed Actor Recovery Algorithm (DARA) [1]; which is considered one of the famous reactive mechanism, works firstly by choosing the appropriate node among the two-hop neighboring nodes of the failed node and then relocates the existing movable nodes in the network to an appropriate location.

Many approaches deal with cut vertex failure only, such as DARA [1], PDARA [7] approach that forms a connected dominating set (CDS), and informs a particular node in advance whether a partition occurs in case of failure. Since cut vertex identification incurs significant overhead in terms of messaging, the Recovery through Inward Motion (RIM) algorithm [31] only needs to maintain a one-hop information table, and all the one-hop neighbors move towards the position of the failed node till the distance is equal to half the communication radius ($Rc/2$). However, the cascaded inward motion increases the moving distances, and total number of the relocated nodes [3]. The Least-Movement Topology Repair algorithm (Le-MoToR) [2] relies on the local information to attempt to minimize the movement distance of each relocated node. A distributed approaches that exploit non cut-vertex actors in the recovery process such as Least Distance Movement Recovery (LDMR) algorithm [8], and distributed autonomous connectivity restoration method based on finite state machine (DCRMF) [33] is based on moving the direct neighbors of the abnormal node toward the position of the abnormal node until it is replaced by the nearest non cut-vertex actor.

Though many heuristic algorithms have been proposed, such as Genetic Algorithm [12,19,22], and Artificial Bee Colony Algorithm [13], it is still extraordinarily time-consuming.

## 3   The Proposed Model

In this research, we propose a Physarum inspired, distributed autonomous fault repair model to relocate the mobile actor nodes in optimal locations (interest points) to restore connectivity while minimising actor movement, and keeping load balance. This model is based on possible heuristics that Physarum use in complex foraging decisions based on trade-offs between their motivation for food, and in the presence of competitors. Each Physarum has its own autonomous behaviors, they react to each other and their own local environment. This kind of interaction will implicitly accelerate the whole system evolving to a global optimum network as in WSAN.

WSANs is an example of graphically expressed problem. A WSAN is represented by an undirected graph $G = (V, E)$ where $V$ is the set of actor nodes, and edges of $E$ established between every two nodes, and $Rc$ is the actors communication radius. Each actor $V_m$ has ID, positions, and energy. Each actor maintains just one-hop neighborhood table, and gets the states of the surrounding nodes through the heartbeat information detection.

Given an initial random deployment of $M$ mobile actor nodes over a 2-D area, we formulated a hexagonal CA reaction diffusion model for dynamic relocation of actors using multiple Physarums as a representation of mobile actors and food sources as interest points. In this model we did not consider the communication influences of MAC layer of the network.

### 3.1   Definitions

Before we discuss the proposed model, some definitions must be clarified.

**Definition 1** (Connectivity). *In an undirected graph G, it is connected if and only if there is at least one path for any two vertices. We are using depth-first search algorithm for traversing the graph starting from the base station as the initial node to check connectivity.*

**Definition 2** (Dominating Nodes). *Are the nodes which constitute the skeleton of a connected network.*

**Definition 3** (1-Hop Neighbors). *An actor $v_j$ is considered to be 1-hop neighbor of another actor $v_i$, if and only if $e_{i,j} \leq R_c$.*

**Definition 4** (Cut-Vertex). *Is the critical node whose failure disrupting network connectivity, and cause the network to be partitioned.*

**Definition 5** (Critical Node). *The node whose failure breaks its directed neighborhood connectivity.*

**Definition 6** (Uncritical Node). *A node is an uncritical node if and only if all its one-hop neighbors form a connected network*

### 3.2    Fault Repair Algorithm

If any active node fails to work, the network connectivity will be prone to be lost. We developed a restoration algorithm; which belongs to the reactive mechanism class, where the location of a failed node will be re-exposed as an interest point (food resource) for actors (Physarum) to compete on. In order to decrease the movement overhead we selected a node actor of its one-hop neighbors to move to replace the failed one. In order to deal with multiple node abnormities, we introduced a distributed regional restoration strategy, where multiple abnormal nodes in different regions will independently execute the restoration algorithm.

**Alternative Node Selection for Fault Repair Algorithm.** To identify a cut vertex requires global information, which is impossible and inefficient in WSANs, also critical nodes failure may not necessarily break network connectivity, and may bring unnecessary restoration. This model avoids any complex mechanism to select the alternative nodes as it is much more efficient and cheaper to restore its direct neighbors connectivity than to identify a cut vertex and restore the network connectivity.

Our mechanism to select the alternative nodes only involves the one-hop information table, and the competition between actors (Physarum) will depends on actor node location (regions of k-influence), residual energy, and the distance from failed node.

*(I) Region of k-influence of base station.* The actors that are closer to the base station can be overloaded and can degrade early. In this model we considered the concept of the regions of k-influence around the base station, where k is an integer denoting the number of hopes needed to reach the base station. The

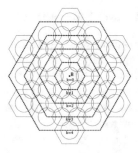

**Fig. 1.** Region of k-influence of base station $B$

actors at distance k from the base station lie on a hexagon termed as ring-k, as shown in Fig. 1. Only the nodes within a region of higher k-influence can compete for failed node, this will keep inward motion to maintain connectivity to base station, and avoid unnecessary repair of leaf nodes.

*(II) Actor residual energy.* In fact, network lifetime is the most important factor which depends on load balancing as an important evaluation metrics for the connectivity restoration algorithms. In our model design the actor energy will decrease by 1% with every movement step, this will give a priority to less used actors to compete over interest points. Actors failed to occupy interest points will not move and stay in stand-by for fault repair. If two actors are competing for the same interest point the actor (Physarum) with more energy will be able to repel the other (Physarum) with less energy. This will give a chance for spare/less used actors with higher energy to have the priority to fill the target point, keeping load balance.

*(III) The distance from failed node.* Physarum is capable of finding the shortest path between two points using simple heuristics [18]. Actor nodes (Physarum) closer to the failed node will have the chance to occupy the location of failed node, this will help to enhance the network lifetime since node movement is considered the major factor of energy exhaustion.

### 3.3 Hexagonal Area Tessellation

Topology control is one of the key issues of WSN, which is important for reducing communication interference and prolonging network lifetime.

In our proposed model, we considered a hexagonal deployments where the area is dynamically tessellated by regular hexagons with its side equal to communication radius ($R_c$). The vertices and the centers of the regular hexagons will be identified as the interest points to attract actors. This node placement will minimize redundancy, and avoid the situations that more than one node senses and processes the same event as much as possible. It has been proved that such node placement technique maximizes the area coverage using a minimum number of nodes [9].

## 3.4   Physarum Inspired Model for WSAN

**Modelling Multiple Physarum and Multiple Food Resources.** Unlike other models based on single Physarum [4,11,15,18,27], we considered a Physarum competitive behavior, where a group of Physarum each has autonomous behaviors react to each other and their own local environment foraging for multiple food sources. We assume that competing Physarum will exert repulsion forces on each other which will affect the evolution of the whole system. We created a new formula to compute two forces (attraction/repulsion) affecting Physarum diffusion, The first is chemo-attraction force to food sources (interest points), and the second is the repulsion negative forces that competing Physarums exert on each other based on its mass (actor energy).

**The Model State of Cellular Automaton (CA).** In order to model mobile-WSN, we considered a CA grid in the two-dimension space, which is divided into a matrix $(X \times Y)$ of identical hexagon cells, in which every cell $c_{(i,j)}$ has six neighbours. In this grid a set of $m$ actors $(V = v_1, v_2, \ldots, v_m)$ are competing on a set of $n$ interest point $(IP = ip_1, ip_2, \ldots, ip_n)$. The state of a cell $c_{(i,j)}^t$ at time $t$ located at position $(i, j)$ is described by its type as in Eq. 1, whether it is an empty cell, an interest point, an actor, a base station, or an obstacle cell (Ex:- physical obstacle, boundary wall).

$$CT_{(i,j)} = \{ \text{"}FREE\text{"}, \text{"}OBSTACLE\text{"}, \text{"}INTEREST\_POINT\text{"},$$
$$\text{"}ACTOR\text{"}, \text{"}BASE\_STATION\text{"} \} \qquad (1)$$

An interest point is defined by its mass, k-influence, while an actor is defined by its energy. Similarly to the original Physarum competition model, where chemical is defined by its mass, quality and Physarum is defined by its mass respectively.

**The Model Rules of Cellular Automaton (CA).** In our model, each actor is a self organized computational unit. Each of them aims to achieve the maximum utility based on its local environment by choosing appropriate behaviors. Models based on CA are fast when implemented on serial computers, because they exploit the inherent CA parallelism [29].

The CA model rules are mainly based on the diffusion equations combined with Physarum heuristics in competition settings, where multiple actors (Physarums) will compete for these interest points (Food resources). Each actor will execute the diffusion (search) process as defined in Eqs. 2, 3, 4 and 5 to explore its neighborhood within its communication radius $(R_c)$ for a number of iteration equaling its $(R_c)$ (every $R_c$ iteration will be defined as one round). Each actor at iteration $(t)$ uses the values of its six neighbours cell to calculate the value of the energy at the next iteration $(t + 1)$.

$$AE_{(i,j)}^{t+1} = AE_{(i,j)}^t + \sum_{(k,l)} \begin{cases} (AF * AD * AE_{(k,l)}^t) - AE_{(i,j)}^t, & \text{if } A\_AA_{(i,j),(k,l)} = 1 \\ 0, & \text{otherwise} \end{cases}$$

$$\forall (k,l) : \quad i-1 \leqslant k \leqslant i+1,$$
$$j-1 \leqslant l \leqslant j+1,$$
$$k \neq l$$

$$AF = 1 + A\_AttForce_{(i,j),(k,l)}^t + A\_RepForce_{(i,j),(k,l)}^t \tag{2}$$

where,

$AE_{(i,j)}^{t+1}$ defines the diffusion of actor energy for the next generation $(t+1)$ at cell $c_{(i,j)}$.

$AE_{(i,j)}^t$ is the current energy of an actor at iteration $(t)$ for cell $c_{(i,j)}$.

$AF$ is the forces affecting an actor.

$AD$ is the actor diffusion coefficient.

$$A\_AA_{(i,j),(k,l)} = \begin{cases} 1, & \text{if } CT_{(k,l)} = \text{``}FREE\text{''} \ OR \ \text{``}INTEREST\_POINT\text{''} \\ 1, & \text{if } CT_{(k,l)} = (\text{``}ACTOR\text{''}) \ AND \ (AID_{(i,j)} = AID_{(k,l)}) \\ 0, & \text{otherwise} \end{cases}$$

$$\tag{3}$$

where,

$A\_AA_{(i,j),(k,l)}$ defines whether an actor at cell $c_{(i,j)}$ is available to diffuse towards a neighbouring cell $c_{(k,l)}$.

$AID_{(i,j)}$ is the ID of an actor.

$$A\_AttForce_{(i,j),(k,l)} = \begin{cases} \dfrac{IPM_{(k,l)}}{Total\_IPM}, & \text{if } IPM_{(k,l)} = MAX(IPM_{(i,j)}) \\ 0, & \text{otherwise} \end{cases} \tag{4}$$

where,

$A\_AttForce_{(i,j),(k,l)}$ defines the value of attraction force of $AE_{(i,j)}$ towards its neighbouring cell $c_{(k,l)}$.

$IPM_{(i,j)}$ is the current mass of the interest point for cell $c_{(i,j)}$.

$Total\_IPM$ is the total sum of all interest points mass on the grid.

$$A\_RepForce_{(i,j),(k,l)} = \begin{cases} \dfrac{AE_{opp(k,l)}^t}{Total\_AE}, & \text{if } AID_{(i,j)} \neq AID_{opp(k,l)}, \\ & \qquad AE_{opp(k,l)}^t > Rep\_Limit \\ 0, & \text{otherwise} \end{cases} \tag{5}$$

where,

$A\_RepForce_{(i,j),(k,l)}$ defines the value of repulsion force of $AE_{(i,j)}$ towards its neighbouring cell $c_{(k,l)}$.
$AE^t_{opp(i,j)}$ is the neighbor actor energy at the opposite direction.
$Rep\_Limit$ is a limit where an actor must reach to repel neighboring actor.

In our model, we have addressed a 1% decrease in actor energy (Physarum mass) with each movement step. Simply, actor superiority in competition is directly proportional to actor energy, a key point for load balancing and will give a priority to less used actors to process messages and replace failed nodes. The process of searching for interest points will be executed for several rounds until connectivity is achieved or other stopping conditions are met. Actors failed to occupy interest points will not move and stay in stand-by for fault repair.

## 4   Experimental Results

The core model was implemented in Java with Processing package https:// processing.org/ being used for graphical simulation.

To validate our model, we have conducted an experiment (connectivity and fault repair) for 50 times using the same parameters of diffusion equation as in [27] (Table 1). All the experiments were statistically analysed to compare average results (mean ± sd) using T-Test (SPSS package). These experiments were executed on a commercial computer (Intel®Core™ i5-6500 CPU @ 3.20 GHz with 8 GB RAM).

**Table 1.** Parameters values for the experiments

| Parameter | Value |
|-----------|-------|
| AD | 0.1 |
| AE | 3000 |
| IPM | 3000 |
| REP_LIMIT | 5 |

In this experiment design, 100 wireless sensor-actor nodes were randomly placed in a 2D (50 × 50) hexagonal grid. All actors were homogeneous with communication radii $R_c = 5$. The base station was located in the center of the grid. After initial random deployment, actors will compete over interest points until we get a fully connected network. Then we will randomly eliminate 10% of actor nodes, the locations of removed nodes will be re-exposed as interest points (food source) for actors (Physarum) to compete on. The connectivity will be updated after elimination of nodes, and the recovery algorithm will be executed until connectivity is restored again or 30 rounds are passed without achieving connectivity. This process of node elimination, and fault repair will be repeated with variable number of actor nodes elimination [10%, 20%, ..., 90%] which provides different densities of network topology.

The outcome of the experiment will be the total number of moves required to be connected again, the number of relocated actor nodes, and the average movement per actor (total moving distance/number of relocated nodes). Each experiment was run in duplicate using the concept of the regions of k-influence around the base station or without using it.

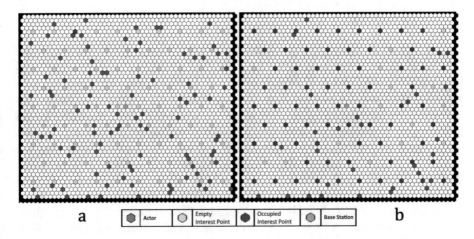

| | Actor | | Empty Interest Point | | Occupied Interest Point | | Base Station |

a                                                                    b

**Fig. 2.** The initial network simulation with a total node of 100 and $R_c = 5$(a), and after algorithm execution to get a connected network (b).

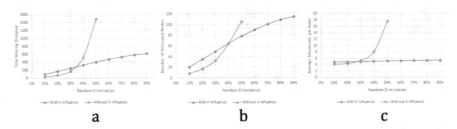

a                              b                              c

**Fig. 3.** Showing the total moving distance (a), the number of relocated nodes (b), and average movement per node (c).

The experimental results showed that network were not connected after initial random deployment (Fig. 2a). And after executing our connectivity algorithm and using the concept of the regions of k-influence around the base station, we get a connected WSANs (Fig. 2b). After we randomly eliminated [10%, 20%, ..., 90%] of actor nodes and applied the fault repair algorithm, it has been shown that using the concept of the regions of k-influence around the base station proved to be efficient while maintaining connectivity even after 90% of nodes elimination (10 actor nodes were only remaining). The algorithm without this concept was only able to restore connectivity up to 40% elimination of actor nodes in all experimental trials, while after eliminating 50% of actor nodes, only 48% of experiments succeeded in achieving connectivity (Fig. 3).

# 5 Conclusion

To the best of our knowledge, this is the first paper attempting to solve connectivity restoration problem using Physarum inspired unconventional computational approach. We have proposed energy aware self-healing autonomous network connectivity restoration algorithm that minimize movement overhead and keep load balance. The location of failed actor will be re-exposed as food source for actors (Physarum) to compete on. This model avoids any complex mechanism to select the alternative nodes, it only involves the one-hop information table, and the competition between actors (Physarum) which depends on actor node location (regions of k-influence), residual energy, and the distance from failed node. The results of our experiment showed that after initial random deployment of actor nodes all the resulting WSANs were unconnected. However, using hexagonal tessellation topology ensured that all actor nodes are connected with minimal overlap. The fault repairing algorithm using the concept of the regions of k-influence around the base station proved to be highly efficient. Only the nodes within a region of higher k-influence can compete for failed node. This algorithm will give a chance for spare/less used actors with maximum energy to have the priority to execute the algorithm and keep load balance.

**Acknowledgement.** Abubakr Awad is supported by Elphinstone PhD Scholarship (University of Aberdeen). Wei Pang and George M. Coghill are supported by the Royal Society International Exchange program (Grant Ref IE160806).

# References

1. Abbasi, A.A., Younis, M., Akkaya, K.: Movement-assisted connectivity restoration in wireless sensor and actor networks. IEEE Trans. Parallel Distrib. Syst. **20**(9), 1366–1379 (2009). https://doi.org/10.1109/TPDS.2008.246
2. Abbasi, A.A., Younis, M.F., Baroudi, U.A.: A least-movement topology repair algorithm for partitioned wireless sensor-actor networks. Int. J. Sens. Netw. **11**(4), 250–262 (2012). https://doi.org/10.1504/IJSNET.2012.047152
3. Abbasi, A.A., Younis, M.F., Baroudi, U.A.: Recovering from a node failure in wireless sensor-actor networks with minimal topology changes. IEEE Trans. Veh. Technol. **62**(1), 256–271 (2013). https://doi.org/10.1109/TVT.2012.2212734
4. Adamatzky, A.: From reaction-diffusion to physarum computing. Nat. Comput. **8**(3), 431–447 (2009). https://doi.org/10.1007/s11047-009-9120-5
5. Adamatzky, A.: Physarum Machines: Computers from Slime Mould. World Scientific (2010). https://books.google.co.uk/books?id=Kbs_AIDbfU8C
6. Afsana, F., Asif-Ur-Rahman, M., Ahmed, M.R., Mahmud, M., Kaiser, M.S.: An energy conserving routing scheme for wireless body sensor nanonetwork communication. IEEE Access **6**, 9186–9200 (2018). https://doi.org/10.1109/ACCESS.2018.2789437
7. Akkaya, K., Senel, F., Thimmapuram, A., Uludag, S.: Distributed recovery from network partitioning in movable sensor/actor networks via controlled mobility. IEEE Trans. Comput. **59**(2), 258–271 (2010). https://doi.org/10.1109/TC.2009.120

8. Alfadhly, A., Baroudi, U., Younis, M.: Least distance movement recovery approach for large scale wireless sensor and actor networks. In: IWCMC 2011 - 7th International Wireless Communications and Mobile Computing Conference, pp. 2058–2063 (2011). https://doi.org/10.1109/IWCMC.2011.5982851

9. Brass, P.: Bounds on coverage and target detection capabilities for models of networks of mobile sensors. ACM Trans. Sens. Netw. **3**(2) (2007). https://doi.org/10.1145/1240226.1240229

10. Goubier, O.N.P., Huynh, H.X., Truong, T.P., Traore, M., Pottier, B., Rodin, V., Nsom, B., Esclade, L., Rakoroarijaona, R.N., Goubier, O., Stinckwich, S., Huynh, H.X., Lam, B.H., Vinh, Udrekh, Muslim, H., Surono: Wireless sensor network-based monitoring, cellular modelling and simulations for the environment. ASM Sci. J. **2017**(Special issue1), 56–63 (2017)

11. Gunji, Y.P., Shirakawa, T., Niizato, T., Haruna, T.: Minimal model of a cell connecting amoebic motion and adaptive transport networks. J. Theor. Biol. **253**(4), 659–667 (2008). https://doi.org/10.1016/j.jtbi.2008.04.017

12. Gupta, S.K., Kuila, P., Jana, P.K.: Genetic algorithm for k-connected relay node placement in wireless sensor networks. Adv. Intell. Syst. Comput. **379** (2016). https://doi.org/10.1007/978-81-322-2517-1_69

13. Hashim, H.A., Ayinde, B.O., Abido, M.A.: Optimal placement of relay nodes in wireless sensor network using artificial bee colony algorithm. J. Netw. Comput. Appl. **64**, 239–248 (2016). https://doi.org/10.1016/j.jnca.2015.09.013

14. Imran, M., Younis, M., Haider, N., Alnuem, M.A.: Resource efficient connectivity restoration algorithm for mobile sensor/actor networks. EURASIP J. Wirel. Commun. Netw. **2012**(1), 347 (2012)

15. Jones, J.: Influences on the formation and evolution of physarum polycephalum inspired emergent transport networks. Nat. Comput. **10**(4), 1345–1369 (2011). https://doi.org/10.1007/s11047-010-9223-z

16. Lam, B.H., Huynh, H.X., Pottier, B.: Synchronous networks for bio-environmental surveillance based on cellular automata. EAI Endorsed Trans. Context-Aware Syst. Appl. **16**(8) (2016). https://doi.org/10.4108/eai.9-3-2016.151117

17. Latty, T., Beekman, M.: Speed-accuracy trade-offs during foraging decisions in the acellular slime mould physarum polycephalum. Proc. R. Soc. B Biol. Sci. **278**(1705), 539–545 (2011). https://doi.org/10.1098/rspb.2010.1624

18. Nakagaki, T., Yamada, H., Tóth, Á.: Maze-solving by an amoeboid organism. Nature **407**(6803), 470 (2000). https://doi.org/10.1038/35035159

19. Ozera, K., Oda, T., Elmazi, D., Barolli, L.: Design and implementation of a simulation system based on genetic algorithm for node placement in wireless sensor and actor networks. In: International Conference on Broadband and Wireless Computing, Communication and Applications, pp. 673–682. Springer, Heidelberg (2016)

20. Qiu, T., Chen, N., Li, K., Qiao, D., Fu, Z.: Heterogeneous ad hoc networks: architectures, advances and challenges. Ad Hoc Netw. **55**, 143–152 (2017). https://doi.org/10.1016/j.adhoc.2016.11.001

21. Qiu, T., Luo, D., Xia, F., Deonauth, N., Si, W., Tolba, A.: A greedy model with small world for improving the robustness of heterogeneous internet of things. Comput. Netw. **101**, 127–143 (2016). https://doi.org/10.1016/j.comnet.2015.12.019

22. Ramezani, T., Ramezani, T.: A distributed method to reconstruct connection in wireless sensor networks by using genetic algorithm. Mod. Appl. Sci. **10**(6), 50 (2016)

23. Reid, C.R., Beekman, M.: Solving the towers of Hanoi - how an amoeboid organism efficiently constructs transport networks. J. Exp. Biol. **216**(9), 1546–1551 (2013). https://doi.org/10.1242/jeb.081158

24. Reid, C.R., Latty, T.: Collective behaviour and swarm intelligence in slime moulds. FEMS Microbiol. Rev. **40**(6), 798–806 (2016). https://doi.org/10.1093/femsre/fuw033
25. Saigusa, T., Tero, A., Nakagaki, T., Kuramoto, Y.: Amoebae anticipate periodic events. Phys. Rev. Lett. **100**(1) (2008). https://doi.org/10.1103/PhysRevLett.100.018101
26. Senturk, I., Yilmaz, S., Akkaya, K.: Connectivity restoration in delay-tolerant sensor networks using game theory. Int. J. Ad Hoc Ubiquitous Comput. **11**(2–3), 109–124 (2012). https://doi.org/10.1504/IJAHUC.2012.050268
27. Tsompanas, M.A.I., Sirakoulis, G.C., Adamatzky, A.: Cellular automata models simulating slime mould computing. In: Advances in Physarum Machines, pp. 563–594. Springer, Heidelberg (2016)
28. Vaidya, K., Younis, M.: Efficient failure recovery in wireless sensor networks through active spare designation. In: DCOSS 2010 - International Conference on Distributed Computing in Sensor Systems, Adjunct Workshop Proceedings: IWSN, MobiSensors, Poster and Demo Sessions (2010). https://doi.org/10.1109/DCOSSW.2010.5593284
29. Wolfram, S.: Computation theory of cellular automata. Commun. Math. Phys. **96**(1), 15–57 (1984)
30. Yan, K., Luo, G., Tian, L., Jia, Q., Peng, C.: Hybrid connectivity restoration in wireless sensor and actor networks. EURASIP J. Wirel. Commun. Netw. **2017**(1) (2017). https://doi.org/10.1186/s13638-017-0921-4
31. Younis, M., Lee, S., Gupta, S., Fisher, K.: A localized self-healing algorithm for networks of moveable sensor nodes. In: GLOBECOM - IEEE Global Telecommunications Conference, pp. 1–5 (2008). https://doi.org/10.1109/GLOCOM.2008.ECP.9
32. Zhang, X., Gao, C., Deng, Y., Zhang, Z.: Slime mould inspired applications on graph-optimization problems. In: Advances in Physarum Machines, pp. 519–562. Springer, Heidelberg (2016)
33. Zhang, Y., Wang, J., Hao, G.: An autonomous connectivity restoration algorithm based on finite state machine for wireless sensor-actor networks. Sensors **18**(1), 153 (2018)

# Mining Unit Feedback to Explore Students' Learning Experiences

Zainab Mutlaq Ibrahim$^{(\boxtimes)}$, Mohamed Bader-El-Den, and Mihaela Cocea

University of Portsmouth, Lion Terrace, Portsmouth PO1 3HE, UK
{zainab.mutlaq-ibrahim,mohamed.bader,mihaela.cocea}@port.ac.uk

**Abstract.** Students' textual feedback holds useful information about their learning experience, it can include information about teaching methods, assessment design, facilities, and other aspects of teaching. This can form a key point for educators and decision makers to help them in advancing their systems. In this paper, we proposed a data mining framework for analysing end of unit general textual feedback using four machine learning algorithms, support vector machines, decision tree, random forest, and naive bays. We filtered the whole data set into two subsets, one subset is tailored to assessment practices (assessment related), and the other one is the non-assessment related data subset, We ran the above algorithms on the whole data set, and on the new data subsets. We also, adopted a semi automatic approach to check the classification accuracy of assessment related instances under the whole data set model. We found that the accuracy of general feedback data set models were higher than the accuracy of the assessment related models and nearly the same value of the non- assessment related modeles. The accuracy of assessment related models were approximated to the accuracy of the assessment related instances under the full data set models.

**Keywords:** Sentiment analysis · Educational data mining
Assessment · Student feedback

## 1 Introduction

Unit feedback is a fundamental part of the learning process for institutions. It can provide data about units in which may hold implicit useful knowledge for researchers and practitioners to understand student learning experiences. This can form a start point for educators to modify or develop units accordingly.

Normally, feedback consist of a simple survey form, most often a combination of Likert scale responses to questions or statements (Responses can be strongly agree, agree, neutral, disagree or strongly disagree), in addition to one or more of open-ended question(s) where students need to write few short sentences.

Quantitative data can be taken from likert scale responses. In fact, these responses have been used for long time to assess the teaching effectiveness [1], due to the fact that they are numerical data and easy to analyse, on the contrary

© Springer Nature Switzerland AG 2019
A. Lotfi et al. (Eds.): UKCI 2018, AISC 840, pp. 339–350, 2019.
https://doi.org/10.1007/978-3-319-97982-3_28

of the Open-ended responses which are not easy to analyse as likert responses. Open ended responses are unstructured data, also they may have keywords that are not included in Likert questions' words or even in their own words in which make them (the open-ended responses) not bias to the trend of a survey. They can be a very important source of any faculty analytic processes to reveal the hidden information in these texts. And they can be a big chance for students to be a real participant of the ongoing studies to advance education.

Students' feedback have been to some extent accepted by researchers and practitioners due to the fact that student ratings and feedback are the most valid source for evaluating teaching effectiveness [2], students are the people who receive the teaching procedures, so their feedback is crucial.

Educational data mining (EDM) is the discipline that focuses on developing methods and algorithms to explore big data that comes from educational databases and sources to better understanding of students and the setting they learn in [3]. In particular, sentiment analysis is the process of analyzing statements and obtaining subjective information from them.

Massive online open courses (MOOC) have been accepted by some institutions since 2012, with such a huge class size, massive blog feedback is being generated which form a big challenge for EDM community to innovate and advance frameworks and techniques to analyze such feedback, hence, this framework can be used to analyse such data.

The rest of the paper is organized as follows: Related work is presented in Sect. 2. The framework and work flow is presented in Sect. 3. The used data set of this study is presented in Sect. 4. Experiments and results are presented in Sect. 5, and finally conclusion and future work are presented in Sect. 6.

## 2   Related Work

Researchers have been encouraged and motivated by EDM community to innovate new frameworks to analyze different educational topics such as assessment, students' emotion, browsing or interaction data, the results of educational research, and many more [4]. EDM community also has urged researchers to apply a previously used frameworks to a new domain or reanalyze an existing data set with a new technique [4].

Data mmining algorithims have been applied, to classify students according to their Moodle usage and the final obtained marks [5], to predict student retention [6], to reduce dropout rates [7], to analyze students' programming assignment [8], and many more.

All of the above studies [5–7] used a structured data sets which were taken from Moodles and databases, structured data sets is easy to search by simple algorithm, example of this is spread sheets, while unstructured data such as tweet is more like human language and searching it is very difficult and needs an advanced and special algorithms.

Analyzing unstructured data (text mining) needs a pre-processing stage to structure it so it becomes easily search-able and manageable. Recently, there is an increasing number of research in utilizing text mining techniques for different educational purposes and applications due to the need of advancing and developing learning process. Abd-Elrahman et al. [9] used WordStat tool to determine the number of positive and negative entries with different teaching aspect categories, they mentioned only spelling errors and nothing about the dimensions of the data, they depends on an algorithm that count the number of occurrence of a specific words. Sliusarenko [10] transformed the qualitative feedback into quantitative by extracting the key-terms, then applied factor analysis to find the most important factors in the feedback, and finally applied regression technique to see which factors have the most impact on student ratings. Pan et al. [11] used SPSS text analysis in a try to strengthen quantitative data with systematic and meaningful qualitative interpretation. Jordan [12] utilized StatSoft Statistica and SPSS, he built a correlation model using text mining methods, he found a weak correlation between the Likert responses and the open-ended written responses. This means there are significant words and patterns within the open-ended responses that can provide additional information to the decision makers. Finally, Pagare, Chen et al. [13,14] analyzed twitter data to understand students' learning experiences. They [14] innovated a new work flow which was a mixed of human efforts and machine learning. Both [13,14] applied Naive Bays(NB). Chen also applied Support Vector Machine(SVM)and Max Margin Multi-label (M3L) classifiers [14]. They [13,14] concluded that NB is very good classifier to use on text data.

## 3 Framework

In this section, we present a general framework for analyzing end of unit students' feedback which was given in text format as shown in Fig. 1. The main aim of this study is to develop a data mining framework to capture students' concern of assessment from general feedback and to investigate the performance of data mining models when they are applied to general feedback verses topic specific. To achieve the above aim, the following objectives are defined:

- Instances classification: Identify the best classification model to automatically detect the class of each instance, to filter and divide the full data set into assessment and non-assessment related sub sets.
- Assessment related instances' sentiment: Identify the best sentiment analysis model that automatically detect the polarity of the assessment related instances, so we can identify issues from negative instances.
- Full data set instances' sentiment: Identify the best sentiment analysis model that automatically detect the polarity of its instances.
- Assessment related instances' sentiment under the full data set model: Detect the polarity of the assessment related instances under the sentiment of whole data set model.
- Non-Assessment related instances' sentiment: Identify the best sentiment analysis model that automatically detect the polarity of the non-assessment related instances.

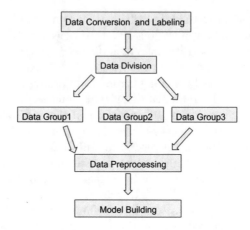

**Fig. 1.** Proposed general framework of students' feedback model

# 4  Data

The used dataset in this study is a hand-written feedback, it was collected from students as end of unit (INDADD) feedback for the years 2012–2016, it consists of 979 instances. It includes responses to the following two Statements: Statement1: The best part of the unit is:, Statement2: The area of this unit that needs improving is:, Statement1 onsidered as positive and statement2 as negative feedback.

## 4.1  Labeling

The data has been labeled using three labels: Assessment related label, where some keywords and their derivatives or synonymous are presented, such as "coursework", "exam", and "quiz", "assessment", "marking", "grading", "test", "feedback" and "evaluation". Assessment not related label, where there is a meaningful text but the assessment related keywords mentioned above are not presented. And irrelevant label to cover the empty instances, misspellings, jokes, or irrelevant statements which we have none of them in our data set.

Three native English speakers reviewed the feedback data and label it according to the suggested labels. Each entry of the feedback has three rating from the three viewers. Although, rules of labeling are very clear, easy to follow, and far to mislead, the raters have a chance to label the entry as irrelevant.

In content analysis literature, statistical measures such as scott's Pi, Fleiss Kappa and Krippendorf's Alpha are used to decide agreement among raters on topics [15,16]. However, Chen and his colleagues [1] used the harmonic mean (F1 measure) to measure how close two label sets are assigned to one entry by two raters as their study were dealing with multi-label classification problem. In our study, the defined labels are mutually exclusive which means that each entry can fall under only one label, so any one of the above measures is applicable to our data set.

In this study Krippendorf's Alpha measure is adopted as a reliability coefficient, it was developed to measure the agreement among raters on a specific topic, it applies to all sort of metrics, any number of raters, any number of categories, incomplete data, and big or small data samples [17].

As this study utilizes Krippendorf's Alpha measure which effectively can deal with the missing data [17], the raters above can have extra chance to skip labeling in tricky text. The Krippendorff's score is 0.9841 which considered to be an optimist value.

### 4.2  Data Distribution

The used data set in this study is anonymous data, which means that students can not be identified but sometimes the contents of the data identified few lecturers, however to make the contents anonymous for the lecturers too, we refer to the persons by index such as L1 (for the first mentioned lecturer), L2, ... and so on. See Fig. 2.

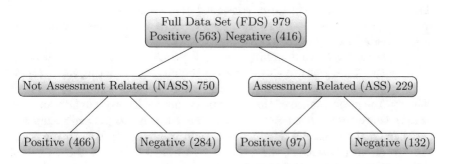

**Fig. 2.** Data set distribution

## 5  Experiments and Results

In this section, we executed five experiments to fulfill the objectives that which are mentioned in the framework section. For each experiment, we used ten fold cross validation method as this method is robust against potential bias to the training data set, in this method we used nine fold for training and one fold for testing, we repeated the test ten times and calculated the average performance for all attempts. We used a PC desktop with quad 2.33 GHZ CPU, 4 GB RAM, and windows 10 operating system. The following section includes the preprocessing components that were executed to all experiments. Followed by a brief description of classifiers that we are going to use and four popular evaluation measures.

## 5.1   Text Pre-Processing

Hand-written feedback can include miss-spelling errors, jokes, irrelevant statements, some special characters such as "@", Punctuation marks, etc. However, it is the open-ended nature of questions that allows students to express what is in their minds and what they feel without the constraint of the carefully worded numerical rating questions.

Pre-processing stage aims at cleaning the data and reducing its dimension, this can contribute positively to more accurate results. In this section the following operations are executed using KNIME analytic platform:

- Punctuation Erasure: Removes all punctuation characters of terms contained in the text, such as exclamation, question marks.
- N Chars Filter: Filters all terms contained in the text data with less than the three characters such as "a", "in", "if" and many more.
- Number Filter: Filters all terms contained in the text data that consist of digits, including decimal separators "," or "." and possible leading "+" or "_".
- Case converter: Converts all terms contained in the text data to lower.
- Stop word Filter: Filters all terms of the text, which are contained stop word such as "because", "again", and "the".
- Snowball Stemmer: Stems terms contained in the text data with the Snowball stemming library to guarantee that each term represent once and only once in the created bag of words (BoW).
- Feature creation: In text mining, text's features are the characters or the words of that text. Feature selection is the process of eliminating the irrelevant and trivial features and keeping the significant features which are genuinely affecting the performance of the constructed model. To explain that more let us have this example from our data set, "That it was 100 percent coursework based on real life seminars which we could easily relate to", in our study which is about assessment, the most significant feature in the above example is the "coursework" term, however, in machine learning we can not eliminate all features and keep the most significant one.

Some of the text pre-processing operations contribute to feature selection process, such as removing stop words, number filter, snowball stemmer, and n-character filter.

One of the most popular feature creation technique is n-grams [18,19]. An n-gram is a sere of n items from a text, it can be letters or words, uni gram, is very popular technique which is selecting single words, bigram is selecting two words at a time for example "the coursework is not clear", "the coursework", "coursework is", "is not", "not clear". In sentiment analysis section (see Table 2), we are going to use uni gram (UNIGRAM) and bi gram (BIGRAM).

## 5.2   Classifiers

The most popular classifiers for text mining are: Support Vector Machines (SVM) [20–22], it is a powerful tool for solving data mining problems such as classification, regression, and feature selection, it has the power to determine an optimal separating point that labels records into different categories [23]; Naive Bayes [1,13,20,22,24], is a probabilistic classifier, it is robust to noisy data and irrelevant attributes, and can cope very well with null values [25]; Decision Tree [22,24], it uses training examples of data to construct the tree, at classification time the tree executed from root to leaf, so the leaf node decides the class of the record [25]; and Random Forest [22,24], it selects attributes randomly and utilizes the decision tree as the base model [25].

### 5.3 Assessment Classification

*Experiment1*
The first experiment aims to build a model that automatically filter the assessment related instances from the whole data set, we labeled the data as mentioned in labeling section and ran the text pre-processing component to build the final model.

Table 1 illustrates the evaluation results of experiment 1 performance.

From the result, we observed that all classifiers' performance in terms of accuracy, precision, recall, and f-score was significant except of the RF classifier which performed poorly. The optimal classifier was DT as it was error free.

**Table 1.** Experiment 1

|  | NB | SVM | DT | RF |
|---|---|---|---|---|
| Accuracy | 0.9640 | 0.9930 | 1 | 0.7640 |
| Prcision | 0.9770 | 0.9855 | 1 | 0.6725 |
| Recall | 0.9255 | 0.9950 | 1 | 0.5450 |
| F-score | 0.9480 | 0.9950 | 1 | 0.5295 |

### 5.4 Sentiment Analysis

In this section we executed four sentiment analysis experiments (2, 3, 4, 5), Fig. 3 shows their proposed framework. Table 2 illustrates the result of Experiments (2, 3, 4, 5) using uni-gram (UNIGRAM) and bi-gram(BIGRAM) features, only experiment 5 used the results of experiment 3 using uni-gram.

**Table 2.** Results of experiment: 2, 3, 4, 5

|     |      |              | Accuracy | Precision | Recall | F-score |
|-----|------|--------------|----------|-----------|--------|---------|
| DT  | Exp2 | ASS-UNIGRAM  | 0.67     | 0.67      | 0.67   | 0.67    |
|     |      | ASS-BIGRAM   | 0.65     | 0.64      | 0.64   | 0.64    |
|     | Exp3 | FDS-UNIGRAM  | 0.71     | 0.70      | 0.69   | 0.70    |
|     |      | FDS-BIGRAM   | 0.71     | 0.70      | 0.69   | 0.70    |
|     | Exp4 | ASS-IN-FDS   | 0.66     | –         | –      | –       |
|     | Exp5 | NASS-UNIGRAM | 0.72     | 0.70      | 0.69   | 0.69    |
|     |      | NASS-BIGRAM  | 0.71     | 0.69      | 0.69   | 0.69    |
| SVM | Exp2 | ASS-UNIGRAM  | 0.69     | 0.70      | 0.70   | 0.69    |
|     |      | ASS-BIGRAM   | 0.72     | 0.72      | 0.73   | 0.72    |
|     | Exp3 | FDS-UNIGRAM  | 0.76     | 0.76      | 0.74   | 0.74    |
|     |      | FDS-BIGRAM   | 0.76     | 0.76      | 0.74   | 0.76    |
|     | Exp4 | ASS-IN       | 0.69     | –         | –      | –       |
|     | Exp4 | NASS-UNIGRAM | 0.76     | 0.75      | 0.73   | 0.74    |
|     |      | NASS-BIGRAM  | 0.75     | 0.74      | 0.72   | 0.73    |
| RF  | Exp2 | ASS-UNIGRAM  | 0.58     | 0.58      | 0.50   | 0.73    |
|     |      | ASS-BIGRAM   | .058     | 0.59      | 0.51   | 0.38    |
|     | Exp3 | FDS-UNIGRAM  | 0.73     | 0.75      | 0.71   | 0.71    |
|     |      | FDS-BIGRAM   | 0.74     | 0.77      | 0.61   | 0.67    |
|     | Exp4 | ASS-IN-FDS   | 0.70     | –         | –      | –       |
|     | Exp5 | NASS-UNIGRAM | 0.62     | 0.31      | 0.50   | 0.38    |
|     |      | NASS-BIGRAM  | 0.63     | 0.31      | 0.50   | 0.38    |
| NB  | Exp2 | ASS-UNIGRAM  | 0.66     | 0.68      | 0.62   | 0.60    |
|     |      | ASS-BIGRAM   | 0.67     | 0.73      | 0.63   | 0.60    |
|     | Exp3 | FDS-UNIGRAM  | 0.60     | 0.59      | 0.56   | 0.55    |
|     |      | FDS-BIGRAM   | 0.61     | 0.60      | 0.57   | 0.55    |
|     | Exp4 | ASS-IN-FDS   | 0.44     | –         | –      | –       |
|     | Exp5 | NASS-UNIGRAM | 0.64     | 0.61      | 0.58   | 0.57    |
|     |      | NASS-BIGRAM  | 0.65     | 0.62      | 0.57   | 0.57    |

## Sentiment Analysis of Assessment Related Instances

*Experiment2*

In this experiment we used 229 of assessment related instances (ASS) to build a model that automatically detect their sentiment. SVM classifier was the best performer in terms of accuracy, it was 72% percent accurate. The poorest performer was the RF classifier, as its error rate was 42%. The recall values show that SVM is the most sensitive of the four classifiers, i.e., it correctly identifies instances of both classes, while RF is the least sensitive. Precision is the highest for SVM and lowest for RF. The best balance between precision and recall is attained by RF which is only 1% higher than the value that attained by SVM.

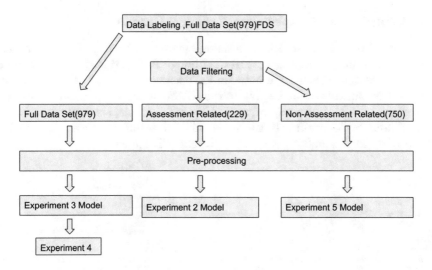

**Fig. 3.** Proposed framework for experiments: 2, 3, 4, 5

## Sentiment Analysis of Full Data Set Instances

*Experiment3*
We used 979 instances to build a model that automatically detect the sentiment of the whole data set. Also, SVM classifier recorded the higher accuracy of 76% followed by DT, then RF. The poorest performer classifier is NB with error rate of about 40%, this opposed to the finding of Chen and Pagare [1,13] that NB is a very good for text classification. The recall values show that SVM is the most sensitive, it has the highest value of correctly identified instances of both classes, while NB is the least sensitive. The highest precision is achieved by SNM, while the lowest is achieved by NB. The best balance between precision and recall is achieved by SVM, while the lowest balance is achieved by NB.

## Sentiment Analysis of Assessment Related Instances Under the Full Data Set Model

*Experiment4*
This experiment was built on the results of experiment 3 which includes a total of 979 instances. The aim was to view how accurate the general feedback (FDS) model in classifying the assessment related instances (ASS). The highest accuracy was scored by RF model, while the least value was scored by NB.

## Sentiment Analysis of Non-assessment Related Instances

*Experiment5*
We used the 750 of non-assessment related (NASS) instances to build a model that automatically detect their sentiment. SVM outperforms all other classifiers in terms of accuracy, precision, recall, and the F-score.

## 5.5    Comparison of Experiment 2, 3, 4, 5 Results

We observed that in all experiments there was no significant difference between using a uni-gram and bi-gram features. Although ASS and NASS are subsets of FDS, there is a notable margin between the accuracy values of FDS models and ASS models, but this did not apply to the NASS models as its accuracy value is approximated to the accuracy value of FDS. For example, the accuracy of SVM for FDS and NASS are nearly the same, while it is different from ASS by 7%. This applies to DT models, but not to RF and NB models.

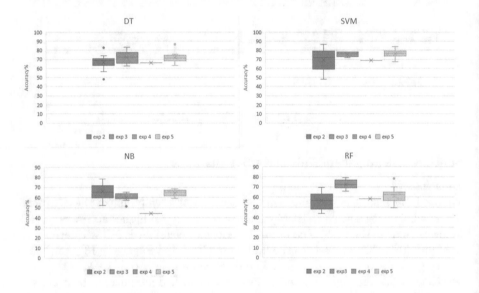

**Fig. 4.** Accuracy range for experiments: 2, 3, 4, 5

The accuracy values of all ASS models are very close to the accuracy values of assessment related instances under FDS models which means it is better to apply classifiers directly on topic domain data set.

# 6   Conclusion and Future Work

In this paper we examined the learning potential of four machine learning classifiers for learning classification and sentiment from students' textual feedback. We ran SVM, DT, NB, and RF classifiers on our data set and its subsets, the FDS was 979 instances, ASS subset was 229, and NASS subset was 750.

Our experiments showed no significant difference between using unigram and bigram features in building our models.

We evaluated each classifier performance. DT classifier was error free for Experiment 1, SVM outperformed DT, NB, and RF in all sentiment analysis experiments 2, 3, 4, 5. Figure 4 illustrates the accuracy range and average of all models, we found that the accuracy of general feedback data set machine learning models were higher than the accuracy of the assessment related machine learning models and nearly the same value of the non- assessment related machine learning models. The accuracy of assessment related machine learning models were approximated to the accuracy of the assessment related instances under the full data set machine learning models.

We used different parameters in each model, bias and gamma in SVM models, default probability in NB models, a static random seed and Gini index in RF models, and attribute selection and no pruning in DT models.

Future work includes more, analyzing of students' sentiment of assessment, recognizing the assessment issues, and using part of speech feature (POS) in building our models.

# References

1. Howsher, L., Chen, Y.: Student evaluation of teaching effectivness: an assessment of student perception and motivation. Assess. Eval. High. Educ. (2003)
2. Schmelkin, L.P., Spencer, K.J.: Student perspectives on teaching and its evaluation. Assess. Eval. High. Educ. **27**(5), 397–409 (2002)
3. Educational Data Mining deffenetion. Accessed 24 Oct 2017
4. Romero, C., Ventura, S.: Data mining in education. Wiley Interdisc. Rev. Data Min. Knowl. Discovery **3**(1), 12–27 (2013)
5. Ventura, S., Espejo, P.G., Hervas, C., Romero, C.: Data mining algorithms to classify students. In: Educational Data Mining 2008 - 1st International Conference on Educational Data Mining, Proceedings (2008)
6. Yadav, S.K., Bharadwaj, B., Pal, S.: Mining education data to predict student's retention: a comparative study. arXiv preprint arXiv:1203.2987 (2012)
7. Pal, S.: Mining educational data to reduce dropout rates of engineering students. Int. J. Inf. Eng. Electron. Bus. **4**(2), 1 (2012)
8. Albrecht, E., Grabowski, J.: Towards a framework for mining students' programming assignments, pp. 1096–1100 April 2016
9. Abbott, T., Abd-Elrahman, A., Andreu, M.: Using text data mining techniques for understanding free-style question answers in course evaluation forms. Res. High. Educ. J. **9**, 12–23 (2010)

10. Ersboll, B.K., Sliusarenko, T., Clemmensen, L.K.H.: Text mining in students' course evaluations relationships between open-ended comments and quantitative scores. In: Proceedings of the 5th International Conference on Computer Supported Education, pp. 564–573 (2013)
11. Pan, D., Tan, G.S.H., Ragupathi, K., Booluck, K., Roop, R., Roop, R., Ip, Y.K.: Profiling teacher/teaching using descriptors derived from qualitative feedback: formative and summative applications. Res. High. Educ. **50**(1), 73–100 (2009)
12. Jordan, D.W.: Re-thinking student written comments in course evaluation: text mining unstructured data for program and institutional assessment. Ph.D. thesis, California State University (2011)
13. Pallavi, P.: Recognizing student's problem using social media data. Int. J. Comput. Sci. Mob. Comput. **4**(6), 440–446 (2015)
14. Madhavan, K., Chen, X., Vornoreanu, M.: Mining social media data for understanding students' learning experiences. IEEE Trans. Learn. Technol. **7**(3), 246–259 (2014)
15. Krippendorff, K.: Reliability in content analysis. Hum. Comm. Res. **30**(3), 411–433 (2004)
16. Bracken, C.C., Lombard, M., Snyder-Duch, J.: Content analysis in mass communication: assessment and reporting of intercoder reliability. Hum. Comm. Res. **28**(4), 587–604 (2006)
17. Krippendorff, K.: Computing Krippendorff's Alpha-Reliability (2011)
18. Agarwal, A., Xie, B., Vovsha, I., Rambow, O., Passonneau, R.: Sentiment analysis of twitter data. In: Proceedings of the Workshop on Languages in Social Media, LSM 2011, Stroudsburg, PA, USA, pp. 30–38. Association for Computational Linguistics (2011)
19. Bhayani, R., Go, A., Huang, L.: Twitter Sentiment Classification using Distant Supervision (2017)
20. Mejova, Y.: Sentiment analysis: an overview. University of Iowa, Computer Science Department (2009)
21. de Groot, R.: Data mining for tweet sentiment classification. Master's thesis (2012)
22. Tian, F., Gao, P., Li, L., Zhang, W., Liang, H., Qian, Y., Zhao, R.: Recognizing and regulating e-learners' emotions basedon interactive chinese texts in e-learning systems. Knowl.-Based Syst. **55**, 148–164 (2014)
23. Pao, H.K., Lee, Y.J., Yeh, Y.R.: Introduction to support vector machines and their applications in bankruptcy prognosis (2012)
24. Lucini, F.R., Fogliatto, F.S., da Silveira, G.J.C., Neyeloff, J.L., Anzanello, M.J., Kuchenbecker, R.D.S., Schaan, B.D.: Text mining approach to predict hospital admissions using early medical records from the emergency department. Int. J. Med. Inf. **100**, 1–8 (2017)
25. Du, H.: Data mining techniques and applications. In: International Series of Monographs on Physics (2010)

# Dimension Reduction Based on Geometric Reasoning for Reducts

Naohiro Ishii[1]($\boxtimes$), Ippei Torii[1], Kazunori Iwata[2], Kazuya Odagiri[3], and Toyoshiro Nakashima[3]

[1] Aichi Institute of Technology, Toyota, Japan
{ishii,mac}@aitech.ac.jp
[2] Aichi University, Nagoya, Japan
kazunori@vega.aichi-u.ac.jp
[3] Sugiyama Jyogakuen University, Nagoya, Japan
{kodagiri,nakasima}@sugiyama-u.ac.jp

**Abstract.** Dimension reduction of data is an important problem and it is needed for the analysis of higher dimensional data in the application domain. Rough set is fundamental and useful to reduce higher dimensional data to lower one for the classification. We develop generation of reducts based on nearest neighbor relation for the classification. In this paper, the nearest neighbor relation is shown to play a fundamental role for the classification from the geometric reasoning. First, the nearest neighbor relation is characterized by the complexity order. Next, it is shown that reducts are characterized and generated based on the nearest neighbor relations based on the degenerate convex cones. Finally, the algebraic operations on the degenerate convex cones are developed for the generation of redŭcts.

**Keywords:** Reduct · Nearest neighbor relation · Characterization of reducts
Convex cones · Degenerate convex cones

## 1 Introduction

By Pawlak's rough set theory [1], a reduct is a minimal subset of features, which has the discernibility power as using the entire features, which shows the dimensionality reduction of features. Skowlon [2, 3] developed the reduct derivation by using the Boolean expression for the discernibility of data, which is a computationally complex task. So, a new concept for the efficient generation of reducts is expected from the point of classification with reduced dimensional data. We have developed a generation of reducts from useful data for the classification in which their partial data computes reducts without using all the data compared to the conventional methods [9, 10]. Nearest neighbor relation with minimal distance between different classes proposed here has a basic information for the dimensionality reduction. For the classification of data, a nearest neighbor method [4, 5, 8] is simple and effective one. In this paper, first, the nearest neighbor relation is characterized by the complexity order from the geometric reasoning. Next, a degenerate convex cones are constructed based on the nearest neighbor relations for reducts. It is shown that reducts are also generated based on the

© Springer Nature Switzerland AG 2019
A. Lotfi et al. (Eds.): UKCI 2018, AISC 840, pp. 351–361, 2019.
https://doi.org/10.1007/978-3-319-97982-3_29

degenerate convex cones. Finally, using nearest neighbor relation, algebraic operations are derived on the degenerate convex cones.

## 2   Nearest Neighbor Relation with Minimal Distance

Nearest neighbor relation with minimal distance is introduced here. The relation with minimal distance plays an important role in the generation of reducts.

**Definition 2.** A nearest neighbor relation with minimal distance is a set of pair of instances, which are described in

$$\{(x_i, x_j) : d(x_i) \neq d(x_j) \wedge |x_i - x_j| \leq \delta\} \tag{1}$$

where $|x_i - x_j|$ shows the distance between $x_i$ and $x_j$ and $\delta$ is the minimal distance.

Then, $x_i$ and $x_j$ in the Eq. (1) are called to be in the nearest neighbor relation with minimal distance $\delta$.

### 2.1   Estimation of Number of Nearest Neighbor Relations

The estimation of nearest neighbor relations is computed as follows. Nearest neighbor relation shows pairs of closest instances in different classes. In Fig. 1, we assume that the data $m_1 (\bullet)$ makes nearest neighbor relations with data ($\times$), where the total data is 7. Then, there are 5 nearest neighbor relations between two partial spherical portions, under the condition of the definition of the nearest neighbor relation.

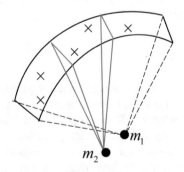

**Fig. 1.** Nearest neighbor relations generated from $\bullet$ data $m_1$ and $m_2$

Further, we assume that the data $m_2(\bullet)$ generates 2 nearest neighbor relations, since the center data $m_2$ is different from $m_1$ as shown in Fig. 1. Thus, the total number of the nearest neighbor relations become 7 ($< n = 9$ relations) + 2($< n$). The complexity order of the nearest neighbor relations becomes $O(n)$. Next, when the data are increased as shown in Fig. 2, the data $\bullet$ becomes $m_1, m_2 \ldots m_s$. Then, assume here the data $m_1$ generates $l_1$ nearest neighbor relations with the data $\times$. Also, the data $m_2$ makes $l_2$ nearest neighbor relation. Finally, the data $m_s$ makes $l_s$ relations. Since the total number of data is $n$,

$$l_1 \leq n \ , \ l_2 \leq n, \ \ldots l_s \leq n \qquad (2)$$

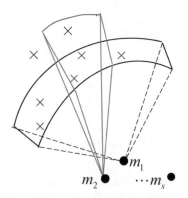

**Fig. 2.** Nearest neighbor relations generated from $\bullet$ data, $m_1, m_2, \ldots, m_s$

The complexity order of $l_1$ becomes $O(n)$. Similarly, that of $l_2$ becomes $O(n)$.

Thus, $l_1 + l_2 + \ldots + l_s \leq sn$ shows that the order of all the nearest neighbor relations becomes $O(n)$.

Thus, the number of the nearest neighbor relations is estimated in the complexity order $O(n)$.

To find minimal nearest neighbor relations, the divide and conquer algorithm is applied, which is well known to search the closest pair [7].

A schematic figure of the search for closest pair of the set (data) is shown in Fig. 3. The closest pair of points is derived as follows. Divide the set (data instances) into equal sized parts by the line $l$ and recursively compute the minimal distance in each part, where $d$ is the minimal of the two minimal distances [7]. Sort the remaining points according to their $y$ coordinates. Merge the two sorted lists into one sorted list. These dividing steps takes $O(\log n)$. Eliminate the points that lie farther than $d$ apart from $l$ and $n$ is the total number of data. Thus, the divide and conquer algorithm derives $O(n \log n)$ for the operations to search the nearest neighbor relations.

**Theorem 1.** When the number of the instances is $n$, the number of the nearest neighbor relations is estimated to be $O(n)$ and the operations finding the nearest neighbor relations are estimated to be $O(n \log n)$ in the complexity order.

## 2.2 Reducts Derived from Boolean Reasoning

As the dimensionality reduction, we adopt the reduct, which is developed from rough set theory. Decision table is necessary for the representation between data and its

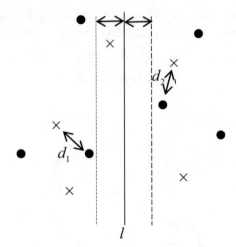

**Fig. 3.** Search of nearest neighbor relations among data

attributes. An example of the decision table is shown in Table 1. The reduct of the decision table is derived based on the Boolean reasoning, which is developed by Pawlak, Skowron and Rauszer [1, 2]. Though the Boolean reasoning method to esti-mate reducts is fundamental, the relations between instances and attributes are not sufficiently interpreted for the dimension reduction. In this paper, nearest neighbor relations is introduced to develop the geometrical reasoning for reducts.

**Table 1.** Decision table of data example (instances)

| Attribute | a | b | c | d | Class |
|---|---|---|---|---|---|
| $x_1$ | 1 | 0 | 2 | 1 | + 1 |
| $x_2$ | 1 | 0 | 2 | 0 | +1 |
| $x_3$ | 2 | 2 | 0 | 0 | −1 |
| $x_4$ | 1 | 2 | 2 | 1 | −1 |
| $x_5$ | 2 | 1 | 0 | 1 | −1 |
| $x_6$ | 2 | 1 | 1 | 0 | +1 |
| $x_7$ | 2 | 1 | 2 | 1 | −1 |

The left side data in the column in Table 1 as shown in, $\{x_1, x_2, x_3, .., x_7\}$ is a set of instances, while the data $\{a, b, c, d\}$ on the upper row, shows the set of attributes of the instance. The gray shaded data in Table 1 shows data of nearest neighbor relations. In the relation $(x_1, x_7)$ in Table 1, the $x_1$ in the class +1 is nearest to the is $x_7$ in the class −1. Similarly, $(x_5, x_6)$ and $(x_6, x_7)$ are nearest relations.

In Table 2, the discernibility matrix of the decision table in Table 1 is shown. In case of instance $x_1$, the value of the attribute a, is $a(x_1) = 1$. That of the attribute b, is $b(x_1) = 0$. Since $a(x_1) = 1$ and $a(x_5) = 2$, $a(x_1) \neq a(x_5)$ holds. In Table 2, Boolean variables of the nearest neighbor relations are shown in the gray elements, which are derived in Table 1. The Boolean product of these four terms in Table 2 becomes [8, 9]

$$(a+b) \cdot (b+c) \cdot (c+d) = b \cdot c + b \cdot d + a \cdot c \tag{3}$$

which becomes a candidate of reducts.

**Table 2.** Discernibility matrix of the decision table in Table 1

|       | $x_1$        | $x_2$        | $x_3$ | $x_4$        | $x_5$ | $x_6$ |
|-------|--------------|--------------|-------|--------------|-------|-------|
| $x_2$ | -            |              |       |              |       |       |
| $x_3$ | a, b, c, d   | a, b, c      |       |              |       |       |
| $x_4$ | b            | b, d         | -     |              |       |       |
| $x_5$ | a, b, c      | a, b, c, d   | -     | -            |       |       |
| $x_6$ | -            | -            | b, c  | a, b, c, d   | c, d  |       |
| $x_7$ | a, b         | a, b, d      | -     | -            | -     | c, d  |

The attributes pair $\{a, c\}$ is found between $x_1$ and $x_4$, also between $x_2$ and $x_4$ in the indiscernibility matrix [9]. The minterm $a \cdot c$ cannot discriminate instances between $x_1$ and $x_4$, also between $x_2$ and $x_4$. Then, the minterm $a \cdot c$ is removed from the Eq. (3). Thus, the reducts

$$b \cdot c + b \cdot d \tag{4}$$

is obtained. The generation of reducts based on the discernibility and the indiscernibility matrices are summarized in the following theorems [9, 10].

**Theorem 2.** Reducts are generated based on the nearest neighbor relations on the discernibility matrix as the necessary conditions of the Boolean forms, which are refined on the indiscernibility matrix as the final reducts.

We assume here the variable {b, c} has not the same values for the different classes.

**Lemma 3.** If only the variable $\xi \in \{a, d\}$ without the variable {b, c} exists with their different values and different classes for the element of in Table 2, {b, c} does not become a reduct.

**Theorem 4.** If both the variable b and c exist with their different values and different classes for the elements in the indiscernibility matrix of Table 2, {b, c} becomes a reduct.

**Corollary 5.** If {b, c} is a reduct, the variable b and c exist with their different values and different classes for the elements in the indiscernibility matrix of Table 2.

## 2.3    Boolean Products Based on Nearest Neighbor Relations

In the previous Sect. 2.2, Boolean products on the indiscernibility matrix is adopted. Since the Boolean products are performed among all the elements in the indiscernibility matrix, the complexity order of the products becomes $O(n^2)$, where $n$ is the number of instances in the matrix. Also, in the Sect. 2.1, the Boolean products on the nearest neighbor relations are performed. Then, the products on the nearest neighbor relations becomes $O(n)$, which is reduced from the $O(n^2)$ without using the nearest neighbor relations.

**Theorem 6.** The total number of Boolean products for reducts on the indiscernibility matrix with $n$ instances become $O(n^2)$, while that of Boolean products on the matrix becomes $O(n)$.

# 3    Geometrical Reasoning by Convex Cones

The geometrical reasoning approach for the generation of reducts is developed by generating convex cones.

## 3.1    Construction of Convex Cone by Nearest Neighbor Relations

The simple convex cones are shown as a point, a line convex cone and a triangular convex cone. The subset $\Omega$ of the vector space V is a convex cone if all vectors

$$\alpha X + \beta Y \tag{5}$$

belongs to $\Omega$, for any positive scalars $\alpha$, $\beta$ and any vector $X$, $Y$ in $\Omega$ [6, 7]. The Eq. (5) is extended to make a convex cone with a higher dimension space. Data set is decomposed to triangular cones, lines and points. To make these convex cones, a reference point is important. The reference point to make a vector from data is assigned to the data in the different class to make the nearest neighbor relation data.

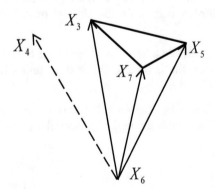

**Fig. 4.** Triangular convex cone by data $\{X_3, X_5, X_7\}$ with nearest neighbor relation

We use capital letter notations $\{X_i\}$ for convex cones instead of small letters $\{x_i\}$ for instances in Table 1. The triangular convex cone is made from the nearest neighbor relation data in Table 1 as shown in Fig. 4. The reference data $X_6$ is assigned, which is the data in the class +1.

The data $\{X_3, X_5, X_7\}$ with nearest neighbor relation in Table 1 generates a triangular convex cone in Fig. 4. This is interpreted as follows. The data $\{X_3, X_5, X_7\}$ exists in the class $-1$, while the data $\{X_6\}$ is in the class +1 as the reference point. The vectors $\vec{X}_3, \vec{X}_5$ and $\vec{X}_7$ are generated as $\vec{X}_3 = X_3 - X_6$, $\vec{X}_5 = X_5 - X_6$ and $\vec{X}_7 = X_7 - X_6$. Any inner vector $\vec{Y}$ in the triangular convex cone in Fig. 4 is indicated as

$$\vec{Y} = \alpha\vec{X}_3 + \beta\vec{X}_5 + \gamma\vec{X}_7 \tag{6}$$

where $\alpha, \beta, \gamma \geq 0$ hold by convex condition. In the vectors in Eq. (6), By removing the reference point data, from vectors in Eq. (6), the following equation holds,

$$Y = \alpha X_3 + \beta X_5 + \gamma X_7 \tag{7}$$

By applying the Eq. (9), $X_4$ does not satisfy the Eq. (9), that is,

$$X_4 \neq \alpha X_3 + \beta X_5 + \gamma X_7 \tag{8}$$

Then, the data $X_4$ constructs other convex cone as in Fig. 4. Next, the reference data in the class +1 is changed to the reference data $X_1$ from that of $X_6$ in Fig. 4. The Boolean variables derived from the nearest neighbor relations $(X_6, X_3), (X_6, X_5)$ and $(X_6, X_7)$ on the triangular convex cone in Fig. 4 are given as Boolean variables $\{b, c\}$, $\{c, d\}$ and $\{a, b\}$, respectively.

### 3.2    Degenerate Convex Cones by Nearest Neighbor Relations

Degenerate convex cone is defined by the nearest neighbor relations. As shown in Fig. 4, the convex is made of the reference data $X_6$ and data $X_3$, $X_5$, $X_7$. From these data, the union of Boolean variables $\{b, c\}$ and $\{c, d\}$ is made, which consists of three variables $(b, c, d)$. Then, a degenerate convex cone is made from three variables $(b, c, d)$, which is removed the variable $(a)$ as shown in Fig. 5. Then, the degenerate convex cone in Fig. 5 consists of $X'_3, X'_5, X'_7$ and $X'_6$, which have components, Boolean variables $(b, c, d)$.

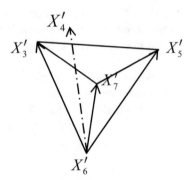

**Fig. 5.** Degenerate convex cone generated from $X_3'$, $X_5', X_7'$ and $X_6'$

### 3.3    Dependence Relations on the Degenerate Convex Cones

Independent vectors of nearest neighbor relations, which consist of the degenerate convex, are useful for the operations of dimensional reduction. Independent vectors in

$$\vec{X}_3' = X_3' - X_6', \vec{X}_5' = X_5' - X_6' \text{ and } \vec{X}_7' = X_7' - X_6' \tag{9}$$

are defined. The dependent relation of these independent vectors removes the item on the indiscernibility matrix.

Since $X_6' = (110)$, $X_3' = (200)$, $X_5' = (101)$, $X_7' = (121)$ and $X_4' = (221)$, the following linear equation is checked whether the $(X_4 - X_6)'$ is dependent to independent vectors $\vec{X}_3'$, $\vec{X}_5'$ and $\vec{X}_7'$.

$$
\overset{(X_4' - X_6')}{\begin{pmatrix} 1 \\ 1 \\ 1 \end{pmatrix}} = \alpha \overset{\vec{X}_3'}{\begin{pmatrix} 1 \\ -1 \\ 0 \end{pmatrix}} + \beta \overset{\vec{X}_5'}{\begin{pmatrix} 0 \\ -1 \\ 1 \end{pmatrix}} + \gamma \overset{\vec{X}_7'}{\begin{pmatrix} 0 \\ 1 \\ 1 \end{pmatrix}} \tag{10}
$$

Then, $\alpha = 1$, $\beta = -(1/2)$ and $\gamma = (3/2)$ hold for the Eq. (10), which shows the $(X_4' - X_6')$ to be dependent to $\vec{X}_3'$, $\vec{X}_5'$ and $\vec{X}_7'$. This shows the difference Boolean term $(X_4 - X_6)$ is removed, since the difference Boolean variables by the nearest neighbor relations, are absorbed by $\{b, c\}$ and $\{c, d\}$. Generally, the following theorem holds.

**Theorem 7.** If a linear equation of the difference Boolean term holds in the degenerate convex made of the nearest neighbor relations, the difference Boolean term is removed by the absorption of Boolean terms of the nearest neighbor relations.

### 3.4    Algebraic Operations Using Degenerate Convex Cones

Boolean absorption is realized by the algebraic operations on the degenerate convex cones. The algebraic operation is made of inequality equations based on the nearest neighbor relations. The inequality shows whether the data difference information between the different classes is unequal to that between the same class.

The data difference equations except the nearest neighbor relations are shown in the case of Fig. 5,

$$(X'_i - X'_j)_k \neq 0 \text{ for } i = 3, 4, 5, j = 1, 2 \text{ and } k = b, c, d \qquad (11)$$

If the Eqs. (11) hold for both $k = b$ and $k = c$, the element $(x_i, x_j)$ in Table 2, is removed by the equality equation of $(X_i = X_j)_{k=b,k=c}$, since it is absorbed from the nearest neighbor relation $\{b, c\}$.

Similarly, if the Eqs. (11) holds for both $k = b$ and $k = c$, the element $(x_i, x_j)$ is removed, since it is absorbed from the nearest neighbor relation $\{c, d\}$.

This is interpreted in the following. The left side of the Eq. (11) is replaced to the following equation using the nearest neighbor relation in Fig. 5,

$$(X'_i - X'_j)_k = (X'_i - X'_6)_k - (X'_j - X'_6)_k \qquad (12)$$

From the Eq. (12), the Eq. (11) becomes

$$(X'_i - X'_6)_k \neq (X'_j - X'_6)_k \qquad (13)$$

The left and right sides of the Eq. (13) is interpreted in two different convex cones as shown in Fig. 6. The left cone is constructed from the different classes, while the right cone is constructed from the same class. Thus, the algebraic operation in the Eq. (13) compares the values between the left convex cone and the right one in Fig. 6.

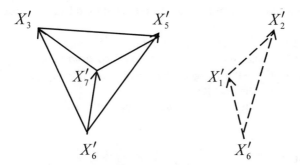

**Fig. 6.** Convex cone (left) for the different classes and cone (right) for the same class

The left side of the Eq. (13) is the difference data between data in the different classes of the nearest neighbor relations, while the right side of the Eq. (13) is the difference data between data in the same class. For $i = 3, j = 1$ and $k = b, c$.
$(X'_3 - X'_6)_b \neq (X'_1 - X'_6)_b$, since $+1 \neq -1$ and $(X'_3 - X'_6)_c \neq (X'_1 - X'_6)_c$ since $-1 \neq +1$.

Thus, the Boolean variables of the cross term $(x_i, x_j) = (x_3, x_1)$ is removed by the absorption of the nearest neighbor relation. Similarly, $(x_i, x_j) = (x_5, x_1)$, $(x_3, x_2)$, $(x_5, x_2)$ and $(x_7, x_2)$ are removed.

These algebraic operations are generally described as the following theorem.

**Theorem 8.** The item $(x_i, x_j)$ with Boolean variables on the Table 2 are removed by the nearest relations with reference data $X'_6$, if the following equations hold in Fig. 6

$$(X'_i - X'_6)_k \neq (X'_j - X'_6)_k, \text{ for } i = 3, 4, 5, j = 1, 2 \text{ and } k = b, c, d \qquad (14)$$

By Theorem 8, when an example of $(X'_3 - X'_1)$, which is in the case of $i = 3$, $j = 1$, is developed as follows,

$$(X'_3 - X'_1) = (X'_3 - X'_6) - (X'_1 - X'_6)$$

Since $(X'_3 - X'_6)_b = +1$ and $(X'_1 - X'_6)_b = -1$ hold, $(X'_3 - X'_6)_b \neq (X'_1 - X'_6)_b$.

Similarly, $(X'_3 - X'_6)_c \neq (X'_1 - X'_6)_c$ and $(X'_3 - X'_6)_d \neq (X'_1 - X'_6)_d$ hold. Thus, The item $(X_3, X_1)$ with Boolean variables $(b, c, d)$ is removed. Similar operations $i = 3, 4, 5$ and $j = 1, 2$ are carried out.

Theorem 8 does not hold for $(X_4, X_1)$, which satisfies the Eq. (14) only for $k = b$. Then, a new Boolean variable $(b)$ is derived except the nearest neighbor relations. By the nearest neighbor relations, the Boolean variable terms $(b + c)$, $(c + d)$ and $(a + b)$ are obtained. By the product of these Boolean terms, reducts $\{bc, bd\}$ are obtained.

The Boolean construction steps based on the geometric reasoning of the nearest neighbor relations are as follows.

1  Convex cone which is based on the nearest neighbor relation is made with reference data point.
2  Boolean variables of nearest neighbor relations is computed from the convex cone. Then, the degenerate convex cones are generated from these variables.
3  Algebraic operations of the degenerate convex cone is computed, which is followed by the removal of Boolean terms.
4  The remained extracted Boolean variables construct reducts.

# 4  Conclusion

In this paper, the dimensionality reduction for reducts is realized based on the geometrical reasoning. We propose a method of the nearest neighbor relations for the data classification which are characterized by the convex cones. First, the number of the nerest neighbor relations are estimated by the complexity order, which shows the number of the Boolean pruducts. Next, it is shown that the nearst neighbor relations generate convex cones. The degenerate convex cones are defined to derive reducts. The algebraic operations for reducts are performed on the degenerate convex cones. Thus, the geometric reasoning for dimensionality reduction for reducts show useful information for the classification and the dimension reduction.

# References

1. Pawlak, Z.: Rough sets. Int. J. Comput. Inf. Sci. **11**, 341–356 (1982)
2. Skowron, A., Rauszer, C.: The discernibility matrices and functions in information systems. In: Intelligent Decision Support - Handbook of Application and Advances of Rough Sets Theory, pp. 331–362. Kluwer Academic Publishers, Dordrecht (1992)
3. Skowron, A., Polkowski, L.: Decision algorithms, a survey of rough set theoretic methods. Fundamenta Informatica **30**(3–4), 345–358 (1997)
4. Cover, T.M., Hart, P.E.: Nearest neighbor pattern classification. IEEE Trans. Inf. Theor. **13** (1), 21–27 (1967)
5. Preparata, F.P., Shamos, M.I.: Computational Geometry. Springer (1993)
6. Prenowitz, W., Jantosciak, J.: Join Geometries, A Theory of Convex Sets and Linear Geometry. Springer (2013)
7. Levitin, A.: Introduction to the Design & Analysis of Algorithms, 3rd edn. Person Publication (2012)
8. Ishii, N., Torii, I., Mukai, N., Iwata, K., Nakashima, T.: Classification on nonlinear mapping of reducts based on nearest neighbor relation. In: Proceedings ACIS-ICIS IEEE Computer Society, pp. 491–496 (2015)
9. Ishii, N., Torii, I., Iwata, K., Nakashima, T.: Generation and nonlinear mapping of reducts-nearest neighbor classification. In: Advances in Combining Intelligent Methods, pp. 93–108. Springer (2017). Chapter 5
10. Ishii, N., Torii, I., Iwata, K., Odagiri, K., Nakashima, T.: Generation of reducts based on nearest neighbor relations and boolean reasoning. In: HAIS2017, LNCS, vol. 10334, pp. 391–401. Springer (2017)

# Anomaly Detection in Activities of Daily Living Using One-Class Support Vector Machine

Salisu Wada Yahaya[✉], Caroline Langensiepen, and Ahmad Lotfi

School of Science and Technology,
Nottingham Trent University, Nottingham NG11 8NS, UK
salisu.yahaya2015@my.ntu.ac.uk,
{caroline.langensiepen,ahmad.lotfi}@ntu.ac.uk

**Abstract.** Different computational methodologies for anomaly detection has been studied in the past. Novelty detection involves classifying if test data differs from the training data. This is applicable to a scenario when there are sufficiently many normal training samples and little or no abnormal data. In this research, a novelty detection algorithm known as One-Class Support Vector Machine (SVM) is applied for detection of anomaly in Activities of Daily Living (ADL), specifically sleeping patterns, which could be a sign of Mild Cognitive Impairment (MCI) in older adults or other health-related issues. Tests conducted on both synthetic and real data shows promising results.

**Keywords:** Novelty detection · Anomaly detection · One-class SVM
Activities of Daily Living (ADL)

## 1 Introduction

Anomaly detection in ADL has been receiving much attention. However, obtaining labeled anomalous data is nearly impossible. The most common approach to anomaly detection in ADL is finding a deviation from a person's usual behaviour routine [1].

In this paper, we aim to detect anomalies in ADL, specifically the sleeping routine. An unhealthy sleeping pattern is detrimental to well-being [2] and may be an early sign of MCI and other health-related issues in old adults [1,2]. As the population of older adults increases [3], identifying this anomaly will improve their well-being and quality of life.

Sleeping habit (start time, end time, duration, wake cycles etc.) varies from one individual to another. Our approach is to model a person's sleeping habit and compare subsequent occurrences to the built model in order to detect deviation (anomaly).

To achieve this, we propose using a novelty detection algorithm - also known as one-class classification. Since the available training data are considered normal

© Springer Nature Switzerland AG 2019
A. Lotfi et al. (Eds.): UKCI 2018, AISC 840, pp. 362–371, 2019.
https://doi.org/10.1007/978-3-319-97982-3_30

and outlier free, and anomalous data is not available during training, one-class classification will be well suited for this scenario.

One-class classification (novelty detection) has to do with recognizing if a test data point differs from the data available during training. This is mostly applicable when there is a sufficiently large normal training sample and little or no abnormal data. This has several real-life applications in the areas of medical diagnosis, industrial processes, finance, sensor network, and IT security [4].

The ideas is, given a set of $n$ training samples $X = \{x_1, ..., x_n\}$ of $d$-dimensional data (i.e. $X \in R^d$), find if a set of $m$ testing samples $X^* = \{x_1^*, ..., x_m^*\}$ of the same dimension (i.e. $X^* \in R^d$) has similar characteristics as X. This is achieved by constructing a model of normality $M_x$ of the training set $X$ to be tested against the test set $X^*$.

This paper is organized as follows: Sect. 2 reviews existing literature in anomaly and novelty detection. Section 3 describes the methodology of this research. Section 4 reports the experimental results, and Sect. 5 contains the conclusion and future plans.

# 2   Related Works

Detecting an anomaly in user's ADL by finding deviations from their usual ADL pattern has been conducted using Random Forest (RF) in [5], Convolutional Neural Network (CNN) in [6], HMM in [7], and different variants of RNN in [1].

A probabilistic approach for novelty detection has to do with estimating the probability density function of the data distribution; regions with higher density indicate normal data, while regions with lower density signify novelty [4]. The parametric approach assumes the data has a certain probability distribution. This is problematic in real-life applications where knowledge of the data distribution is not known a priori. Non-parametric approaches such as the Parzen Window have been proposed such that the data distribution is estimated from the training sample. A probabilistic approach has been applied in [8] for detecting abnormality in jet engine vibration, network intrusion in [9], and to identify cancerous mass in mammographic images [10]. A variant of GMM novelty detection with confidence value has been proposed for selecting suitable density thresholds in [11] and "SmartSifter" based on statistical learning for outlier detection in [12].

Approaches based on the distance between data points such as nearest neighbours are studied in [13]. This approach involves searching for the nearest neighbour of a data point and classifying the data point as normal if the neighbour is close or as an outlier if the nearest neighbour is far. This is computationally expensive, especially when finding the nearest neighbour in a high dimensional space [4]. This has been applied for detection of disease outbreak using social media streams [14], and detection and removal of outliers in audio streams for improved speaker recognition [15].

K-Means clustering has been applied for detection of faulty building equipment (e.g. heating, air conditioning) [16], while an ensemble of novelty detection

364     S. W. Yahaya et al.

algorithms including k-means clustering is applied for the detection of faults in semiconductors production [17]. Principle component analysis (PCA) and its non-linear extension known as kernel PCA are applied for novelty detection in [18,19] and Self-Organizing Map (SOM) in [19].

Neural Networks (NN) are used for novelty detection in image sequence analysis [20] and in combination with Support Vector Data Description (SVDD) for filtering data for adaptive learning systems [21]. Augusteijn and Folkert [22] however, argue that back-propagation NN is not suitable for novelty detection and assert that Probabilistic Neural Network (PNN) has better performance. This assertion is supported by Yadav and Devi [23] and uses PNN on different data sets with good novelty detection accuracy.

The one-class SVM proposed by Scholkopf et al. has been used to detect outliers in medical diagnosis of Melano Prognosis of over 270 patients with a ROC curve of 0.71 [24]. Syed et al. [25] use one-class SVM and other novelty detection algorithms on medical data to identify high-risk patients after surgery. Similarly, one-class SVM is applied for the detection of seizure in humans using electroencephalogram (EEG) data [26], for vehicle diagnosis [27], and for time series novelty detection [28].

## 3   Methodology

### 3.1   Data Description

Both synthetic and real data are used for this research. The real data is obtained from the UCI Machine Learning repository. It contains the daily ADL of two (2) individuals living in their separate homes. 14 days ADL data is available for the first individual while 21 days ADL data is available for the second individual. The activities recorded include eating, sleeping, toileting, watching TV etc. Each activity has a start time, end time and the activity type as shown in Table 1. For this research, the data is filtered and only sleeping ADL is utilized.

**Table 1.** Sample ADL data

| Start time | End time | Activity |
|---|---|---|
| 2011-11-28 02:27:59 | 2011-11-28 10:18:11 | Sleeping |
| 2011-11-28 10:21:24 | 2011-11-28 10:23:36 | Toileting |
| 2011-11-28 10:25:44 | 2011-11-28 10:33:00 | Showering |
| ... | ... | ... |

The synthetic data is generated in a similar format as the real data for two individuals. However, different start time, end time, and duration are used in order to achieve generalized testing across the various data. The duration of the synthetic data is within the range of the lowest and highest value of the real data. The data is generated for a 99 day period for the two (2) individuals. For easy reference, the real data will be referred to as R1 and R2, while synthetic data will be referred to as S1 and S2 as shown in Table 2.

**Table 2.** Experiment data set

| Data type | Size of data (Days) | Assigned ID |
|---|---|---|
| Real data for subject 1 | 14 | R1 |
| Real data for subject 2 | 21 | R2 |
| Synthetic data 1 | 99 | S1 |
| Synthetic data 2 | 99 | S2 |

Figure 1 shows the combined 3D scatter plot of both the synthetic and real data. It can be seen that there are four (4) separate data groups with some overlap.

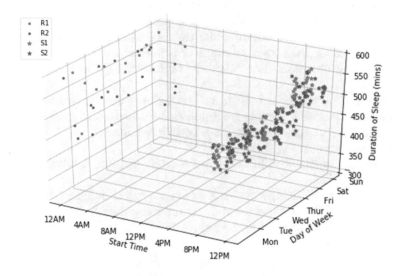

**Fig. 1.** 3D scatter plot of sleeping activity

### 3.2 Data Preprocessing

Feature selection is a crucial aspect of machine learning. It is necessary to select the features with the ability to discriminate between normal and abnormal pattern. The extracted features are:

1. The duration of the activity - by subtracting the start time from the end time.

2. The starting hour and minute.
3. The day of the activity.
4. And if the activity falls on weekday or weekend.

Machine Learning algorithms can be sensitive to data scaling (especially algorithms based on SVM); extracted features at different scales can easily mislead the classifier and may result in wrong predictions. In order to normalize the different features and convert them to a similar scale, Min-Max Scaler is applied to the selected features.

$$x' = \frac{x - x_{min}}{x_{max} - x_{min}} \tag{1}$$

where $x_{min}$ is the minimal data value, $x_{max}$ is the maximal data value and $x$ is the data point to be scaled.

### 3.3   One-Class SVM

One class SVM is an unsupervised novelty detection algorithm. Unlike unsupervised outlier detection algorithms where the training data contains both normal and anomalous samples, One-Class SVM assumes that the training data contains no or very few outliers. This is mostly applicable in a situation where abnormal data is hard to obtain while training data is readily available. For example, obtaining the sensor readings of a normal working machine is easy while obtaining abnormal data may involve sabotaging the machine in every possible way.

A model of normality is built using the training data (presumably outliers free) and subsequent data points are tested against the trained model. If the data falls within the region of normality, it is considered normal; else, it is marked as an outlier as shown in Fig. 2.

This section briefly introduces the one-class SVM. More detailed explanation can be found in [29].

Given a set of $n$ training samples $X = \{x_1, ..., x_n\}$ of $d$-dimensional data (i.e. $X \in R^d$), let $\Phi : R^d \rightarrow F$ be a non-linear mapping from a data space $R^d$ to a feature space $F$, The support vector separating the data is computed as:

$$\min_{w \in F, \xi \in R^d, \rho \in R} \frac{1}{2}||w||^2 + \frac{1}{vn} \sum_{i=1}^{n} \xi_i - \rho \tag{2}$$

subject to

$$(w \cdot \Phi(x_i)) \geq \rho - \xi_i, \quad \xi_i \geq 0, \quad i = 1, ..., n$$

where $n$ is the number of training samples, $v \in (0,1)$ is a trade-off parameter characterizing the expected fraction of outliers, $\xi$ is a slack variable, $w$ and $\rho$ are the hyper-plane parameters in the feature space $F$ with $\rho$ being the distance to the origin and $w$ as a parameter of the hyper-plane.

Since the training data for this research are considered outlier free, a small value (i.e. 0.01) is used for $v$ which is obtained using a grid search.

Solving the above equation, the decision function $f(x)$ which determines if $x_i$ is an outlier is obtained as:

$$f(x) = \sum_{i=1}^{n} \alpha_i k(x_i, x) - \rho \tag{3}$$

$$f(x) = \begin{cases} +1 & x \quad is \quad Inlier \\ -1 & x \quad is \quad Outlier \end{cases}$$

where $\alpha_i$ is a Lagrange multiplier for each vector $x_i$ and the kernel function for the non-linear mapping $\Phi$ is $k(x_i, x) = \Phi(x_i) \cdot \Phi(x)$

We applied the widely used kernel function for One-Class SVM which is Radial Basis Function (RBF):

$$K(x_i, x) = exp\left(-\gamma ||x_i - x||\right) \tag{4}$$

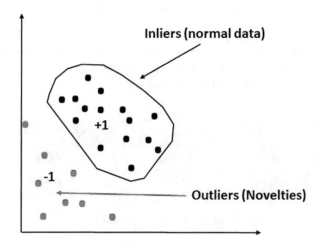

Fig. 2. One-Class SVM

## 4    Experiment

The experiment is conducted on both the synthetic and real data shown in Fig. 1 to compare the novelty detection performance of the model on ADL sleeping data.

### 4.1    Testing the Model on Synthetic Data (S1 and S2)

The One-Class SVM is tested on the two synthetic data sets S1 and S2. It is first trained with the synthetic data for the first individual (S1) and then tested against the synthetic data of the second individual (S2) and the two real data sets (R1 and R2).

Due to significant variation in the start time between the real and synthetic data, all the real data (R1 and R2) are identified as outliers. However, some of the synthetic data points for the second individual (S2) are identified as normal since the data overlaps with some data points of the first individual (S1).

Similarly, on training the model with the second synthetic data (S2), it identifies all the real data (R1 and R2) as outliers while some of the first synthetic data (S1) as inliers due to their overlap.

| S1 | Predicted Inliers | Predicted Outliers |
|---|---|---|
| Actual Inliers | 94.31% | 5.69% |
| Actual Outliers | 2.68% | 97.32% |

Fig. 3. Result for synthetic data 1 (S1)

Figure 3 shows the result obtained when the model is trained on the synthetic data of the first individual (S1). It can be seen that the model achieved above 90% accuracy in detecting outliers and inliers.

| S2 | Predicted Inliers | Predicted Outliers |
|---|---|---|
| Actual Inliers | 95.36% | 4.64% |
| Actual Outliers | 9.76% | 90.24% |

Fig. 4. Result for synthetic data 2 (S2)

Figure 4 shows the result obtained when the model is trained on the synthetic data of the second individual (S2). An accuracy of above 90% is also achieved.

## 4.2 Testing the Model on Real Data (R1 and R2)

The model is first trained on the real data for the first individual (R1) and tested against the real data of the second individual (R2) and the two synthetic data sets (S1 and S2).

It is then trained on the second real data (R2) and tested against the first real data (R1) and the two synthetic data sets (S1 and S2).

In both cases, the synthetic data (S1 and S2) are both detected as outliers because of their significant variation from the real data. The real data (R1 and R2) have overlapping data points and therefore some of the data points of R2

| R1 | Predicted Inliers | Predicted Outliers |
|---|---|---|
| Actual Inliers | 76.19% | 23.81% |
| Actual Outliers | 0% | 100% |

**Fig. 5.** Result for real data 1 (R1)

are classified as normal when the model is trained on R1 data and vice versa. Figures 5 and 6 show the obtained results.

Figure 5 shows the result obtained when the model is trained on the real data for the first individual (R1). The result shows that the model achieved 100% accuracy in detecting outliers. However, some data points that are inliers are misclassified as outliers thereby result in a lower accuracy of about 76% for inliers detection.

| R2 | Predicted Inliers | Predicted Outliers |
|---|---|---|
| Actual Inliers | 77.24% | 22.58% |
| Actual Outliers | 0% | 100% |

**Fig. 6.** Result for real data 2 (R2)

Figure 6 shows the result obtained when the model is trained on the real data for the second individual (R2). 100% accuracy is achieved for outlier detection while 77% accuracy is achieved for inlier detection.

In comparison, the model performs better on the synthetic data (S1 and S2) than on the real data (R1 and R2). This cannot be unconnected to the insufficient amount of real data which may affect the model's ability to build an accurate decision boundary.

## 5 Conclusion

In this paper, we have shown the potential of One-Class SVM in the detection of anomaly in ADL, specifically in sleeping patterns. Despite the inadequate amount of real data, the model is able to distinguish the data set it is trained on and classify the other data as outliers for both the synthetic and real data with promising results.

The approach, however, lacks the ability to detect the parameter which qualifies the data as an outlier. Future work will address this by looking at trend analysis as well as application to a large collected data set. Raw EEG data collected from a wearable sleep monitoring sensor will also be considered as well as a self-progressive learning approach in which the novelty detection algorithm can learn behavioural changes and dynamically adjust its decision boundary while taking data ageing into consideration by giving more priority to new data points.

# References

1. Arifoglu, D., Bouchachia, A.: Activity recognition and abnormal behaviour detection with recurrent neural networks. Procedia Comput. Sci. (2017)
2. Borazio, M., Berlin, E., Kucukyildiz, N., Scholl, P., Van Laerhoven, K.: Towards benchmarked sleep detection with wrist-worn sensing units. In: 2014 IEEE International Conference on Healthcare Informatics (2014)
3. Chernbumroong, S., Cang, S., Atkins, A., Yu, H.: Elderly activities recognition and classification for applications in assisted living. Expert Syst. Appl. (2013)
4. Pimentel, M.A., Clifton, D.A., Clifton, L., Tarassenko, L.: A review of novelty detection. Sign. Process. (2014)
5. Lundstrom, J., De Morais, W.O., Cooney, M.: A holistic smart home demonstrator for anomaly detection and response. In: 2015 IEEE International Conference on Pervasive Computing and Communication Workshops (PerCom Workshops) (2015)
6. Tasfi, N.L., Higashino, W.A., Grolinger, K., Capretz, M.A.M.: Sampling for electrical anomaly detection deep neural networks with confidence sampling for electrical anomaly detection. In: 2017 IEEE International Conference on Internet of Things (iThings) and IEEE Green Computing and Communications (GreenCom) and IEEE Cyber, Physical and Social Computing (CPSCom) and IEEE Smart Data (SmartData) (2017)
7. Tonchev, K., Koleva, P., Manolova, A., Tsenov, G., Poulkov, V.: Non-intrusive sleep analyzer for real time detection of sleep anomalies. In: 2016 39th International Conference on Telecommunications and Signal Processing (TSP) (2016)
8. Nairac, A., Corbett-Clark, T.A., Ripley, R., Townsend, N.W., Tarassenko, L.: Choosing an appropriate model for novelty detection. In: Fifth International Conference on Artificial Neural Networks (1997)
9. Yeung, D.-Y., Chow, C.: Parzen-window network intrusion detectors. In: Object Recognition Supported by User Interaction for Service Robots (2002)
10. Tarassenko, L., Hayton, P., Brady, M., Cerneaz, N.: Novelty detection for the identification of masses in mammograms. In: 1995 Fourth International Conference on Artificial Neural Networks (1995)
11. Ilonen, J., Paalanen, P., Kamarainen, J.-K., Kälviäinen, H.: Gaussian mixture pdf in one-class classification: computing and utilizing confidence values. In: 18th International Conference on Pattern Recognition (ICPR 2006) (2006)
12. Yamanishi, K., Takeuchi, J.I., Williams, G., Milne, P.: On-line unsupervised outlier detection using finite mixtures with discounting learning algorithms. Data Min. Knowl. Discov. (2004)
13. Ghoting, A., Parthasarathy, S., Otey, M.E.: Fast mining of distance-based outliers in high-dimensional datasets. Data Min. Knowl. Discov. (2008)
14. Dai, X., Bikdash, M.: Distance-based outliers method for detecting disease outbreaks using social media. In: SoutheastCon (2016)
15. Ali, I., Saha, G.: A distance metric based outliers detection for robust automatic speaker recognition applications. In: 2011 Annual IEEE India Conference (2011)
16. Habib, U., Zucker, G., Blochle, M., Judex, F., Haase, J.: Outliers detection method using clustering in buildings data. In: IECON 2015 - 41st Annual Conference of the IEEE Industrial Electronics Society (2015)

17. Kim, D., Kang, P., Cho, S., Lee, H.J., Doh, S.: Machine learning-based novelty detection for faulty wafer detection in semiconductor manufacturing. Expert Syst. Appl. (2012)
18. Hoffmann, H.: Kernel PCA for novelty detection. Pattern Recogn. (2007)
19. Modenesi, A.P., Braga, A.P.: Analysis of time series novelty detection strategies for synthetic and real data. Neural Process. Lett. (2009)
20. Markou, M., Singh, S.: A neural network-based novelty detector for image sequence analysis. IEEE Trans. Pattern Anal. Mach. Intell. (2006)
21. Liu, Y., Cukic, B., Fuller, E.: Novelty detection for a neural network-based online adaptive system. In: 29th Annual International Computer Software and Applications Conference (2005)
22. Augusteijn, M.F., Folkert, B.A.: Neural network classification and novelty detection. Int. J. Remote Sens. (2010)
23. Yadav, B., Devi, V.S.: Novelty detection applied to the classification problem using probabilistic neural network. In: 2014 IEEE Symposium on Computational Intelligence and Data Mining (CIDM) (2014)
24. Dreiseitl, S., Osl, M., Scheibböck, C., Binder, M.: Outlier detection with one-class SVMs: an application to melanoma prognosis. In: Proceedings of the AMIA Annual Fall Symposium (2010)
25. Syed, Z., Saeed, M., Rubinfeld, I.: Identifying high-risk patients without labeled training data: anomaly detection methodologies to predict adverse outcomes. In: AMIA Annual Symposium Proceedings (2010)
26. Gardner, A., Krieger, A., Vachtsevanos, G., Litt, B.: One-class novelty detection for seizure analysis from intracranial EEG. J. Mach. Learn. Res. (2006)
27. Theissler, A.: Multi-class novelty detection in diagnostic trouble codes from repair shops. In: 2017 IEEE 15th International Conference on Industrial Informatics (INDIN) (2017)
28. Ma, J., Perkins, S.: Time-series novelty detection using one-class support vector machines. In: Proceedings of the International Joint Conference on Neural Networks (2003)
29. Schölkopf, B., Platt, J.C., Shawe-Taylor, J., Smola, A.J., Williamson, R.C.: Estimating the support of a high-dimensional distribution. Neural Comput. (2001)

# Towards Active Muscle Pattern Analysis for Dynamic Hand Motions via sEMG

Jiahan Li[1], Yinfeng Fang[2], Yongan Huang[5], Gongfa Li[1,3,4],
Zhaojie Ju[2(✉)], and Honghai Liu[2]

[1] Key Laboratory of Metallurgical Equipment and Control Technology,
Wuhan University of Science and Technology, Ministry of Education,
Wuhan 430080, China
[2] School of Computing, University of Portsmouth, Portsmouth PO1 3HE, UK
zhaojie.ju@port.ac.uk
[3] Research Center for Biomimetic Robot and Intelligent Measurement and
Control, Wuhan University of Science and Technology, Wuhan 430081, China
[4] Institute of Precision Manufacturing, Wuhan University of Science
and Technology, Wuhan 430081, China
[5] State Key Lab of Digital Manufacturing Equipment and Technology,
Huazhong University of Science and Technology Wuhan, Wuhan, China

**Abstract.** Surface Electromyographys (sEMG) as a widespread human-computer interaction method can reflect the activity of human muscles. When the human forearm finishes different hand motions, there will be strong sEMG signals in different regions of the skin surface. This paper investigates the mapping relationship between sEMG signal patterns and the dynamic hand motions. Four different hand motions are studied based on the extracted signal with mean absolute value (MAV) features and the shape-preserving piecewise cubic interpolation method. In the experiments, a 16-channel electrode sleeve is used to collect 9-subject EMG signals. According to the distribution of electrodes in the forearm, the forearm surface is divided into 8 different muscle regions. The preliminary experimental results show that different hand motions can cause different distribution of sEMG signals in different regions. It confirms that different subjects show similar patterns for the same motions. The experimental results can be applied as new sEMG features with a higher computational speed.

**Keywords:** sEMG · MAV · Shape-preserving piecewise cubic interpolation
Local maximum · Muscle regions

## 1 Introduction

sEMG has been widely used as the effective method of controlling prosthetic. sEMG-based hand movement classification and recognition are the main research direction. In order to get high classification rates, there are several methods and algorithms have been developed such as multichannel surface EMG system [1], adaptive directed acyclic graph [2], ANNs [3], neuro-fuzzy classifiers [4] and other methods.

© Springer Nature Switzerland AG 2019
A. Lotfi et al. (Eds.): UKCI 2018, AISC 840, pp. 372–382, 2019.
https://doi.org/10.1007/978-3-319-97982-3_31

sEMG is a kind of biological signal with complex noise. That is caused by inherent equipment and environmental noise, electromagnetic radiation, motion artifacts, and the interaction of different tissues [5]. Many algorithms and features are proposed to optimize hand motion-classification and recognition based on EMG signals. We adopted multiple features commonly used in previous studies such as mean absolute value, zero crossing, slope sign change, and waveform length, in the algorithm for extracting hand-posture features, and the k-nearest-neighbors (KNN) algorithm as the classifier to perform hand-posture recognition [6]. Feature extraction and feature classification are two important tasks after sEMG data acquisition [7]. In recent decades, numerous studies have investigated the features of EMG signals as well as machine learning methods [8]. Many advanced algorithms have also been introduced into the field of pattern recognition based on EMG signals such as GMM to handle the dimension-reduction problem for EMG missing-data compensation [9].

Many studies have focused on processing EMG signals only. Many researchers hope to apply sEMG to prosthetic hand control. In recent years, some researcher proposed a synchronous robot control system based on the sEMG signals of human upper limb motions [10]. Although sEMG technology improvements have been made over the years, the sEMG control is usually limited to open and closing. The main reason for this is the lack of robustness of the prosthetic hand control method. The position of the sensors is one of the main factors influencing the sEMG signals and, as a consequence, control robustness [11]. In order to better get good motions classification results. Some researchers fuse EMG signals with other signals. Xiangxin Li et al. research a motion-classification strategy based on sEMG-EEG signal combination for upper-limb amputees [12]. Ying Sun et al. proposed a gesture recognition method based on kinect and sEMG signal fusion [13].

Recently, some researchers began to apply sEMG signals to biological and medical research. S. Moore and M.P. McGuigan research the functional wavelet resolution of the sEMG frequency spectrum to represent high and low frequency motor unit recruitment in human lower limb muscles. This suggests the same force was produced using different patterns of neuromuscular recruitment [14]. Jose E. Cavazos and Jonathan Halford use of EMG signal to inform GTC seizure semiology. They prove the duration of tonic and clonic phases of GTCS may be quantified from sEMG data recorded by the Brain Sentinel device [15]. Some researchers aim at creating create patient-specific models to improve and objectify these predictions. The relationship between 3D lip movement and sEMG activities was accommodated in a state-space model [16]. Surface electromyography (sEMG) data acquired during lower limb movements have the potential for investigating knee pathology. Some researchers propose an ICA-EBM-Based sEMG classifier for recognizing lower limb movements in individuals with and without knee Pathology [17]. Some researchers develop a diagnostic evaluation system of facial nerve paralysis using sEMG [18]. These studies extend the field of EMG applications.

EMG is a technique used for evaluating and recording the activation signal of muscles and also utilized for the electrical manifestation of the contractions of muscles [19]. This paper based on sEMG studies the relationship between hand movements and muscle activity. At the same time, this paper proves the mapping relationship between the active muscle areas and the hand motions through the experimental results. Based on the experimental results, new EMG features can be developed later. The rest of this paper is organized as follows. Section 2 discusses the experimental data collection methods and experimental data analysis methods. Section 3 proves the mapping relationship between the active muscle areas and the hand motions through the experimental results. Section 4 draws the conclusion, gives the discussion of future work.

## 2   Data Collection and Processing

### 2.1   sEMG Signals Collection

Nine subjects took part in the experiment. All of them were healthy able-bodied subjects, without any prior neuromuscular disorders (eight males and one female, aged 20–30, denoted as P1 to P9).

Experiments use a 16-channel electrode sleeve to collect experimental data. The electrodes are modified from standard disposable Ag/AgCl ECG electrodes by means of gel removing. Electrodes are evenly fixed on an elastic fabric in advance. The procedure of arranging the electrodes on the forearm is as simple as wearing a sleeve. An empty sleeve (without electrodes on it) covers the previous one to guarantee a proper electrode to skin impedance. Before wearing the electrodes sleeve, subjects are obliged to wash their forearm for the better electrode to skin impedance as well [1].

Electrode sleeve is worn in the subject's right hand to collect data. Subjects use the ulna as a reference to wearing the 16-channel electrode sleeve. The ulnar coincides with the midline of the connection of electrode 8 and electrode 16 and the connection of electrode 1 and electrode 9. Sleeve wear position is shown in Fig. 1.

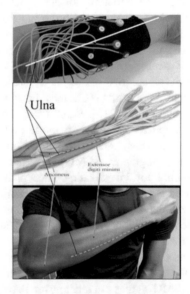

**Fig. 1.** The reference position to wear the 16-channel electrode sleeve.

Each subject wears the electrode sleeve for about 2 min. This operation can make the electrode full contact with the skin to reduce signal noise. The experiment collects the sEMG signals of 4 different hand motions of each subject. The four motions were fist firstly and then turn left wrist, turn right wrist, turn the wrist up and down the wrist. Experimental motions are shown in Fig. 2. The subjects repeated each action 12 times. The experimenter needs to complete the hand motions within 3 s.

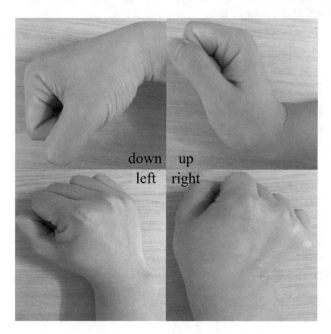

down  up

left  right

**Fig. 2.** The experimental motions.

## 2.2 Processing of sEMG Signals

In this paper, sEMG signal sampling frequency is set to 1 kHz, which is twice higher than the maximum frequency of sEMG signals. The newly received data with 16 channels are put into a buffer under a customized format. Once the buffer is full, a data parcel would be copied to a special buffer in the USB unit and prepared to be read by the USB host to a PC.

The original sEMG signals are filtered skin surface voltage. The subject's experimental motions are completed in 3 s. Every 3 s intercepts a sEMG signal segment. Mean absolute value (MAV) is computed for each signal segment. MAV is the average rectified value (ARV) and can be calculated using the moving average of full-wave rectified EMG. More specifically, it is calculated by taking the average of the absolute value of the EMG signal. Since it represents the simple way to detect muscle contraction levels, it becomes a popular feature for myoelectric controlled applications. It is defined as (1). Where N denotes the length of the signal and $x_n$ represents the EMG signal in a segment [5].

$$MAV = \frac{1}{N} \sum_{N=1}^{N} |x_n| \tag{1}$$

Each channel in the subject's experimental motion will correspond to a MAV value. In this paper, we use the MAV value as the ordinate and the electrode number as the abscissa. This will get a lot of data points in the coordinate system. A subject data points of a motion shown in Fig. 3.

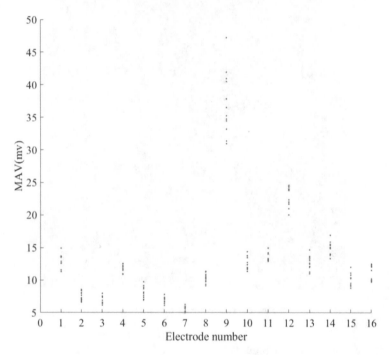

**Fig. 3.** A subject data points of a motion.

This paper uses shape-preserving piecewise cubic interpolation to fit data points. After fitting the data, a fitting curve is obtained. Shape-preserving piecewise cubic interpolation uses a piecewise cubic polynomial $P(x)$. On each subinterval (2), the polynomial $P(x)$ is a cubic Hermite interpolating polynomial for the given data points with specified derivatives (slopes) at the interpolation points. $P(x)$ interpolates y, that is (3), and the first derivative $dp/dx$ is continuous. The second derivative $d^2p/dx^2$ is probably not continuous so jumps at the $x_j$ are possible. Cubic interpolant $P(x)$ is shape preserving. The slopes at the $x_j$ are chosen in such a way that $P(x)$ preserves the shape of the data and respects monotonicity. Therefore, on intervals where the data is monotonic. So is $P(x)$, and at points where the data have a local extremum, so does $P(x)$ [20]. One of the subjects of a motion to the curve shown in Fig. 4.

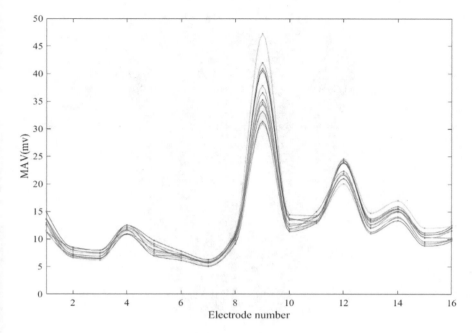

**Fig. 4.** One of the subjects of a motion to the curve.

$$X_k \leq X \leq X_{k+1} \tag{2}$$

$$P(x_j) = y_j \tag{3}$$

## 3 Pattern Analysis and Results

A total of 36 datasets were collected in the experiment. The MAV is calculated for each dataset, and the corresponding curve is fitted using shape-preserving piecewise cubic interpolation. This paper calculates the local maximum on the curve. Each fitting curve represents the intensity of a sEMG between electrodes.

According to the distribution of the electrode sleeve and the human forearm surface muscle distribution, the mapping between the electrode and the muscle can be obtained. This paper will be placed on the surface of the forearm muscle to the two-dimensional plane. This method can intuitively show the positional relationship between the muscle and the electrode. The positional relationship between the muscle and the electrode shown in Fig. 5. The positional relationship is not completely accurate, and the experimental positional relationship will have a small deviation from the positional relationship with the Fig. 5. MAV represents the simple way to detect muscle contraction levels. According to the local maximum of the curve and the positional relationship between the muscle and the electrode, it concludes that each subject has an active muscle area for each type of hand motion.

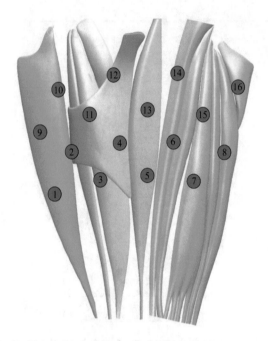

**Fig. 5.** Positional relationship between the muscle and the electrode.

In this paper, 36 datasets are calculated in turn. Active electrodes for each activity of each experimenter are counted. According to the statistical results, a total of 149 local maximum points were collected in the 4 experimental hand motions. Each local maximum point corresponds to one active electrode number. Statistical results (active electrodes number) are shown in Table 1.

**Table 1.** Active electrodes number.

| Motion Subject | Up | Down | Left | Right |
|---|---|---|---|---|
| P1 | 4,12,14 | 4,6,8,13 | 2,5,10,13 | 4,9,12,14 |
| P2 | 2,4,10,12,14 | 4,6,8,10,14 | 2,5,10,13 | 4,6,9,12,14 |
| P3 | 4,6,10,12,14 | 4,6,12,14 | 2,5,10,12,14 | 3,10,12,14 |
| P4 | 4,8,10 | 5,8,11,14 | 5,9,11,13 | 4,8,11,14 |
| P5 | 4,7,10,12,15 | 4,6,7,10,12,15 | 7,10,12,14 | 4,7,9,12,14 |
| P6 | 5,8,10,12,15 | 4,8,10,12,15 | 4,6,10,15 | 5,9,13 |
| P7 | 4,10,11,14 | 4,7,10,14 | 5,10,14 | 9,10,14 |
| P8 | 5,7,10,14 | 2,10,4 | 5,9,11,13,14 | 5,9,11,14 |
| P9 | 5,10,14 | 3,5,10,14 | 5,9,11,14 | 5,9,11,14 |

Of the 149 local maximum points, there are 27 local maxima belonging to the up motion, 39 local maxima for the down motion, 37 local maxima for the left motion, and

36 local maxima for the right motion. The active electrode reflects the major role of the corresponding muscle in the motion. Combining the statistical results in Table 1 with the positional relationship between the muscle and the electrode map can give an active muscle area during the motion.

In this paper, because the electrodes are distributed in pairs, the human upper arm muscle is divided into 8 regions. The area covered by a pair of longitudinal electrodes divides the muscle. Each muscle area covers 1 to 2 muscles. This paper will be the 8 areas number R1–R8. As showed in Fig. 6. The active electrodes of each action are matched to the positional relationship map.

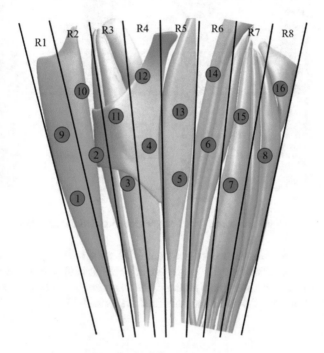

**Fig. 6.** R1–R8 muscle areas.

Due to the hand motions are done by muscle synergies. This paper selected 3 active muscle regions for each experiment motion. Based on preliminary observations of active electrode data. R2, R3, and R4 are the active area of up-hand motion. R4, R5, and R6 is the active area of down-hand motion. R2, R5 and R6 are the active area of left-hand motion. R1, R4, and R6 are the active area of right-hand motion. 21 up motion active electrodes were located in R2, R3 and R4 region. The amount of data in the R2, R3, and R4 region accounted for 57% of the total data for up motion. 22 down motion active electrodes were located in R4, R5, and R6 region. The amount of data in the R4, R5, and R6 region accounted for 56% of the total data for down motion. 26 left motion active electrodes were located in R2, R5, and R6 region. The amount of data in the R2, R5, and R6 region accounted for 70% of the total data for left motion. 24 left motion active electrodes were located in R1, R4, and R6 region. The amount of data in the R1, R4 and R6 region accounted for 67% of the total data for the right motion. Active electrodes that are not in the active muscle area are also mostly located around the area. The matching rate of the active electrode and the muscle region is shown in Fig. 7.

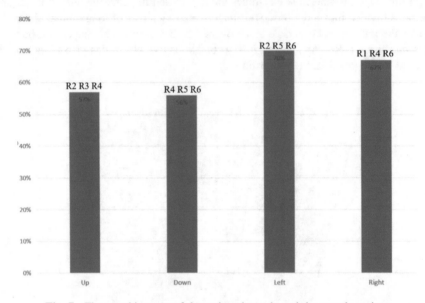

**Fig. 7.** The matching rate of the active electrode and the muscle region.

In this paper, the above results are combined with the support vector machine (SVM) classification algorithm. The active muscle region is used as the motion feature to classify the corresponding hand motion. The experimental results showed that the classification accuracy of 4 hand motions reached 87%, 92%, 87% and 83% respectively (see Fig. 8).

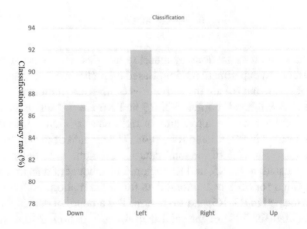

**Fig. 8.** The classification accuracy of the 4 hand motions

# 4    Conclusion and Future Work

## 4.1    Conclusion

The research work has proved that the forearm has different areas of muscle activity when performing different motions. There are some clear mappings between the active electrode areas and motions. When the subject finishes up motion, R2, R3, R4 muscle areas are more active. When the subject finishes the down motion, R4, R5, R6 muscle areas are more active. When the subject finishes left motions, R2, R5, R6 muscle areas are more active. When the subject finishes the right motion, R1, R4, R6 muscle areas are more active. Subject motions and muscle activity areas showed good mapping relations. These mappings can be used to improve the classification of the forearm motions as a new sEMG feature. The accuracy rate of hand motions recognition also proves it. On the other hand, this mapping can help medical researchers to research the muscular synergy problems through the sEMG signals.

## 4.2    Future Work

There are many factors in the experiment that are unfavorable to the experimental results such as positional relationship between the muscle and the electrode map is not completely accurate; each subject's forearm thickness will affect positional relationship between the muscle and the electrode; subject sleeve wear position is not exactly accurate; the signal acquisition process will be influenced by some uncertainties. Experiments in the later stage can be improved to obtain more accurate experimental results. And more motions added to the experiment.

Improve experimental results as a sEMG feature, using this feature to complete the classification of different hand motions. The computational method in this paper is concise, so it is expected that the classification speed of this feature will be superior to the traditional sEMG feature. Based on this experimental result, a real-time visualization system for muscle activity can be constructed. This visualization system is of great importance for the study of muscle activity and EMG-based hand motion recognition.

**Acknowledgment.** The authors would like to acknowledge the support from the Natural Science Foundation of China under Grant No. 51575412, 51575338 and 51575407, the EU Seventh Framework Programme (FP7)-ICT under Grant No. 611391, the Grants of National Defense Pre-Research Foundation of Wuhan University of Science and Technology (GF201705) and the Research Project of State Key Lab of Digital Manufacturing Equipment & Technology of China under Grant No. DMETKF2017003. And this paper is funded by Wuhan University of Science and Technology graduate students' short-term study abroad special funds.

# References

1. Yinfeng, F., Honghai, L., Gongfa, L., Xiangyang, Z.: A multichannel surface EMG system for hand motion recognition. Int. J. Humanoid Robot. **12**(2), 1550011 (2015)
2. Yaxu, X., Zhaojie, J., Jing, C., Honghai, L.: Multiple sensors based hand motion recognition using adaptive directed acyclic graph. Appl. Sci. **7**(4), 358 (2017)

3. Rui, S., Rong, S., Kai-yu, T.: Complexity analysis of EMG signals for patients after stroke during robot-aided rehabilitation training using fuzzy approximate entropy. IEEE Trans. Neural Syst. Rehabil. Eng. **22**(5), 1013–1019 (2013)
4. Khezri, M., Jahed, M., Sadati, N.: Neuro-fuzzy surface EMG pattern recognition for multifunctional hand prosthesis control. In: 2007 IEEE International Symposium on Industrial Electronics, Spain, pp. 269–274 (2007)
5. Rezwanul, M., Ahsan, M., Ibrahimy, I., Othman, K.: Electromygraphy (EMG) signal based hand gesture recognition using artificial neural network (ANN). In: 4th International Conference on Mechatronics, Malaysia, vol. 4, pp. 1–6 (2015)
6. Wan-Ting, S., Zong-Jhe, L., Shih-Tsang, T., Tsorng-Lin, C., Chia-Yen, Y.: A bionic hand controlled by hand gesture recognition based on surface EMG signals: a preliminary study. Biocybern. Biomed. Eng. **38**(1), 126–135 (2018)
7. Manea, S., Kamblib, R., Kazic, F., Singhc, N.: Hand motion recognition from single channel surface EMG using wavelet & artificial neural network. Procedia Comput. Sci. **49**(1), 58–65 (2015)
8. An-Chih, T., Jer-Junn, L., Ta-Te, L.: A novel STFT-ranking feature of multi-channel EMG for motion pattern recognition. Expert Syst. Appl. **42**(7), 3327–3341 (2014)
9. Qichuan, D., Jianda, H., Xingang, Z., Yang, C.: Missing-data classification with the extended full-dimensional gaussian mixture model: applications to EMG-based motion recognition. IEEE Trans. Ind. Electron. **62**(8), 4994–5005 (2015)
10. Boyang, Z., Erwei, Y., Jun, J., Zongtan, Z.: A synchronous robot control system based on the sEMG signals of human upper limb motions. In: Proceedings of the 36th Chinese Control Conference, China, pp. 5136–5140 (2017)
11. Francesca, P., Matteo, C., Arjan, G., Henning, M.: Repeatability of grasp recognition for robotic hand prosthesis control based on sEMG data. In: 2017 International Conference on Rehabilitation Robotics (ICORR), UK, pp. 17–20 (2017)
12. Xiangxin, L., Oluwarotimi, W., Samuel, X., Hui, Z.: A motion-classification strategy based on sEMG-EEG signal combination for upper-limb amputees. J. NeuroEng. Rehabil. **14**(1), 2–5 (2017)
13. Ying, S., Cuiqiao, L., Gongfa, L., Guozhang, J., Du, J., Honghai, L., Zhigao, Z.: Gesture recognition based on kinect and sEMG signal fusion. Mobile Netw. Appl. 1–9 (2018)
14. Moore, S., McGuigan, M.: Functional wavelet resolution of the sEMG frequency spectrum to represent high and low frequency motor unit recruitment in human lower limb muscles. Int. J. Sci. Med. Sport **20**, 73–75 (2017)
15. Jose, E., Cavazos, J., Halford, M.: Use of sEMG to inform GTC seizure semiology. Neurology **88**(2), 2–11 (2017)
16. Merijn, E., Maarten, A., Ludi, S., Dieta, B., Alfons, B., Ferdinand, H.: Predicting 3D lip movement using facial sEMG: a first step towards estimating functional and aesthetic outcome of oral cancer surgery. Med Biol Eng Comput **55**(4), 1–11 (2016)
17. Ganesh, N., Easter, S., Sridhar, P., Arjunan, A., Acharyya, D., Kumar, A.: Predicting 3D lip movement using facial sEMG: an ICA-EBM-based sEMG classifier for recognizing lower limb movements in individuals with and without knee pathology. IEEE Trans. Neural Syst. Rehabil. Eng. **26**(3), 675 (2018)
18. Shogo, O., Misaki, S., Hiroki, T., Takahiro, N.: Development of diagnosis evaluation system of facial nerve paralysis using sEMG. In: The 2017 International Conference on Artificial Life and Robotics, Japan, pp. 11893–11908 (2017)
19. Yinfeng, F., Nalinda, H., Dalin, Z., Honghai, L.: Multi-modal sensing techniques for interfacing hand prostheses: a review. IEEE Sens. J. **15**(11), 6065–6076 (2015)
20. Fritsch, F., Carlson, R.: Monotone piecewise cubic interpalation. SIAM J. Numer. Anal. **17**(2), 238–246 (1980)

# A Novel Crossings-Based Segmentation Approach for Gesture Recognition

Dario Ortega Anderez$^{(\boxtimes)}$, Ahmad Lotfi, and Caroline Langensiepen

School of Science and Technology,
Nottingham Trent University, Nottingham NG11 8NS, UK
dario.ortegaanderez2013@my.ntu.ac.uk

**Abstract.** Human activity recognition (HAR) has mainly been directed to the recognition of static or quasi-periodic activities like sitting, walking or running, typically for fitness applications. However, activities like eating or drinking are neither static nor quasi-periodic. Instead, they are composed of sparsely occurring motions or gestures in continuous data streams. This paper presents a novel adaptive segmentation technique based on crosses of moving averages to identify potential eating or drinking gestures from accelerometer data. The novel crossings-based segmentation approach proposed is able to identify all eating and drinking gestures from continuous accelerometer data including different activities. A posteriori, potential gestures are classified as food or drink intake gestures using a combination of Dynamic Time Warping (DTW) as signal similarity measure and a k-Nearest Neighbours (KNN) classifier. An outstanding classification rate of 100% has been achieved.

**Keywords:** Adaptive signal segmentation · Gesture recognition
Wearable sensors

## 1 Introduction

Despite the increasing attention given to Human Activity Recognition (HAR) with the use of inertial sensors, efforts are principally applied to fitness applications, where quasi-periodic activities like walking or running are studied. In contrast, activities like eating or drinking cannot be considered quasi-periodic. Instead, they are composed of sparsely occurring motions in continuous data streams. For convenience in discussion, we refer to these activities as "sequential activities". Figure 1 illustrates an example of a quasi-periodic activity (walking), a static activity (sitting) and a sequential activity (eating).

Most activity classification studies employ a rough segmentation technique, namely the sliding window, whereby signals are typically divided into consecutive (often overlapping) windows of equal length [2,3,8]. Once split, a feature vector is calculated from those windows and fitted into supervised machine learning classifiers for the final activity classification. Alternatively, template-based

© Springer Nature Switzerland AG 2019
A. Lotfi et al. (Eds.): UKCI 2018, AISC 840, pp. 383–391, 2019.
https://doi.org/10.1007/978-3-319-97982-3_32

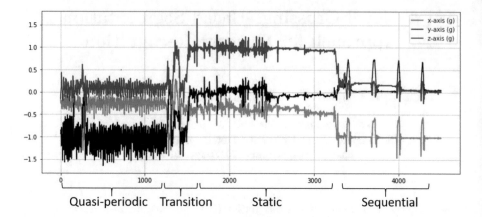

**Fig. 1.** Example of a quasi-periodic (walking), static (sitting) and sequential (eating) activities. Data was collected with a wrist-mounted accelerometer at 100 Hz.

approaches, through which activities are decomposed into fundamental motion segments, have also been investigated [10].

Both window-based and template-based approaches have been proven to be accurate on identifying static or quasi-periodic activities. However, limitations are found when applied to activities which are neither static nor quasi-periodic, like eating or drinking.

Firstly, the window length is typically fixed for all activities, either selected from previous work or treated as a hyper-parameter of the classification problem. Secondly, the window length is dependent on the activity set chosen. That is, windows must be sufficiently long to capture fundamental characteristics of the signal but sufficiently short to avoid capturing signal from multiple activities. Considering these limitations, the inclusion of a sequential activity to an activity set formed by quasi-periodic and static activities could compromise the performance of the model considerably, since the use of a fixed window will no longer be appropriate given the sporadic and irregular nature of the gestures occurring in sequential activities.

To overcome this problem, this paper presents a novel adaptive segmentation technique based on cross-overs and cross-downs between a slow and a fast moving average. The reasoning behind this is based on the fact that to eat or to drink something, one has to grasp the corresponding tool, take it to the mouth and finally take it back to the rest position. Such a sequence of motions suggests that a fast moving average will react faster to the gesture of taking food or drink to the mouth, crossing over a slower moving average. A cross-down will follow when moving the hand back to the rest position.

Given the variability in length of segmented gestures, Dynamic Time Warping (DTW) was used for signal alignment. A posteriori, a k-Nearest Neighbours (KNN) classifier was trained to classify between eating and drinking gestures using the DTW distances from the different segments. Outstanding results reveal a 100% segmentation recall and a 100% classification accuracy.

The rest of the paper is structured as follows: Sect. 2 discusses previous work on gesture recognition and time series segmentation. Section 3 presents the methodology proposed for gesture recognition. Section 4 presents the preliminary results. Section 5 presents the conclusions and future work.

## 2    Previous Work

As mentioned above, activities like eating or drinking are neither static nor quasiperiodic. Instead, they are formed by sparsely occurring motions in continuous data streams. Despite the significant progress achieved in activity recognition, gesture recognition remains a challenge. Traditional methods for HAR commonly use a fixed window length; however the sporadic nature of gestures motivates the development of adaptive signal segmentation techniques.

Among time series segmentation algorithms, Piecewise Linear Representation (PLR) approaches are the most frequently used. Three major PLR segmentation techniques, namely Top-Down, Bottom-Up and Sliding-Window, have been outlined by previous work [6]. Additionally, the Sliding Window and Bottom-Up (SWAB) segmentation technique was proposed by [5] as a combination of the Bottom-Up and Sliding Window segmentation techniques. On top of these, different customized approaches have been proposed for gesture recognition.

The authors in [7] developed an adaptive sliding window segmentation technique from waist-worn accelerometer data for the detection of transitional activities (sit-to-stand, stand-to-sit, sit-to-lie and lie-to-sit) among static and dynamic activities. First, a fixed length window (over which thirteen features were extracted) was used to classify between transitional, dynamic and static activities. Once a window was classified as a transitional activity, each such window was extended by a factor proportional to the original window size, until the Gaussian probability density function indicated a decrease in likelihood for a particular transitional activity. An overall classification recall of 93.0% was achieved, improving on the 89.9% achieved without adaptive segmentation.

The authors in [1] developed a drink intake recognition algorithm based on a set of 20 features after reduction via the Mann-Whitney-Wilcoxon test, and a feature similarity search (FSS) with an equidistant segmentation of 0.5 Hz and Euclidean distance as similarity measure. They achieved recognition rates of 93.8% and 89.2% for "fetch" and "sip" motions respectively using tri-axial accelerometer and tri-axial gyroscope data from a wrist-mounted inertial unit.

The recognition of seven hand gestures (up, down, left, right, tick, circle, cross) using bi-axial data from an accelerometer held horizontally in the hand was studied by [9]. Before the recognition phase, signals were segmented using a set of five features, whereby potential starting and end points for independent gestures were identified. A maximum recognition accuracy of 95.6% was achieved by a template matching model using sequences of sign changes in acceleration.

The authors in [4] implemented a Hidden Markov Model (HMM) using data collected from five wearable sensor units (two on each arm and one on the trunk) to recognize a set of four dietary gestures. Signals were divided into motions using the Sliding-Windows and Bottom-Up (SWAB) segmentation approach proposed by [5], and a posteriori a similarity search was developed to identify potential motion segments to be further examined by the HMM. Performance metrics of 0.73 and 0.79 were achieved for precision and recall respectively.

## 3 Methodology

Given the sporadic occurrence of eating and drinking gestures, as opposed to those of quasi-periodic or static activities, rough segmentation techniques like the traditional sliding window, by which time series are divided into windows of same length, seem inappropriate. The solution to gesture recognition proposed in this paper incorporates a comprehensive approach, whereby computer efforts are increasingly augmented as the difficulty of the problem rises. First, a novel adaptive segmentation approach based on crossings of a slow and a fast moving average is proposed to identify potential segments that can include an eating or drinking gesture. Second, the similarity between those segments is measured. Given the variability in length of different segmented motions, a motion sequence may be "warped" non-linearly, thusly requiring stretching or shrinking along the time axis to be compared to another motion sequence. To overcome this problem, Dynamic Time Warping (DTW) is used for alignment. After the signal alignment, distances between motions are computed and a k-Nearest Neighbours (KNN) classifier is trained for the classification of the gestures as "drinking" or "eating".

### 3.1  Signal Segmentation

Adaptive segmentation of inertial signals is a challenging task, since it needs to adapt to the nature of the signal itself. First, the same gesture can have different lengths in time, this implying the segmentation has to adapt to length variability to produce the best representation of each segment. Second, the segmentation should happen online, that is, it needs to adjust to new incoming data. To solve these constraints, a novel segmentation technique based on crossing of moving averages is proposed. Given a sequence of accelerometer readings $[A_{x-(n-1)}, A_{x-1}, \ldots, A_x]$, the moving average for the last n readings is calculated as follows:

$$MvAvg_n = \frac{A_x, A_{x-1}, \ldots, A_{x-(n-1)}}{n} = \frac{1}{n}\sum_{i=0}^{n-1} A_{x-i} \qquad (1)$$

The reasoning behind is based on the fact that to eat or to drink something, one has to grasp the corresponding tool, take it to the mouth and finally take it back to the rest position. Such a sequence of motions suggest that a fast moving average will react faster to the gesture of taking food or drink to the mouth, crossing over a slower moving average. A cross down will follow when moving the hand back to the rest position. This sequence is illustrated in Fig. 2.

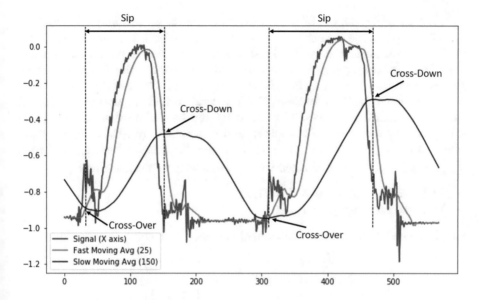

**Fig. 2.** Signal segmentation using crosses of two moving averages.

To filter out either too short or too long unwanted gestures, it was assumed that the duration of drinking and eating gestures are no shorter than 1 s and no longer than 8 s. Once a window or segment for a potential drinking or eating gesture is identified (limited by a cross-over on the left and a cross-down on the right), that window is widened by 1 s on each side.

A comprehensive search was made to select the optimal pair of moving averages based on two factors. First, each segment should accurately identify the sought gestures without losing information which can be crucial for the subsequent classification. Second, the aim at this preliminary step should be the optimization of the classification recall; that is, avoiding missing any true positives.

### 3.2   Gesture Similarity

Once the segmentation process is completed, the similarity of the different gestures is studied to determine whether they are an eating or a drinking gesture. Segments were manually annotated and those segments which were not an eating or a drinking gesture were deleted from the dataset.

Generally, the distance between two sequences $S = [s_1, s_2, ... s_n]$ and $Q = [q_1, q_2, ... q_n]$ with values at every time instant $t = [1, ..., n]$, is measured using the Euclidean distance:

$$d(S, Q) = \sqrt{\sum_{t=1}^{n}(S_t - Q_t)^2} \tag{2}$$

Unfortunately, the use of the Euclidean distance has two major limitations: (1) The length of the sequences must be the same for both sequences, that is, $|S| = |Q|$. (2) It measures the vertical distance between pairs of points according to their index in their respective sequences. This implies that the distance between two same gestures occurring at different speeds can be undesirably large.

To overcome these limitations, DTW is used to find the optimal alignment between temporal sequences. The alignment can be explained as follows: Let x and y be two sequences of lengths m and n, respectively. DTW finds a mapping path $\{(p_1, q_1), \dots, (p_j, q_j)\}$ such that the distance on the mapping path $\sum_{i=1}^{j} |x(p_i) - y(q_i)|$ is minimized with the following two constraints:

- Anchored beginning: $(p1, q1) = (1, 1)$
- Anchored end: $(p_j, q_j) = (m, n)$

where the cost of the optimal alignment can then be recursively computed as:

$$Di, j = |x(i) - y(j)| + min \left\{ \begin{array}{c} D(i\text{-}1,j) \\ D(i\text{-}1,j\text{-}1) \\ D(i,j\text{-}1) \end{array} \right\} \tag{3}$$

Figures 3 and 4 show a signal alignment of two drinking gestures using DTW and the computed warping path respectively. It can be seen that DTW overcomes the limitations of the Euclidean distance. First, it can compare signals with different lengths, since one point of the sequence x can be aligned to more than one point of the sequence y and vice versa. Second, it captures flexible similarities by aligning the coordinates of both sequences.

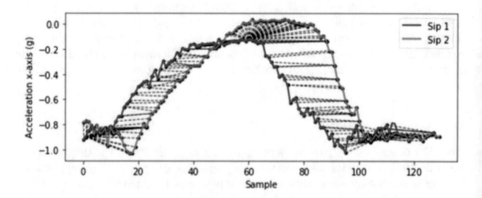

**Fig. 3.** Alignment of two drinking gestures using Dynamic Time Warping.

## 3.3   Classification

The final step of the proposed methodology is gesture classification. Once the signals are segmented and a suitable similarity measure is found, the next step

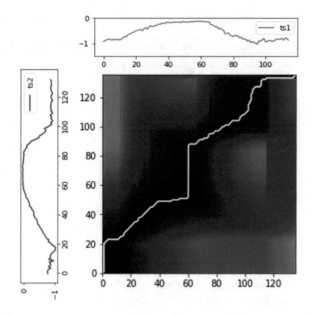

**Fig. 4.** Warping path of two drinking gestures using Dynamic Time Warping.

is to train a classifier to model the similarity between the different segments. A k-Nearest Neighbours is employed in this case. Thus, unseen segments are classified by a majority vote of its k nearest neighbours, being thus assigned to the most common class among them. The robustness of the methodology proposed is studied across the hyper-parameter k.

## 4   Preliminary Results

This section shows the results achieved by the combination of the novel segmentation technique and the employment of DTW as a distance measure between gestures. First, different pairs of moving averages were compared to find the optimum pair to capture drinking and eating gestures in continuous data streams. The performances of each pair are shown in Table 1.

It is noticed the classification recall decreases when the slow moving average has a value higher than 50. Additionally, when two moving averages are close to each other, the technique produces low precision values, since the moving averages tend to cross more frequently. As mentioned in Sect. 3.1, the desired output should present a high classification recall as well as good accuracy in terms of determining the limits of the gestures. The (pair 25/150) is found to present the best behaviour in terms of these requirements.

The classification performance was studied using (k = number of neighbours) as a classification hyper-parameter. For all values a of k from 0 to 10, the classification accuracy achieved is 100%, proving the robustness of the approach proposed.

**Table 1.** Precision and recall on eating and drinking gestures identification

| Moving averages | Drinking precision (%) | Drinking recall (%) | Eating precision (%) | Eating recall (%) |
|---|---|---|---|---|
| 25/50 | 50.00 | 100.00 | 61.29 | 100.00 |
| 25/100 | 66.67 | 100.00 | 76.00 | 100.00 |
| 25/150 | 90.91 | 100.00 | 86.36 | 100.00 |
| 25/200 | 100.00 | 100.00 | 95.00 | 100.00 |
| 50/100 | 62.50 | 100.00 | 86.36 | 100.00 |
| 50/150 | 86.96 | 100.00 | 60.87 | 73.68 |
| 50/200 | 100.00 | 100.00 | 82.61 | 100.00 |
| 50/250 | 100.00 | 100.00 | 85.00 | 89.47 |
| 75/100 | 60.61 | 100.00 | 70.83 | 89.47 |
| 75/150 | 90.91 | 100.00 | 22.73 | 26.32 |
| 75/200 | 100.00 | 95.00 | 60.00 | 63.16 |
| 75/250 | 100.00 | 90.00 | 42.11 | 42.11 |
| 100/150 | 73.91 | 85.00 | 22.73 | 26.32 |
| 100/200 | 50.00 | 50.00 | 17.39 | 21.05 |

Despite the need to include additional data from different users and a greater variety of scenarios, the promising results achieved suggest that the comprehensive methodology implemented in this paper has the potential to be further developed. Not only is the crossings-based approach able to detect 100% of the eating and drinking gestures in continuous data streams, but DTW is also proven to be an accurate similarity measure able to perfectly classify eating and drinking gestures which in principle look extremely similar.

## 5   Conclusions and Future Work

The combination of appropriate adaptive signal segmentation alongside the employment of DTW for signal alignment exhibits outstanding results. As shown above, the crossings-based segmentation technique is able to recognize eating and drinking gestures in continuous data streams including different activities with a classification recall of 100%. Given the flexibility of this technique, it could be used to detect other kinds of gesture by experimentally adapting the pair of moving averages. In addition, DTW is proven to be an accurate similarity measure on temporal sequences of accelerometer data, achieving 100% classification accuracy when used with a KNN classifier.

Future work will be directed to the collection of additional data from several users in different scenarios and testing the methodology against a benchmark dataset. More challenging gestures will be added to the classification model. Further investigation on similarity measures for temporal sequences will be also carried out.

# References

1. Amft, O., Bannach, D., Pirkl, G., Kreil, M., Lukowicz, P.: Towards wearable sensing-based assessment of fluid intake. In: 2010 8th IEEE International Conference on Pervasive Computing and Communications Workshops (PERCOM Workshops), pp. 298–303. IEEE (2010)
2. Bayat, A., Pomplun, M., Tran, D.A.: A study on human activity recognition using accelerometer data from smartphones. Procedia Comput. Sci. **34**, 450–457 (2014)
3. Casale, P., Pujol, O., Radeva, P.: Human activity recognition from accelerometer data using a wearable device. In: Pattern Recognition and Image Analysis, pp. 289–296 (2011)
4. Junker, H., Amft, O., Lukowicz, P., Tröster, G.: Gesture spotting with body-worn inertial sensors to detect user activities. Pattern Recogn. **41**(6), 2010–2024 (2008)
5. Keogh, E., Chu, S., Hart, D., Pazzani, M.: Segmenting time series: a survey and novel approach. In: Data mining in time series databases, pp. 1–21. World Scientific (2004)
6. Lovrić, M., Milanović, M., Stamenković, M.: Algoritmic methods for segmentation of time series: an overview. J. Contemp. Econ. Bus. Issues **1**(1), 31–53 (2014)
7. Noor, M.H.M., Salcic, Z., Kevin, I., Wang, K.: Adaptive sliding window segmentation for physical activity recognition using a single tri-axial accelerometer. Pervasive Mob. Comput. **38**, 41–59 (2017)
8. Ravi, N., Dandekar, N., Mysore, P., Littman, M.L.: Activity recognition from accelerometer data. In: AAAI, pp. 1541–1546 (2005)
9. Xu, R., Zhou, S., Li, W.J.: Mems accelerometer based nonspecific-user hand gesture recognition. IEEE Sens. J. **12**(5), 1166–1173 (2012)
10. Zhang, M., Sawchuk, A.A.: Motion primitive-based human activity recognition using a bag-of-features approach. In: Proceedings of the 2nd ACM SIGHIT International Health Informatics Symposium, pp. 631–640. ACM (2012)

# Author Index

© Springer Nature Switzerland AG 2019
A. Lotfi et al. (Eds.): UKCI 2018, AISC 840, pp. 393–394, 2019.
https://doi.org/10.1007/978-3-319-97982-3